PLANT PHYSIOLOGY

PLANT PHYSIOLOGY

Editor

Edgar Ibbot

scitus academics

Plant Physiology

Edited by **Edgar Ibbot**

This Book is Printed on Acid-Free Paper.

ISBN: 9781681179452
Library of Congress Control Number: 2017949083

SCITUS Academics LLC,
616, Corporate Way, Suite 2, 4766,
Valley Cottage, NY 10989

www.scitusacademics.com

Notice

Printed in United States of America.

Preface

Plant physiology includes the study of plant form and function. As plants developed to life on land they were obligatory to evolve methods to continue with the separation of CO_2 and water. Plant physiology looks at the different parts of plants and how they work, either individually or collectively. In recent years our knowledge of the metabolic pool in plants, with its numerous eddies, cross currents and vortexes, has steadily become wider and deeper. The metabolic pathways in the oxidation of carbohydrates, as previously indicated, are quite well understood. The biochemistry of photosynthesis, although not, parenthetically, its energetics, is approaching a comparable status of knowledge. Similarly, knowledge of the metabolism of organic acids, fats, and nitrogenous compounds has progressed substantially in the last decade or two. The general advance in our understanding of the overall dynamics of plant metabolism has been one of the most outstanding achievements of plant physiology during the last two decades. Research in plant hormones is another field of investigational activity which has retained its prominence in recent years. Work on other kinds of plant hormones besides the auxins has; continued in recent years, and some entirely new ones have been discovered. Investigations have continued on the roles of thiamin, niacin, pyridoxine and other plant hormones of the B vitamin group. The truly grand challenges in plant physiology are those that will yield important insights about how plants live and function, and also provide at least partial solutions to critical societal needs.

The book provides general principles and new insights of some plant physiology aspects covering abiotic stress, biochemistry and metabolism, transport and translocation, plant water relations, mineral nutrition and reproduction. This book may serve as source of methods, theories, ideas and tools for students, researchers and experts in the areas of plant physiology.

Table of Contents

CHAPTER 1

The Biological Pigments in Plants Physiology

Arturo Solís Herrera

Human Photosynthesis® Study Center, Aguascalientes, Mexico

ABSTRACT

The physiology of the plants has come to be a formidable challenge. Despite best efforts, the mysteries remain, most of the processes identified cannot be played in vitro, and to date it cannot be said that knowledge about the complex biochemical processes of the plant is understood properly. The need to produce food is an emergency in the world, so any knowledge that will allow us to advance in this sense is important. Our discovery about the intrinsic property of melanin transform light energy into chemical energy by means of the dissociation of the molecule of water, such as chlorophyll in plants, represents a turning point in relation to the chemical reaction so far considered the most important in the world: photosynthesis.

Keywords: Chlorophyll, Melanin, Lignin, Water Dissociation, Hydrogen, Oxygen, Energy, Pigments

1. INTRODUCTION

A pigment is defined as any substance capable of absorbing light [1] , so it could be considered pigments almost all substances. Usually, substances absorb very specific wavelengths, and emit a characteristic electromagnetic signal that allows us to identify them. Color is the property of electromagnetic radiation with a wavelength between 300 - 400 and 700 - 800 nm. The color of individual chemical substances derived from their interaction with white light. If the total radiation falling on a chemical compound passes through or is reflected from it, the substance has a white color. If white light on contact with a substance is completely absorbed, appears black. Most often chemical compounds selectively absorb within certain radiation ranges and the color of a given compound is complementary to the absorbed radiation.

This allows us to infer that the energy inherent in the light produced a change in substance or molecule that it absorbed, so far in accordance with the law of conservation of energy, whose first postulate says that energy, defined as everything that produces a change is not created or destroyed, only is thus transformed.

Therefore, the light energy absorbed by the compound or molecule, brought about a change in it, and the new molecule issued part of the absorbed energy, in the form of electromagnetic signal, for example, light with different wavelength to which was originally absorbed.

Nature is full of examples in this regard, and the plants are no exception. And also in them it happens that the wavelengths of light can be, wholly or partly, absorbed, transmitted or reflected.

And many compounds containing a plant that they can classified as pigments, for example: porphyrins, carotenoids, anthocyanins, etc. However, there are two that have powerfully drawn attention: the chlorophyll and recently the melanin. And the reason is that both molecules have very peculiar behaviors, we could call them unique in nature. This is: both molecules are able to dissociate the water molecule.

And it is the attention, because we are against the way in which nature transforms light energy into chemical energy. It is possible that there are other molecules that can do so, but the two examples most known or at least shown are chlorophyll, which until a decade ago was considered the only molecule with such capacity, and recently, melanin.

Chlorophyll, absorbs the ends of the visible spectrum, and with that power separates the liquid water in its gaseous components, hydrogen and molecular oxygen; and it carried out at room temperature and in an irreversible manner, then ejects the oxygen into the atmosphere. The chemical reaction can be written in the following way.

$$2H_2O_{(liquid)} \rightarrow 2H_{2\ (gas)} + O_{2\ (gas)}$$

It is the attention that the reaction happens so quickly and at room temperature inside the chlorophyll's leaves, as in the laboratory we can dissociate water, but requires warm water at 2000°C and the products that we get, they are not so ordered as in chlorophyll and melanin. This reaction is impelled by the extremes of visible light (violet and red)

Another interesting detail is that part of the energy that is released when water dissociates, is transported by molecular hydrogen, which was to be expected since it is the main carrier of energy in the entire universe.

On the other hand, molecular oxygen, which is very stable and toxic, is expelled into the atmosphere by the plant, and since it has always been present in the equation, the plant optimizes it, given its high toxicity.

In melanin (Greek: melanos; dark), the reaction is similar, but quicker and thousands of times more efficient, because the melanin absorbs electromagnetic full spectrum and in addition is capable not only of separate water into its components but also re-shape the molecule, this is: gaseous components takes her to the liquid state again [2] . This reaction happens inside melanin, and can be located both intra and extracellularly. The chemical reaction is written thus:

$$2H_2O_{(liquid)} \rightarrow 2H_{2\,(gas)} + O_{2\,(gas)} \rightarrow 2H_2O_{(liquid)} + 4e^-$$

Or so:

$$2H_2O_{(liquid)} \leftrightarrow 2H_{2\,(gas)} + O_{2\,(gas)} + 4e^-$$

For every two molecules of water re-formed, are generated 4 high energy electrons, which are electrons whose probability cloud is located farthest from the atomic nucleus in question as a low-energy electron or stable state. High energy electrons or excited are exchanged easily, making it almost impossible to identify the atom or molecule to which they belong.

2. MELANIN, THE META-CHLOROPHYLL

Melanin is what lies behind the chlorophyll, which explains the misunderstood chlorophyll, which remedied the deficiencies; in few words, is that even explains the origin of life.

Chemicals Processes are known in plants, which the mere dissociation of water by chlorophyll cannot explain, could tell that the physiology of the plant is full of mysteries until now misunderstood. The energy that emanates from the chlorophyll does not provide a satisfactory explanation for so many and so many observed phenomena, both in the root, stem and leaves. Only the circulation of the SAP not has been explained to date.

The so-called cohesion-tension theory that try to explain SAP transport by the xylem, from the roots of the plant to the leaves through the wood tubules, is to date; a theory.

The mysterious forces that drive the viscous solutions through the trunk, from the root to the leaves, in order, in a sequence that is not understood, no doubt require energy, cannot be explained by the chlorophyll, glucose or ATP, but the energy that emanates from the melanin comes to fill a hole that will allow us to advance in the understanding of the intricate biochemical processes that affect the life of the plant.

Melanin in humans had passed as unnoticed as the lignin in plants, the main interest at present was to modify the composition of lignin, which is the melanin in the plants; as a basic strategy to increase the digestibility of cell walls in order to produce biodiesel [3] .

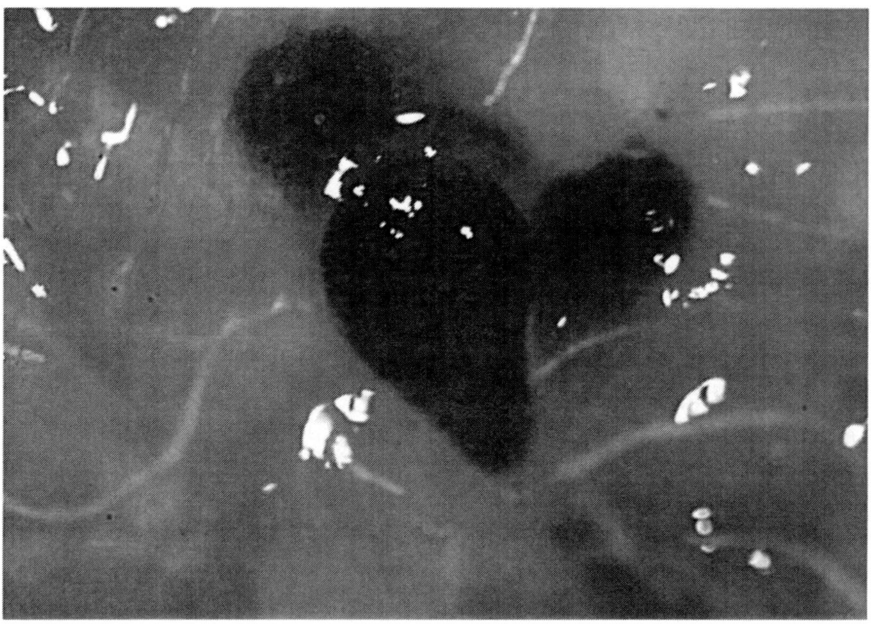

Figure 1. Melanin is present in practically all of the seeds, and the pitaya is no exception. When the amount of water is scarce, power that can produce melanin also is, by what the mechanisms leading to the emergence of seed not can be driven, when the amount of water is adequate, then the energy that emanates from the melanin will be adequate to promote each and every one of the mechanisms that lead to the germination of the seeds (16×).

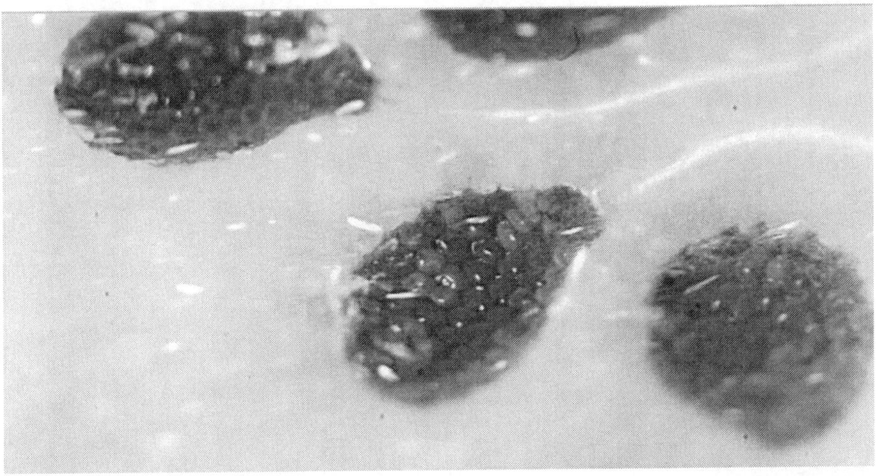

Figure 2. Seed color depends on the amount of melanin, granule size, the orientation of the molecule, the combination with other compounds, as well as the nature of the structures that are in, but, either way, its main function is the same: production of chemical energy (16×).

It is expected that as you go knowing the unsuspected intrinsic property of melanin transform light energy into chemical energy by means of the water molecule dissociation, the interest becomes more reasonable, because the omnipresence of melanin explains the germination of the seeds, as it was a mystery as the seed detected level appropriate water to flourish (Figures 1-3).

The similarity of the surface of the iris of the human eye (Figure 4, Figure 5) with the surface of some seeds, for example, the peach, are given by the notable content of melanin in both structures.

The concept that photosynthetic membrane is the microscopic powerhouse of our biosphere, now is broken into small pieces.

Molecular hydrogen and high energy electrons coming from the dissociation of water, due to the melanin molecule, constitute the true universal biological engine. The oxidation of water as the basis of molecular biology is more true than ever, but as the melanindone, chlorophyll only is an additional mechanism that optimizes some biochemical processes, as does the hemoglobin in human body, as it is a molecule that also has the amazing ability to dissociate the water molecule irreversibly, and was to be expected, because the biochemical differences between chlorophyll and hemoglobin are really minimal.

Figure 3. The same principle applies in all living organisms, apparently is the only or at least the best system to convert light energy into chemical energy. In Sunflower seed we can see melanin. The magnification is 25×.

Because of the huge differences between melanin and chlorophyll, it is that, based on this last, it has not been possible to explain the metabolism of plants. To date, the questions are many, despite the best efforts of researchers and Governments that support research.

3. BRIEF HISTORY OF THE DISCOVERY

Since melanin absorbs all kinds of energy and dissipates it separating the water molecule, their behavior is unique in nature, because it does not emit any electromagnetic, visible or invisible signal that will allow us to characterize it. As a result, the study of melanin in the laboratory has been the date; a formidable challenge [4] . So it is understandable that the finding of its intrinsic property of transform light energy into chemical energy, by means of the water molecule dissociation, came from another type of study, in this case, of a clinical trial; because we observed incidentally melanin acting in one of its multiple location that is in nature. Why is that we consider important to a brief description of the how it came to knowledge.

In 1990, we started an observational, descriptive, study about the three main causes of blindness in Mexico, and that is no coincidence that is the same around the world: Glaucoma, diabetic retinopathy, and macular degeneration related to age.

Our working hypothesis was trying to characterize the morphology of the blood vessels of the optic nerve, very small structure in humans, whose average diameter is equivalent to twelve human hair together. Our aim was to try to find vascular anatomical changes that could eventually serve as indicators of early disease and therefore allowed us early treatments.

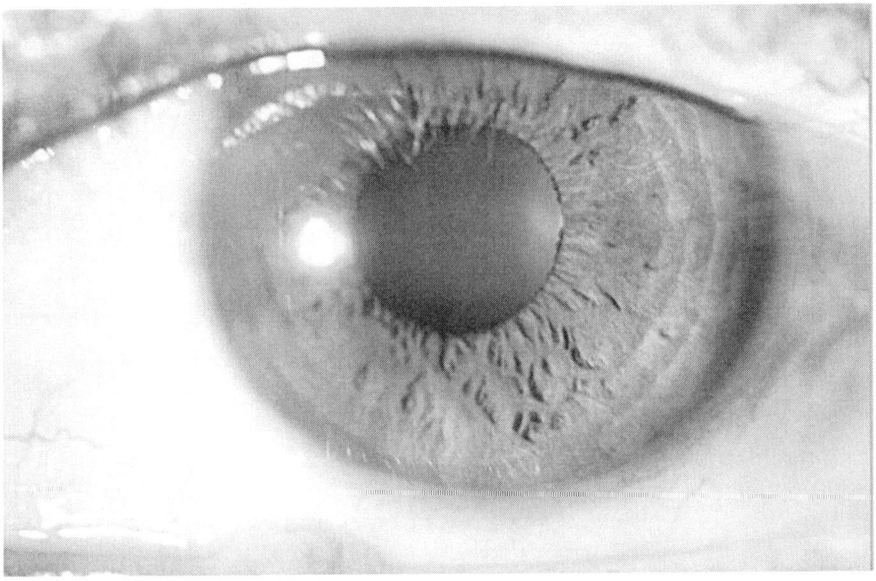

Figure 4. Melanin, wherever you are or whatever that he is called, has the same function, namely the transformation of light, visible and invisible power; in energy chemistry through dissociation and further re-formating of the water molecule. The effect of melanin on the nearest atoms and molecules is surprisingly consistent, as evidenced the huge resemblance between the surface of the iris of the eye and the surface of some seeds, bark of trees, etc.

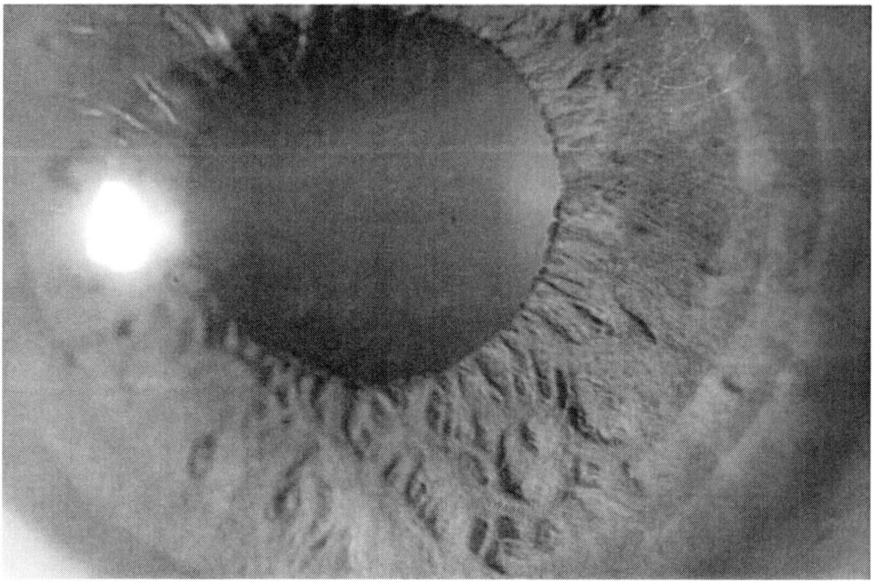

Figure 5. To greater amplification, the resemblance is even greater between the surface of the iris and some seeds, including with the bark of some trees. Is conceivable that at higher magnifications, come a time that would be difficult to discern which is which.

However, the magnifications required to carry out properly the characterization of vascular anatomical pattern, are in the order of 20× to 40×. And to these amplifications, we started to notice the omnipresence of the melanin in the vicinity of the optic disc (Figure 6).

And the insistence of nature in place melanin in the entirety of the almost 6000 patients who checked in the span of 12 years which lasted the study (1990-2002), struck us mightily, because nature just insists on important things.

So a few months of starting the study, instead of one, there were two main variables in study: blood vessels and melanin.

We were sure that melanin should play an important biological role, but we didn't understand that. Eventually we began to detect an antagonism between the blood vessels and the melanin that was becoming more and more evident. Some time later, we had a conclusion: more melanin, fewer blood vessels and vice versa (Figure 7).

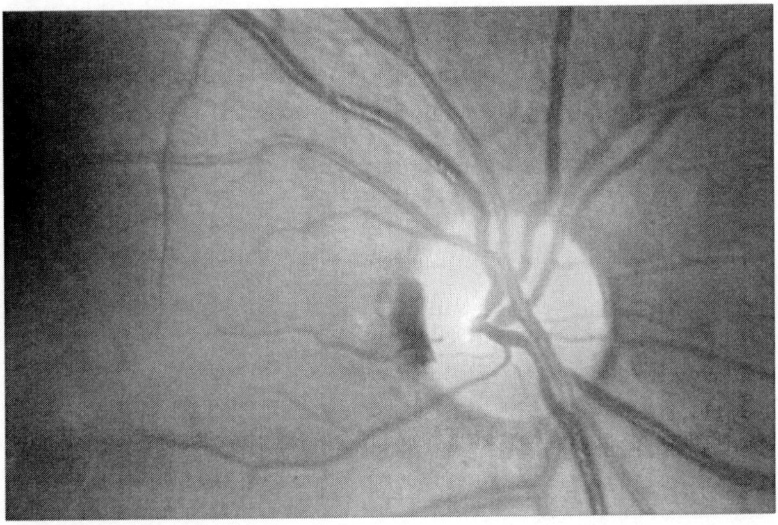

Figure 6. The picture shows the optic nerve of the right eye, with its characteristic orange colour, its oval shape, vessels entering and leaving; and in the meridian of the 9 there is a patch of dark color which corresponds to melanin.

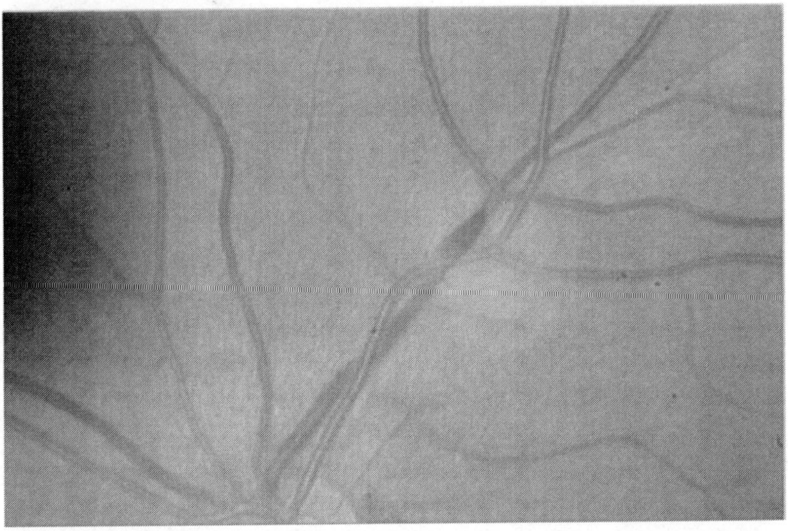

Figure 7. PICTURED are the retinal vessels, and a dark spot that surrounds them partially and which corresponds to the melanin. The effect on the vessels is somewhat subtle, because in this case only observed a light pale around the melanin. In people with fair skin angiogenesis inhibitor effect of melanin is still more subtle, and in dark-skinned people is almost impossible to detect by the amount of melanin.

What remained was to try to explain it. In principle, we did not know which predominated on which. On the one hand, melanin does not possess something like organelles that could produce some kind of transmitters, e.g. peptides, neither nor has something that could be interpreted as receivers. In relation to the blood vessels, these have rough endoplasmic reticulum, so it can synthesize factors that function as transmitters of information, but melanin has no receptors.

And something that was surprising was that the melanin antiangiogenic effect is very powerful, because it is present all the time, which was at odds with some kind of factor, for example: Pigment epithelium derived factor or Vascular endothelial growth factor; because the biological variability in the same factor and the corresponding receiver significantly reduces its effect, perhaps in 60%.

And the effect of melanin is surprisingly consistent, then it had to be something different from the known factors. By then, we realized that molecular oxygen levels were different between tissues, according to the amount of pigment. A greater amount of melanin, higher levels of molecular oxygen and vice versa. So the oxygen was the answer, because it is the best anti-angiogenic agent referred to.

We have been solving the unknowns, but now appeared a new: where does oxygen came from? The difference in oxygen levels between the pigmented tissues and the less pigmented, in our laboratory conditions, in the human eye, it goes from 34% to 54%, a huge difference.

We review several molecules, various tissues, looking for changes consistent with a constant oxygen donation, and did not find anything, it seemed that not passing anything, but a molecule or tissue that had to submit so much oxygen, , sooner or later, will has any perceptible change, and nothing.

Finally, we had only one option: the water containing the eye and that is something more than 95% of the body. And it was the perfect candidate, because in water is hardly notice the changes. But the water not releasing oxygen for free, have to pay it, i.e. energy is required.

So that left us with an unknown: where does the energy so the water yield their oxygen atoms? And after several unsuccessful attempts, the answer appeared in front of a Blackboard full of cluttered biochemical equations, was February 2002; "He is dissociating water" I thought, and to write the equation on the Blackboard, suddenly appeared the order in the rest of the equations.

$$2H_2O \rightarrow 2H_2 + O_2$$

She was the equation written, but as we did laboratory tests, we found that apart from the elevated oxygen levels, also had flow of electrons; which was easy to explain because the reaction is reversible in appropriate conditions. So, eventually we perfect the reaction in the following manner:

$$2H_2O \rightarrow 2H_2 + O_2 \rightarrow 2H_2O + 4e^-$$

Melanin releases energy in the form of symmetrical in all directions. Something like increased energy spheres (Figure 8).

The position of the granules of melanin into the cell eucarionte is strategic since it is located mainly in the perinuclear space, allowing him to be the source of energy of the nucleus, which does not possess neither mitochondria nor ATP; and at the same time it floods the cell cytoplasm with these growing spheres of energy, which are distributed throughout the entire cell, following the laws of simple diffusion (Figure 9).

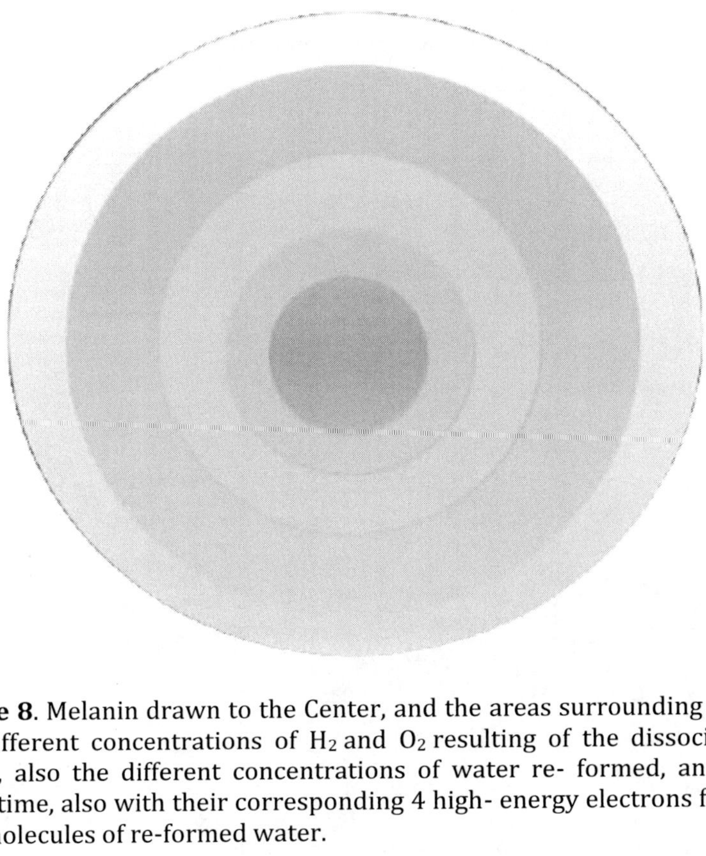

Figure 8. Melanin drawn to the Center, and the areas surrounding it, mean the different concentrations of H_2 and O_2 resulting of the dissociation of water, also the different concentrations of water re- formed, and at the same time, also with their corresponding 4 high- energy electrons for every two molecules of re-formed water.

Figure 9. The scheme represents the cell membrane, mitochondria, Golgi apparatus, rough endoplasmic reticulum, which, like the granules of melanin (melanosomes) surrounds completely to the cell nucleus.

Figure 10. Melanin is also present in the common moth, and its main function is also transform light energy into chemical energy through dissociation and re-forming of the water molecule.

The process of transformation of light energy into chemical energy seems to be the same in all living organisms, including insects (Figure 10).

4. MELANIN AND PLANTS

The downward movement of sugary plant fluids in phloem requieres energy, indeed, and melanin is the explanation. Mistery about how plants move fluids long distances in their phloem is easiest of understand taking in account the energy that comes from melanin.

That sucrose is either used as an energy source for its own metabolism is an idea very difficult to explain. So far, there are only hypothesis, in example, hypothesis of how water rises up in the xylem to the tops of tall trees, how sucrose and a proton (H*) are pumped into the cell; how water comes and build a pressure, how the sap can move a very significant distance; how sucrose is pumped out, how water is pumped out, how sucrose can be splitted to its monomers and thence to other hexoses.

Let us remember that any chemical reaction involves exchange of energy. The metabolism of plants as animals is poorly understood, since the power source was not known. Glucose is a special molecule that organisms use as a source of biomass as a source of carbon chains, but even there, and both animals and plants take the necessary energy from water. Energy can be defined as everything that produces a change, and the word metabolism means continuous change.

The metabolic needs of the plant are continuous, incessant, both day and night; and it is the way in which melanin delivered energy, in the form of incessant, continuous, both day and night. Therefore, the mysteries about life in the plants will have to look in the trunk, not so much in the leaves (Figure 11).

5. CONCLUSIONS

Both plants and humans are enveloped in a dense cloud of mysteries, at least in regards to your metabolism. It is no coincidence that of the 7000 cellular reactions described in the literature, only about 200 are described in the same way in the different sources of information, there are significant disputes in the remaining 6800. And it is that cell biology has puzzled trying to concatenate the glucose as a source of energy and at the same time of carbon chains; which is to say that a wonderful molecule not only brings

Figure 11. The chlorophyll in plants, as well as the hemoglobin in humans, possess the ability to dissociate the water molecule in an irreversible manner, using the light energy it contains. But both molecules are only an extension of the master molecule that is melanin [5] .

carbon atoms to organisms but also carries the necessary energy for its own metabolism.

This wrong concept is deeply rooted in both the scientist and public in general. But glucose just provides the building-blocks of the 99% of the biomolecules that mold our organism. Thereby the body knows glucose very well, so is able to split it, to combine it with other elements or molecules, and in a very precise way; even more, the cell is able to metabolize glucose to CO_2 and water; but the energy that supposedly glucose contains is a myth. Therefore, that biology at the molecular level has been stagnant. It is not possible to extract more energy from glucose containing that.

However, if from now forward we take into account the unexpected role of melanin as an important source bioenergetics, the role of the glucose will refine and allows us to re-order our concepts in a manner more consistent with nature, where everything depends on the generation and distribution of energy.

The discovery that the main source of energy comes from lignin [6] and chlorophyll, represents a substantial change in the study of photosynthesis in plants. We will have to reconsider many metabolic pathways so far accepted dogmatically but remain without plausible explanation, whereas to date theories. A significant advance in the knowledge will be able to be here, and it is a sharp change of direction to redirect the attention toward the trunk of the plants, rather than the leaves; but this will allow us to better understand the mysteries of the physiology of plants.

CITE THIS PAPER

Arturo Solís Herrera (2015) The Biological Pigments in Plants Physiology. *Agricultural Sciences*,**06**,1262-1271.

REFERENCES

1. Hoffman, J. and Puszynski, A. (2009) Chemical Engineeering and Chemical Process Technology. Vol. V. Pigments and Dyestuffs. Encyclopedia of Life Support Systems (EOLSS).
2. Solis Herrera, A., Arias Esparza, M.C., Solís Arias, R.I., et al. (2010) The Unexpected Capacity of Melanin to Dissociate the Water Molecule Fill the Gap between the Life before and after ATP. Biomedical Research, 21, 224-227.
3. Anderson, A.N., Tobimatsu, Y., Ciesielki, P.N., Ximenes, E., Ralph, J., Donohoe, B.S., Ladisch, M. and Chapple, C. (2015) Manipulation of Guaiacil and Syringyl Monomer Byosynthesis in an Arabidopsis Cinnamyl Alcohol Dehydrogenase Mutant Results in Atypical Lignin Byosynthesis and Modified Cell Wall Structure. The Plant Cell, 27, 2195-2209.
4. Hill, H.Z. (1992) The Function of Melanin or Six Blind People Examine an Elephant. BioEssays, 14, 49-56.
5. Solís-Herrera, A. And Solís-Arias, M.P. (2013) The Odyssey of Atmospheric Oxygen in Their Futile Attempt to Reach the Interior of the Cell. International Journal of Health Research and Innovation, 1, 37-52.

6. Solís-Herrera, A. And Arias-Esparza, M.C. and Solís Arias, M.P. (2015) Photosynthesis in Humans, Chapter 12 of the Book: Photosynthesis, Functional Genomics, Physiological Processes and Environmental Issues. Edited by Nafees Khan. Nova Publishers, New York, 257-279.

CHAPTER 2

Biochemical And Physiological Constituents and Their Correlation in Wheat (Triticum Aestivum L.) Genotypes Under High Temperature At Different Development Stages

H. R. Ramani[1*], M. K. Mandavia[1], R. A. Dave[1], R. P. Bambharolia[2], H. Silungwe[3] and N. H. Garaniya[1]

[1]Department of Biochemistry, Junagadh Agricultural University, Junagadh-362001, Gujarat, India.

[2]College of Agriculture, Navsari Agricultural University, Waghai-394730, Gujarat, India.

[3]Department of Food science and Technology, University of Venda, Limpopo province, P/bag X5050, Thohoyandou 0950, South Africa

ABSTRACT

Wheat is the most popular and staple food for millions of people. It is severely affected by heat stress in many countries. Vegetative growth and reproductive phases in wheat differ in their sensitivity to temperature. Heat tolerant (GW-190) and Heat susceptible (J-2010-11) genotypes grown up to tillering and grain filling stages and Heat treatments (40°C and 45°C for 2h and 4h) were given using Heating House. After heat treatment samples were collected, the biochemical and physiological analysis such as Protein, Proline, Glycine betaine, Membrane stability, Relative water content, Germination percent, Seed vigour and Heat tolerant index were performed. Protein, proline and glycine betaine were found significantly highest 34.06 mg/g Fr. Wt., 13.70 mg% and 902.24 μg/F. wt respectively in heat tolerant genotype GW-190 at 45°C for 4 h at tillering stage. Membrane stability and relative water content were found significantly highest 56.83% and 86.15 % respectively in heat tolerant genotype GW-190 at tillering stage. Germination percent, Seed vigor and Heat tolerance index were found higher in control group of Heat tolerant GW-190 genotypes. Where, all the biochemical and physiological contents were found lower in heat susceptible J-2010-11 genotype. From the above results it was concluded that GW-190 was heat tolerant genotypes which is suitable for grown in area of high temperature and J-2010-11 was found heat susceptible.

Keywords: Relative water content, membrane stability, seed index, heat tolerant index, correlation.

INTRODUCTION

Wheat is one of the world's most popular crops. It is the staple food for millions of people and forms an important part of many people's daily diet (Curtis et al., 2002). The yield and quality of cereals are severely affected by heat stress in many countries. Wheat genotypes express a differential response to chronic heat as well as a heat shock (Kamal et al., 2010). Temperature requirements and temperature extremes varies widely for different cultivars of the same species, and among species. The reproductive phase of many crop species is relatively more sensitive to heat stress than the vegetative phase (Hall, 1992; Martiniello et al., 2011).

Vegetative growth and the reproductive phase in wheat, differ in their sensitivity to temperature (Almeselmani et al., 2011; Chakrabarti et al., 2011; Karmanenko et al., 2011; Khakwani et al., 2011; Al-Karaki, 2012; Hakim et al.,2012; Hossain et al., 2012; Hossain et al., 2012; Noorka and Teixeira, 2012; Noorka et al., 2013). The total area of wheat production affected by some or other form of heat stress is estimated to be 65 to 70 million ha. Of these, 7 million ha are grown under continual heat stress

(Reynolds, et al., 2001). High temperature stress is a major growth restraining factor for most crop plants. Long term or even a temporary exposure to high temperature can change different metabolic functions, thereby affecting various plant parts including leaves, flower buds and roots (Tsukaguchi et al., 2003; Iwaya-Inoue et al., 2004). Most of the changes are noticeable in the cellular membranes. The metabolic changes include changes in the activities or structures of enzymes and biochemical contents. Heat stress affects the enzymes from metabolic compartments such as mitochondria (Nash et al.,1982), cytoplasm (Laurie et al.,1990) or chloroplasts (Dionisio et al., 1999). Thus, high temperature is believed to produce an array of changes in plants. Under heat stress, biochemical and physiological constitutes changed in heat tolerant and heat susceptible genotypes.

MATERIALS AND METHODS

Heat tolerant (GW-190) and heat susceptible (J-2010-11) wheat genotypes were selected from fourteen genotypes for experiment (Ramani et al., 2015). One set was grown up to tillering stage (35 days) and divided into two groups, control and heat treatments. Other set was grown up to Grain filling stage (around 100 days) and also divided into two sets, control and heat treatments. Heat treatments (40°C and 45°C for 2 h and 4 h) were given using Heating House at tillering and grain filling stages and samples were collected under ice condition. After heat treatments the leaf samples were taken out and leaves were weighed and then transferred immediately to the respective extracting medium for biochemical and physiological analysis.

ESTIMATION OF TOTAL PROTEIN

True protein was estimated by the standard method of Lowry et al., 1951. The samples (0.1 ml) and standard BSA (0.5 - 3 ml from 0.1 mg/ml BSA stock) were taken in a series of test tubes and the volume was made up to 3.0 ml with distilled water. Then 5 ml solution-C was added, mixed well and incubated at room temperature for 10 mins. Then 0.5 ml Folin - Ciocalteau reagent was also added mixed immediately and incubated in dark for 30 mins. A reagent blank was prepared by taking only reagents and volume made up with distilled water. Light blue color was observed after incubation which measured at 660 nm in a spectrophotometer. The protein content was calculated by taking Bovine serum albumin as standard.

ESTIMATION OF FREE PROLINE

Proline was estimated by using acid ninhydrin method of Bates et al., 1973. The samples (0.3 ml) and standard proline (0.1 to 0.6 ml from 0.05 mg/ml proline stock) were taken in a series of test tubes and the volume was made up to 1.0 ml with distilled water. Then 2 ml gacial acetic acid and 2 ml acid ninhydrin reagent were added. Then tubes were kept in boiling water bath for 1 hour. The tubes were cooled in running water at room temperature. After that 4 ml toluene was added. The absorbance was recorded from toluene phase at 520 nm in spectrophotometer. The free proline was calculated by taking Proline as standard.

Estimation of glycine betaine

Glycine betaine was estimated by Grieve and Grantta, 1983, 0.5g leaf powder was shaken in 20 ml distilled water for 24h at 25 °C. The samples were filtered and supernatant was used for analysis. 0.5 ml of supernatant and equal volume of 2N H_2SO_4 were kept in ice water for 1 h. 0.2 ml of cold KI-I_2 was added and reactants were gently vortex. The tubes were kept at 0 to 4°C for 16 h and centrifuge at 10,000 rpm for 15 minute at 0°C. Supernatant were discarded. A periodical crystal were dissolved in 9 ml of 1, 2- dichromoethane. After 2h, the absorbance measured at 365nm using glycine betaine as standard.

ESTIMATION OF MEMBRANE STABILITY

0.2 g of leaves of uniform size from control and treated plants were taken in tubes containing 30 ml of distilled water. The treatment tubes were then covered with plastic wrap and incubated in a thermostatically controlled water bath for 15 min at 53°C. The control tubes were maintained at 25 °C for the same period. After the elevated temperature treatment, the treatment tubes were quickly cooled down to 25°C. Both tubes were then stored in an incubator overnight for 18 h at 10°C to allow the diffusion of electrolytes from the discs. The tubes were then brought to 25 °C, inverted several times to mix the contents and initial conductance was measured using an electrical conductivity meter. After completing the measurement, tubes were covered and kept at 95 °C for 10 min to completely kill the leaf tissue. Then tubes were cooled down to 25 °C, mixed thoroughly, and the final conductance was measured. Membrane stability was calculated as described in Martineau et al., 1979 method. Membrane Stability Index= (1-T1/T2) X 100, Where, T1 = Initial conductance of sample kept at 53°C, T2 = Final conductance of sample kept at 95°C.

ESTIMATION OF RELATIVE WATER CONTENT

0.2 g fresh leaves were weighed to record fresh weight (FW), followed by dipping half of their portion in Petri dish containing 30 ml distilled water for 12 h. The leaves were blotted to wipe off excess water, weighed to record fully turgid weight (TW), and Subjected to oven drying at 70°C for 8 h to record the dry weight (DW). Relative water content (RWC) was calculated as per formula and expressed as: RWC = [FW-DW] × 100/ [TW-DW] of standard method of Turner, 1986.

ESTIMATION OF GERMINATION PERCENT

Seed were surface sterilized with 0.1% HgCl solution for 2 min and then washed with distilled water. Twenty five seeds each were sown on Whatman filter paper No. 1 bed kept in petri dishes and watered with distilled water. These petri dishes were kept for germination at 22°C (Control), 40°C and 45°C in seed germinator for 7 days to germinate. The filter paper beds were moistened whenever necessary. The germination was recorded after 48 hours and after 7 days. The germination percentage and rate of germination was calculated as per formula given by Almaghrabi, (2012).

Germination Percentage (GP) = Ng/Nt × 100,

Where, Ng = Total number of germinated seeds, Nt = Total number of seeds evaluated.

ESTIMATION OF SEED VIGOR

Seeds were allowed to germinate for 7 days after 7 days, the seedling length (shoot and root length) was recorded. The seedling vigor was calculated as per the formula described by Buriro et al., 2011.

Seedling Vigor = Germination percentage × Seedling length at 7 days.

ESTIMATION OF HEAT TOLERANT INDEX

Heat tolerance index (HTI) was calculated as per the formula given by Fernandez (1992). Heat tolerance index (HTI) = $(G_{si} \times G_{pi})/Gp^2$, Where, G_{si} = germination of each genotype under stress condition, G_{pi} = Germination of each genotype under control condition, Gp^2= Average of germination in all genotypes under control condition.

Correlation of biochemical and physiological constitutes

Mean values of biochemical and physiological constitutes were used for correlation analysis using SPSS 16.0 version statistical software.

RESULTS AND DISCUSSION

Results shown in Table 1 to 3 revealed different levels of the biochemical and physiological constitutes and their correlation between heat tolerant and heat susceptible genotypes under high temperature stress at different development stages.

Table 1. Biochemical and physiological constitutes of heat tolerant and heat susceptible wheat genotypes at tillering and grain filling stages

Sr. No	Genotypes	Heat treatment (ċ)	Protein(mg/g.f.wt)				Proline (mg %)				Glycine betaine (µg/f.wt)			
			Tillering		Grain filling		Tillering		Grain filling		Tillering		Grain filling	
			Time (h)		Time (h)		Time (h)		Time (h)		Time (h)		Time (h)	
			2	4	2	4	2	4	2	4	2	4	2	4
		Control	19.49	19.57	11.86	11.66	1.88	1.93	1.59	1.61	185.55	185.93	156.35	156.12
1	GW-190	40	24.91	27.07	13.74	18.44	3.58	6.17	3.07	5.22	335.49	644.93	267.09	449.49
		45	29.01	34.06	19.65	20.84	11.07	13.70	8.32	11.00	439.72	902.24	342.01	771.96
		Control	17.88	18.06	11.23	11.66	1.66	1.73	1.30	1.38	150.22	149.83	117.32	117.26
2	J-2010-11	40	20.77	25.68	12.60	13.31	2.86	4.92	2.55	4.52	211.72	556.98	188.92	361.55
		45	28.03	29.05	15.14	16.92	8.59	10.19	7.24	9.29	315.95	765.44	227.23	635.16
CD@ 5%			0.78		N.S.		N.S.		0.16		N.S.		N.S.	
CV %			1.91		1.90		0.98		0.21		5.15		6.45	

Sr. No	Genotypes	Heat treatment (ċ)	Membrane stability (%)				Relative water content (%)			
			Tillering		Grain filling		Tillering		Grain filling	
			Time (h)		Time (h)		Time (h)		Time (h)	
			2	4	2	4	2	4	2	4
		Control	56.83	55.65	53.49	53.23	86.15	86.07	56.53	56.42
1	GW-190	40	52.68	48.84	43.29	37.26	80.01	74.79	46.82	44.61
		45	43.55	31.58	35.80	33.04	67.37	58.41	37.51	29.29
		Control	51.23	51.37	45.29	46.72	79.19	79.37	51.73	51.67
2	J-2010-11	40	42.53	41.22	35.26	31.43	66.79	62.91	45.00	31.24
		45	37.92	37.32	36.96	35.33	53.45	46.14	28.21	23.45
CD@ 5%			N.S		N.S.		N.S		3.31	
CV %			3.95		4.6		2.21		4.69	

*table show the mean values of three replication, N.S. = none significant.

Table 2. Germination per cent, Seed vigor and heat tolerant index of heat tolerant and heat susceptible wheat genotypes under high temperature.

Sr. No	Genotypes	Heat treatment (ċ)	Germination (%)	Seed vigor	Heat tolerant index
		Control	100	1945	1.00
1	GW-190	40	92.67	1190	0.97
		45	90.67	980	0.89
		Control	100	1871	1.00
2	J-2010-11	40	70.33	820	0.87
		45	62.67	658	0.67

PROTEIN CONTENT

Protein was found significantly highest 34.06 mg/g in heat tolerant genotype GW-190 at 45°C for 4 h at tillering stage. The heat susceptible genotype J-2010-11 recorded the lowest (11.23 mg/g. wt.) at grain filling stage. When plants were kept for 2 h, protein content was found lower in both genotypes at both stages as compared to 4 h of heat treatment. All the control plants of heat tolerant GW-190 and susceptible genotypes J-2010-11 showed lower protein as compared to treated plants at both stages. Compared to grain filling stage the protein content was found higher at tillering stage in heat tolerant and heat susceptible genotypes. As the temperature and duration of temperature treatment increased, the protein was increased in both genotypes at tillering and grain filling stages (Table 1). Plants cope with heat stress in a complex manner, where heat shock proteins (HSPs) which might play a central role in the complex cellular network (Baniwal et al., 2004).

Differential expression of carbohydrate biosynthesis and metabolic pathway enzymes was also inferred in plants that need high energy to cope with heat stress. The responses of rice seedlings to different high-temperature stresses at 35, 40 and 45°C for 48 h. At 35°C, some protective mechanisms were activated to maintain the photosynthetic capability. At 40°C, antioxidative pathways were also active so, the overall protein content increased with temperature (Han et al., 2009). When seven day old rice seedlings encountered high-temperature stress at 45°C, in addition to those induced at 35°C and 40°C, heat shock proteins were effectively induced so the protein content also increased as temperature increased. The appearance of new proteins in wheat shoots subjected to the high temperature stress are heat shock proteins of molecular weights 111, 90, 70, 45, 32, 24 and 8 KDa and they were accumulated under high temperature(Khalid and Devaraj, 2011).

PROLINE CONTENT

Proline was found significantly highest 13.70 mg% in heat tolerant genotype GW-190 at 45°C for 4 h at tillering stage. The heat susceptible genotype J-2010-11 recorded the lowest (1.30 mg%) at grain filling stage. When plants were kept for 2 h, proline content was found lower in both genotypes at both stages as compared to 4 h of heat treatment. All the control plants of heat tolerant GW-190 and susceptible genotypes J-2010-11 showed lower proline as compared to treated plants at both stages. Compared to grain filling stage, the proline content was found higher at tillering stage in heat tolerant and heat susceptible genotypes. As the temperature and duration of temperature treatment increased the proline content was also increased in both genotypes at tillering and grain filling stages (Table 1). Accumulation of certain low molecular mass organic compounds, generally called compatible solutes or osmoprotectants, is a key adaptive mechanism in many plants grown under abiotic stress including extreme temperature (Hare, et al., 1998). A variety of osmolytes may accumulate in different plant species under stress. These include sugars and sugar alcohols (polyols), proline, tertiary and quaternary ammonium compounds and tertiary sulphonium compounds (Sairam and Tyagi, 2004). At high temperature (35°C) the heat tolerant genotypes produced more than double (> 200%) proline than that of 25°C but the heat susceptible genotypes produced less quantity of proline at 35°C when compared to that in heat susceptible genotypes at 25°C (Ahmed and Hasan, 2011). Proline content was remarkably declined at different stages of growth with lowest accumulation at seed hardening stage (Kumar et al., 2012). Compared to heat sensitive genotypes, proline content were higher in tolerant genotypes under late sown conditions (Dhyani et al., 2013).

GLYCINE BETAINE

Glycine betaine was found significantly highest 902.24 µg/F. wt in heat tolerant genotype GW-190 at 45°C for 4 h at tillering stage. The heat susceptible genotype J-2010-11 recorded the lowest (117.32 µg/F. wt) at grain filling stage. When plants were kept for 2 h glycine betaine was found lower in both genotypes at both stages as compared to 4 h of heat treatment. All the control plants of heat tolerant GW-190 and susceptible genotypes J-2010-11 showed lower glycine betaine as compared to treated plants at both stages. Compared to grain filling stage the glycine betaine content was found higher at tillering stage in heat tolerant and heat susceptible genotypes. As the temperature and duration of temperature treatment increased the glycine betaine content was also increased in both genotypes at tillering and grain filling stages (Table 1). Glycine betaine is an amphoteric quaternary amine that plays an important role as osmoprotectant in wheat plants and accumulated under a range of abiotic stresses including high temperature. Stress, caused a significant increase in Glycine betaine content of wheat genotypes at tillering and anthesis stages. The responses of wheat genotypes to drought, heat stress condition and analyzed mechanism of glycine betaine were involved in an improvement of wheat tolerance (Wang et al., 2010).

MEMBRANE STABILITY

Membrane stability was found significantly highest 56.83% in heat tolerant genotype GW-190 at tillering stage. The heat susceptible genotype J-2010-11 recorded the lowest (31.43%) at grain filling stage. When plants were kept for 2 h, membrane stability was found higher in both genotypes at both stages as compared to 4 h of heat treatment. All the control plants of heat tolerant GW-190 and susceptible genotypes J-2010-11 showed higher membrane stability when compared to treated plants at both stage. Compared to grain filling stage the membrane stability was found higher at tillering stage in heat tolerant and heat susceptible genotypes. As the temperature and duration of temperature treatment increased the membrane stability was also decreased in both genotypes at tillering and grain filling stages (Table 1). Based on membrane thermostability (MT) test, varieties took maximum heat killing time and were classified as heat tolerant, three varieties are moderately tolerant and the rest took the shortest heat killing time and are considered as heat sensitive (Skider, et al., 2001). Thermo tolerant wheat genotypes showed higher Membrane stability index when compared to thermo sensitive wheat genotypes at different development stages (Mohammadi, et al., 2007). The Membrane thermostabilty was significantly affected by different growth stages. Membrane stability of wheat genotypes decreased during the later developmental stages. Membrane stability of flag leaf at the early milk stage was significantly correlated with grain yield (Yildirim et al., 2009).

RELATIVE WATER CONTENT

Relative water content was found significantly highest 86.15% in heat tolerant genotype GW-190 at tillering stage. The heat susceptible genotype J-2010-11 records the lowest (23.45%) at grain filling stage. When plants were kept for 2 h, relative water content was found higher in both genotypes at both stages as compared to 4 h of heat treatment. All the control plants of heat tolerant GW-190 and susceptible genotypes J-2010-11 showed higher relative water content compared to treated plants at both stage. Compared to grain filling stage the relative water content was found higher at tillering stage in heat tolerant and heat susceptible genotypes. As the temperature and duration of temperature treatment increased the relative water content was also decreased in both genotypes at tillering and grain filling stages (Table 1). The rise in temperature during late sowing significantly decreased leaf relative content at 8 and 23 days after anthesis in wheat (Sairam et al., 2000). This is in accordance with present results. Wheat genotypes had lower relative water content values at early sowing and late sowing date compared to normal sowing date (Bhesaniya, 2005). Relative water content in root was not affected by higher temperature at all stages of wheat seedling. While coleoptiles relative water content decreased at all stages of seedling development (Savicka and Skute, 2012).

GERMINATION

The mean germination percentage recorded significantly higher 100% germination in control of GW-190 followed by (92.67%) and (90.67%) at 40°C and 45°C respectively while significantly lowest value was observed with genotype J-2010-11 (62.67%) at 45°C (Table 2). As the temperature increased the germination of seeds decreased in both genotypes but the ratio of decreased in heat susceptible genotype J-2010-11 was more. Germination characters increased with increasing temperature from 15 to 35°C in all wheat genotypes but heat tolerant ability was not based on speed of germination (Hasan et al., 2004). Germination characteristics in heat tolerant cultivars increased with increasing temperature as compared to heat sensitive cultivars (Sikder and Paul, 2010). However, high temperature treatment at 30 to 40°C, germination was reduced in rice (Prasanth et al., 2012) and at 27°C and 30°C temperature treatments, germination was reduced in barley (Ghazi, et al., 2007).

SEED VIGOR

The mean seed vigor was recorded significantly higher (1945) in control of GW-190 followed by (1190) and (980) at 40°C and 45°C respectively while significantly lowest value was observed with genotype J-2010-11 (658) at 45°C. As the temperature increased the seed vigor decreased in both genotypes but the ratio of decreased in heat susceptible genotype J-2010-11 was more (Table 2). Seedling vigor index was increased with increasing temperature (10 to 30°C) and maximum was recorded at 20°C and 30°C (Buriro et al., 2011). The seedling vigor is adversely affected by heat stress (36°C) which resulted in reduced shoot and root length (Grass and Burris, 1995).

Generally shoot and root growth of heat susceptible genotype in relation to length is affected more than thoseof heat tolerant genotype at higher temperature-35°C (Hasan et al., 2004).

Table 3. Pearson correlations between biochemical and physiological constitutes in heat tolerant and susceptible wheat genotypes under high temperature stress.

	Protein	Proline	Glycine betaine	Membrane stability	Relative water content	Germination percent	Seed vigor	Heat tolerance index
Protein	1.000							
Proline	0.960**	1.000						
Glycine betaine	0.994**	0.950**	1.000					
Membrane stability	-0.750*	-0.808*	-0.803*	1.000				
Relative water content	-0.721	-0.818*	-0.768*	0.962**	1.000			
Germination percent	-0.368	-0.452	-0.454	0.801*	0.861*	1.000		
Seed vigor	-0.735	-0.732*	-0.804*	0.944**	0.920**	0.868*	1.000	
Heat tolerance index	-0.531	-0.661	-0.580	0.811*	0.934**	0.913**	0.817*	1.000

*correlation is significant at 0.05, ** correlation is significant at 0.01 (1-tailed).

HEAT TOLERANT INDEX

The mean Heat tolerant index was recorded significantly higher (1.00) in control of GW-190 followed by (0.97) and (0.89) at 40°C and 45°C respectively while significantly lowest value was observed with genotype J-2010-11 (0.67) at 45°C. As the temperature increased the seed vigor decreased in both genotypes but the ratio of decreased in heat susceptible genotype J-2010-11 was more (Table 2). Stress tolerance index (STI) and stress susceptible index (SSI) were more accurate criteria for the selection of heat tolerant and high yielding genotypes by growing under normal and stress condition (Khodarahmpouret et al., 2011).

CORRELATION ANALYSIS

One heat tolerant and one heat susceptible genotypes GW-190 and J-2010-11 respectively were used for correlation analysis (Pearson correlation test) including biochemical and physiological parameters as listed earlier (Table 3). Results of correlation analysis under heat stress condition showed that, protein had highly significant and positive correlation with proline and glycine betaine content whereas negatively and significant correlation with membrane stability. Also protein content was negatively and non- significantly correlated with relative water content, germination percentage, seed vigor and heat tolerant index. Proline content had high significant positive correlation with glycine betaine and membrane stability, while relative water content and seed vigor were negatively and significantly correlated whereas, germination percent and heat tolerant index were negatively and non-significantly correlated. Glycine betaine had significant negative correlation with membrane stability, relative water content and seed vigor, Membrane stability had highly significant positive correlation with relative water content and seed vigor, and positive significant correlation with germination and heat tolerant index. Relative water content was positively and significantly correlated with germination percent, seed vigor and heat tolerant index. Seed vigor was positively correlated with heat tolerant index (Table 3). Result clearly indicated that relative water content and membrane stability was positively correlated and more effective indicator for screening of heat tolerant genotypes. Protein content, proline and glycine betaine content were positively correlated, they were higher in heat tolerant genotypes. Membrane stability index was good indicators of stress tolerance (Sairam and Srivastava, 2001).

CONCLUSION

Some abiotic constraints such as high tempe-rature, cold, drought, high concentrations of toxic minerals and salinity results in, severe loss of crop yield and quality. At both tillering and grain filling stages, Protein, proline, glycine betaine content were higher in tolerant genotype as compared to susceptible genotypes. Also these content were increased with increase in temperature in both genotypes. But the amount of increase was higher in tolerant genotype when compared to susceptible genotypes. Membrane stability, relative water content, heat tolerant index, seed vigor and germination were higher in tolerant genotypes when compared to susceptible genotypes but as temperature increased, they were decreased in both the genotypes at both stages. Also biochemical and physiological content was significantly co-related.

CONFLICT OF INTERESTS

The authors have not declared any conflict of interest.

REFERENCES

1. Ahmed JU, Hasan MA (2011). Evaluation of seedling proline content of wheat genotypes in relation to heat tolerance. Bangladesh J. Bot. 40(1):17-22.
2. Al-Karaki GN (2012). Phenological development-yield relationships in durum wheat cultivars under terminal high temperature stress in semiarid conditions. Agron. pp 7.
3. Almeselmani M, Teixeira da Silva JA, Deshmukh P (2011). Stability of different physiological characters, yield and yield components under high temperature stress in tolerant and susceptible wheat genotypes. Fruit Veg. Cereal Sci. Biotechnol. 5:86-92.
4. Baniwal SK, Bharti K, Chan KY, Fauth M, Ganguli A, Kotak S, Mishra SK, Nover L, Port M, Scharf KD, Tripp J, Weber C, Zielinski D, Von Koskull-Doring P (2004). Heat stress response in plants: a complex game with chaperones and more than twenty heat stress transcription factors. J. Biol. Sci. 29:471-487.
5. Bates LS, Waldren RP, Teare ID (1973). Rapid determination of free proline for water stress studies. Plant Soil 39:205-207.
6. Bhesaniya SV (2005). Effect of high temperature on biochemical changes in different genotypes of wheat (Triticum aestivum L.). M. Sc. Thesis (Unpublished). Junagadh Agricultural University, Junagadh. Gujarat, India.
7. Buriro M, Oad FC, Kerio MI, Tunio S, Gandahi AW, Hassan SWU, Oad SM (2011). Wheat seed germination under the influence of temperature regimes. Sarhad J. Agric. 27(4):539-543.
8. Chakrabarti B, Singh SD, Nagarajan S, Aggarwal PK (2011). Impact of temperature on phenology and pollen sterility of wheat varieties. Aust. J. Crop Sci. 5(8):1039-1043.
9. Curtis BC, Rajaram S, Macpherson HG (2002). Bread wheat: FAO. Plant Production and Protection Series 30:12-15.

10. Dhyani K Ansari MW, Rao YR, Verma RS, Shukla A, Tuteja N (2013). Comparative physiological response of wheat genotypes under terminal heat stress. Plant Signal. Behav. 8(6):1-6.
11. Dionisio ML, Shono M, Tobita S (1999). Effect of prolein and betain on heat inactivation of ribulose 1,5- bisphosphate carboxylase/oxygenase in crude extract of rice seedling. Photosynthetica 36(4):557-563.
12. Fernandez GCJ (1992). Effective selection criteria for assessing stress tolerance. In: Kuo, C. G. eds. Proceedings of the International Symposium on Adaptation of Vegetables and Other Food Crops in Temperature and Water Stress. Publication, Tainan, Taiwan.
13. Ghazi NA, Asadulla A, Yahia O (2007). Seed germination and early growth of three barley cultivars as affected by temperature and water stress. Am. Euro. J. Agric. Environ. Sci. 2(2):112-117.
14. Grass L, Burris JS (1995). Effect of heat stress during seed development and maturation on wheat (Triticum durum L.) seed quality. I. Seed germination and seedling vigor. Can. J. Plant Sci. 75:821-829.
15. Grieve CM, Grattan SR (1983). Rapid assay for determination of water soluble quartnary ammonium compounds. Plant Soil 70:303-307.
16. Hakim MA, Hossain A, Teixeira da Silva JA, Zvolinsky VP, Khan MM (2012). Yield, protein and starch content of 20 wheat (Triticum aestivum L.) genotypes exposed to high temperature under late sowing conditions. J. Sci. Res. 4(2):477-489.
17. Hall AE (1992). Breeding for heat tolerance. Plant Breed. Rev. 10:129-168.
18. Han F, Chen H, Li XJ, Yang MF, Liu GS, Shen SH (2009). A comparative proteomic analysis of rice seedling under various high temperature stresses. Biochem. Biophys. Acta 1794:1625-1634.
19. Hare PD, Cress WA, Staden JV (1998).Dissecting the roles of osmolyte accumulation during stress. Plant Cell Environ. 21:535-553.
20. Hasan MA, Ahmed JU, Hossain T, Hossain MM, Ullah MA (2004). Germination characteristics and seed reserve mobilization during germination of different wheat genotypes under variable temperature regimes. J. Nation Sci. Foundation Sri Lanka 32(3, 4):97-107.
21. Hossain A, Teixeira da Silva JA, Lozovskaya MV, Zvolinsky VP, Mukhortov I (2012). High temperature combined with drought affect rainfed spring wheat and barley in south-eastern Russia: Yield, relative performance and heat susceptibility index. J. Plant Breed. Crop Sci. 4(11):184-196.
22. Iwaya-Inoue M, Matsui R, Fukuyama M (2004). Cold- or heat-tolerance of leaves and roots in perennial ryegrass determined by 1H-NMR. Plant Prod. Sci. 7(2):118-128.
23. Kamal AHM, Kim KH, Shin KH, Choi JS, Baik BK, Tsujimoto H, Hwa Young Heo HY, Park CS, Woo SH (2010). Abiotic stress responsive proteins of wheat grain determined using proteomics technique. Aust. J. Crop Sci. 4(3):196-208.
24. Karmanenko NM, Osipova LV, Nilovskaya NT (2011). Acid, cold and drought tolerance in cereal. Russ. Acad. Sci. 37(5):354-357.
25. Khakwani AA, Dennett MD, Munir M (2011). Drought tolerance screening of wheat varieties by inducing water stress conditions. J. Sci. Technol. 33(2):135-142.
26. Khalid MN, Devaraj VR (2011). Antioxidant and other biochemical defense responses of Macrotyloma uniflorum (Lam.) Verdc. (Horse gram) induced by high temperature and salt stress. Braz. J. Plant Physiol. 23(3):187-195.
27. Khodarahmpouret Z, Choukan R, Bihamta MR, Hervan EM (2011). Determination of the best heat stress tolerance indices in maize (Zea mays L.) inbred lines and hybrids under khuzestan province conditions. J. Agric. Sci. Technol. 13:111-121.
28. Kumar RR, Goswami S, Sharma SK, Singh K, Gadpayle KA, Kumar N, Rai GK, Singh M, Rai RD (2012). Protection against heat stress in wheat involves change in cell membrane stability, antioxidant enzymes, osmolyte, H2O2 and transcript of heat shock protein. Int. J. Plant Physiol. Biochem. 4(4):83-91.

29. Laurie S, Stewart GR (1990). The Effects of Compatible Solutes on the Heat Stability of Glutamine Synthetase from Chickpeas Grown under Different Nitrogen and Temperature Regimes. J. Exp. Bot. 41(11):1415-1422.
30. Lowry OH, Rose rough AL, Farr AL, Randall RJ (1951). Protein measurement with the Folin Phenol reagent. J. Biol. Chem. 193(1):265-275.
31. Martineau PM, Williams JH, Speght JE (1979). Temperature tolerance in soybeans-II. Evaluation of segregating population for membrane thermo-stability. Crop Sci. 19:79-81.
32. Martiniello P, Teixeira da Silva JA (2011). Physiological and bio–agronomical aspects involved in growth and yield components of cultivated forage species in Mediterranean environments: A review. Eur. J. Plant Sci. Biotechnol. 5(2):64-98.
33. Mohammadi V, Bihamta MR, Zali AA (2007). Evaluation of screening techniques for Heat tolerance in wheat. Pak. J. Biol. Sci. 10(6):887-892.
34. Nash D, Paleg LG, Wiskich JJ (1982). Effect of Proline, Betaine and Some Other Solutes on the Heat Stability of Mitochondrial Enzymes. Aust. J. Plant Physiol. 9(1): 47-57.
35. Noorka IR, Teixeira da Silva JA (2012). Mechanistic insight of water stress-induced aggregation in wheat (Triticum aestivum L.) quality. Note Sci. Biol. 4(4):32-38.
36. Noorka IR, Batool A, Rauf S, Teixeira da Silva JA, Ashraf E (2013). Estimation of heterosis in wheat (Triticum aestivum L.) under contrasting water regimes. Int. J. Plant Breed. 7(1):55-60.
37. Prasanth VV, Chakravarthi DVNT, Kiran V, Rao YV, Panigrahy M, Mangrauthia SK, Viraktamath BC, Subrahmanyam D, Voleti SR, Sarla N (2012). Evaluation of rice germplasm and introgression lines for heat tolerance. Annals Biol. Res. 3(11):5060-5068.
38. Ramani HR, Mandavia MK, Khunt AG, Golakiya BA (2015). Evaluation of wheat genotypes (Triticum aestivum L.) against heat stress based on their biochemical and physiological responses. India J. Agric. Biochem. 27(2):133-137.
39. Reynolds MP, Nagarajan S, Razzaque MA, Ajeeb OOA (2001). Heat tolerance. In: M.P. Reynolds, J. I. Ortiz-Monasterio and A. McNab, Eds, Application of physiology in wheat breeding, CIMMYT, Mexico, pp. 124-135.
40. Sairam RK, Srivastava GC (2001). Water stress tolerance of wheat (Triticum aestivum L.): variations in hydrogen peroxide accumulation and antioxidant activity in tolerant and susceptible genotypes. J. Agron. Crop Sci. 186:63-70.
41. Sairam RK, Tyagi A (2004). Physiology and molecular biology of salinity stress tolerance in plants. Curr. Sci. 86(3):407-421.
42. Sairam RK, Srivastva GC, Saxena DC (2000). Increased antioxidant activity under elevated temperature: a mechanism of heat stress tolerance in wheat genotypes. Biologia-Plantarum (Czech Republic.) 43(2):245-251.
43. Savicka M, Skute N (2012). Some morphological, physiological and biochemical characteristics of wheat seedling Triticum aestivum L. organs after high-temperature treatment. Ekologija 58(1):9-21.
44. Sikder S, Paul NK (2010). Study of influence of temperature regimes on germination characteristics and seed reserves mobilization in wheat. Afr. J. Plant Sci. 4(10):401-408.
45. Skider S, Ahmed JU, Hossain T (2001). Heat Tolerance and Relative Yield Performance Of Wheat Varieties Under Late Seeded Conditions. Indian J. Agric. Res. 35(3):141-148.
46. Tsukaguchi TY, Kawamitsu H, Takeda K, Suzuki K, Egawa Y (2003). Water status of flower buds and leaves as affected by high temperature in heat-tolerant and heat-sensitive cultivars of snap bean (Phaseolus vulgaris L.). Plant Prod. Sci. 6:24-27.
47. Turner NC (1986). Crop water deficit: a decade of progress. Advance Agronomy, 39, Almaghrabi, O. A. (2012). Impact of drought stress on germination and seedling growth parameters of some wheat cultivars. Life Sci. J. 9(1):590-598.

48. Wang GP, Zhang XY, Li F, Luo Y, Wang W (2010). Over accumulation of glycine betaine enhances tolerance to drought and heat stress in wheat leaves in the protection at photosynthesis. Photosynthetica 48:117-126.
49. Yildirim M, Bahar B, Mujde KOC, Barutcular C (2009). Membrane Thermal Stability at Different Developmental Stages of Spring Wheat Genotypes and Their Diallel Cross Populations. J. Agric. Sci. 15(4):293-300.

CHAPTER 3

Growth Analysis and Yield of Two Varieties of Groundnut (*Arachis Hypogaea* L.) as Influenced by Different Weed Control Methods

Bolaji U. Olayinka[1] • Emmanuel O. Etejere[1]

Department of Plant Biology, P.M.B. 1515, University of Ilorin, Ilorin, Nigeria

ABSTRACT

Field trials were carried out to evaluate the effects of seven weed management strategies on the growth and yield of two groundnut varieties (Samnut 10 and MK 373) for two successive seasons (2010–2011). The experimental layout was a split plot complete randomized block design with three replications. The two groundnut varieties showed identical pattern of results for leaf area index, dry matter accumulation, relative growth rate, net assimilation rate and crop growth rate as well as yield. All the weed control treatments significantly enhanced the growth and yield compared with the weedy check. The weed free check had the highest growth but the highest yield was recorded from rice straw mulch at 0.1 m depth + one hand weeding at 6 weeks after sowing (WAS) due to increase in number of matured pods per plant, seed weight per plant and 100-seed weight. The results showed that rice straw mulch at 0.1 m depth + one hand weeding at 6 WAS was better agronomical practice for enhancing growth and yield of groundnut. This enhancement could be as a result of its positive influence on physiological parameters such as leaf area index, dry matter accumulation, relative growth rate, net assimilation rate and crop growth rate. Its use is also ecofriendly as it limits the need for synthetic herbicide.

Keywords: Leaf area index Net assimilation rate Relative growth rate Straw mulching Yield

INTRODUCTION

Groundnut is a major oil seed crop grown in tropics and sub-tropic parts of the world. Nigeria is the fourth largest producer of groundnut after China, India and USA (USDA 2009). In United States of America groundnut yield is as high as 3000 kg ha^{-1}, while the average yield in tropical Africa is about 800 kg ha^{-1}, which is traceable to weed infestation (Akobundu 1987). In Nigeria, groundnut yield loss could be as high as 51 % (Etejere et al. 2013). This is because in groundnut, less crop canopy during the first 6 weeks of growth favours strong competition with weeds causing significant reduction in yield (Akobundu 1987).

The degree of damage caused by the weeds has been found to be a function of their leaf area index (LAI) as compared to the crop they are competing with (El-Naim and Ahmed 2010). Vijay Kumar (1992) found that groundnut growth in the absence of weeds attained maximum LAI. LAI, therefore, is an important determinant of dry matter accumulation and grain yield, and it can be adjusted by various agronomical practices (Rahman et al. 1994).

Nakaseko et al. (1979) had observed that high crop productivity can be achieved by exploring the pattern of dry matter accumulation and partitioning, which helps in adjusting proper crop management practices. However, studies of dry matter accumulation and partitioning in relation to yield of groundnut under organic mulch treatment in comparison to hand weeding and chemical weed control are very limited. By exploring the pattern of dry matter partitioning in relation to LAI, the present investigation was carried out to compare and evaluate the effects of rice straw mulch either sole or integrated with one handing weeding with other weed control methods on physiological performance of two groundnut varieties.

MATERIALS AND METHODS

The field trials reported here was carried out at Lafiagi—a Guinea savanna zone of Nigeria (latitude 8°50′N and longitude 5°25′E) in 2010 and 2011. The soil of the experimental field was sandy-loam, with low organic matter (0.74 %), moderately high in available nitrogen (0.25 %), and slightly acidic pH (6.35). The total rainfall received during the cropping seasons at the site was less in 2010 (844.6 mm) than 2011 (1030.3 mm). The mean relative humidity and average temperature were 35.7 °C and 80.9 %; 32.4 °C and 84.1 % in 2010 and 2011, respectively. Predominant weed species were *Brachiaria deflexa, Cleome viscosa, Cochlospermum planchoni, Dactyloctenium aegyptium and Daniellia oliveri.*

Description of the varieties and field operation

Two varieties of groundnut Samnut 10 and MK 373 were used for the study. Samnut 10 is erect with small pod containing 1–2 seeds. It matures between 98–105 days. MK 373 is decumbent type. Pod is large with moderate constriction. Pod contains 2–4 seeds. It matures earlier than Samnut 10 (84–92 days). The seeds of the two varieties used in this study were obtained from College of Agriculture, Mokwa, Niger State. Before sowing seeds were treated with Seedrex (33 % permethrin + 15 % carbonderzine + 12 % chlorothalonil) at the rate of 10 g per 4 kg of seeds to prevent soil borne diseases.

Experimental layout and treatment details

The experiment was laid out in split plots design, with varieties in the main plot and the weed control treatments as the subplot. The various weed control treatments investigated were arranged to fit a complete randomized block design with three replications. Plot size was 3 m × 3 m consisting of four rows spaced 0.4 m apart with 0.1 m inter-crop spacing. Data on weeds and crop were collected within the two inner rows of 4.8 m^2 leaving the outer rows as buffer. The treatments consisted of seven weed control strategies: T_1: Pendimethalin at 1.5 l ha^{-1} as pre-emergence (water at 600 l ha^{-1} was used as carrier after proper calibration with the aid of Knapsack sprayer of 20 liters capacity) T_2: Pendimethalin at 1.5 l ha^{-1} as pre emergency + one hand weeding at 6 weeks after sowing (WAS), T_3: Rice straw mulch at 0.1 m depth (straw was laid in the inter-row) T_4: Rice straw mulch at 0.1 m depth + one hand weeding at 6 WAS, T_5: Two hand weeding at 3 and 6 WAS, T_6: Weed free check (continuous weeding was done at interval of 2 weeks) and T_7: Weedy check (plots were infested with weeds throughout the period of the experiment.

Data collection

Weed count was achieved by placing three quadrates of 1 m^2 at random in each subplot within the weed management zone of 4.8 m^2. Weeds rooted within this zone in each treatment were counted at harvest. The harvested weeds were oven dried at 80 °C to a constant weight to determine the weed biomass. The weed control efficiency and weed index were calculated following the formula of Gill and Kumar (1966).

Plant samples were harvested at interval of 2 weeks within the weed management zone. The above-ground portion of the harvested crop was first washed in running water and segmented into leaves and stems. Thereafter, these plant parts were oven dried at 80 °C to constant weight. The dry weight of different plant parts were then measured using Electronic 101 Balance with precision of 0.1 g. LAI was estimated as (LAI = surface area of sampled leaf/ground area occupied by the sampled plants). Physiological parameters such as relative growth rate (RGR), net assimilation rate (NAR) and crop growth rate (CGR) were estimated according to Gardner (1985). In each experimental plot, data on seed and pod yield were recorded on 10 randomly selected plants harvested within weed management zone of 4.8 m^2after drying to 12 % moisture. Harvest index and yield attributes such as number of matured pod per plant, seed weight per plant and 100-seed weight were also determined.

Data analysis

Data recorded in 2010 and 2011 cropping seasons were pooled together on account of non-significant interaction between years, treatment and variety. The data were then subjected to Univariate Analysis of Variance. Treatment means were separated using least significant difference (LSD) at 0.01 probability level.

RESULTS AND DISCUSSION

Weed control effect

Various weed control methods significantly ($p \leq 0.01$) influenced weed biomass, weed control efficiency and weed index (Table 1).Weed biomass was significantly higher in weedy check in Samnut 10 (662.1 g m^{-2}) and

MK 373 (686.26 g m^{-2}) as compared to other weed control treatments. Significantly lowest weed biomass was recorded in weed free check in both Samnut 10 and MK 373 with values of 68.51 and 71.71 kg ha^{-1}, respectively. In both varieties, sole pendimethalin at 1.5 l ha^{-1} and rice straw mulch had significantly higher weed biomasses as compared to rice straw mulch + one hand weed at 6 WAS, pendimethalin at 1.5 l ha^{-1} + one hand weeding at 6 WAS and two hand weeding at 3 and 6 WAS (Table 1). The results of weed control efficiency were significantly higher in weed free check (94–95 %), and followed an inverse relationship with weed biomass (Table 1). The weed index followed the same trend as weed biomass (Table 1). The results showed that all the weed control treatments were effective in reducing weed biomass compared to the weedy check. However, sole application of rice straw mulch and pendimethalin at 1.5 l ha^{-1} were not very effective in checking weed growth, as evident from the higher weed index and lower weed control efficiency as compared to other weed control treatments. Straw mulch is known to suppress weeds to some extent (Singh 2009). With regard to sole application of pendimethalin at 1.5 l ha^{-1}, the results corroborates the earlier studies of Olorunmaiye and Olorunmaiye (2009), who reported that pre-emergence herbicide treatments without supplementary hoe-weeding could not provide season long weed control because of their short persistence.

The reduction in weed biomass by rice straw mulch + one hand weeding at 6 WAS, two hand weeding at 3 and 6 WAS and pendimethalin at 1.5 l ha^{-1} + one hand weeding at 6 WAS was because these treatments prevented further fresh flush of weeds besides suppressing the already emerged ones. Evidence in support of this was observed in weed control efficiency, which was as high as 72–82 %, when either two hand weeding were carried out at 3 and 6 WAS or rice straw mulch or pendimethalin were integrated with onc hand weeding at 6 WAS compared to the sole pendimethalin and sole rice straw mulch (46–63 %). Similar results have been reported by Agasimani et al. (2010) and El-Naim et al. (2011), where integrated weed control methods or two hand weeding had been practiced in groundnut.

Table 1. Effect of different weed control methods on weed biomass, weed control efficiency, weed index and leaf area index in Samnut 10 and MK 373 (pooled data of cropping seasons 2010 and 2011)

Variety	Treatment	Weed biomass ($g\ m^{-2}$)	Weed control efficiency (%)	Weed index	Leaf area index (%) Week after sowing		
					8	10	12
Samnut 10	T_1	248.22^c	63.54^c	13.79^b	2.93^a	3.20^{ab}	2.60^{ab}
	T_2	128.34^d	77.75^b	-2.03^c	3.43^a	3.93^{ab}	2.90^{ab}
	T_3	335.79^b	46.92^d	18.58^b	3.20^a	3.30^{ab}	2.63^{ab}
	T_4	122.54^e	81.11^a	-20.47^e	3.97^a	4.63^a	3.73^{ab}
	T_5	132.26^d	82.58^a	-12.54^d	3.63^a	4.13^{ab}	3.60^{ab}
	T_6	32.88^f	94.94^a	0.000^e	3.83^a	4.93^a	3.93^a
	T_7	662.11^a	0000^e	47.03^a	2.47^a	2.63^b	1.93^b
	Mean	237.45	63.83	44.36	3.35	3.82	3.05
MK 373	T_1	258.16^c	61.09^c	11.30^{bc}	3.93^a	3.17^{bc}	2.63^a
	T_2	151.36^d	79.83^b	-18.17^{cd}	4.50^a	3.83^{ab}	3.53^a
	T_3	370.22^b	46.94^d	18.17^b	4.00^a	3.33^{abc}	2.87^a
	T_4	133.67^e	78.84^b	-14.10^d	4.63^a	4.10^a	3.83^a
	T_5	108.16^f	80.78^b	-6.78^{cd}	4.37^a	3.90^{ab}	3.63^a
	T_6	34.42^g	95.11^a	0000^{bcd}	4.70^a	4.10^a	3.97^a
	T_7	686.26^a	0000^e	40.53^a	3.43^a	2.87^c	2.47^a
	Mean	248.89	63.17	30.95	4.22	3.62	3.28
	Grand mean	243.17	63.50	37.66	3.79	3.72	3.17
LSD ($p < 0.01$)	Variety	36.85	11.98	99.43	0.94	0.74	4.00
	Treatment	16.32	25.80	10.74	0.63	0.95	0.56
	V × T	23.60	12.23	43.33	0.57	0.57	0.21

Within column means followed by the same superscripts are not significantly different at $p \leq 0.01$

T_1: Pendimethalin at 1.5 l ha^{-1}, T_2: Pendimethalin at 1.5 l ha^{-1} + one hand weeding at 6 WAS, T_3: Rice straw mulch, T_4: Rice straw mulch + one hand weeding at 6 WAS, T_5: Two hand weeding at 3 and 6 WAS, T_6: Weed free check (positive control), T_7: Weedy check (negative control)

Leaf area index (LAI)

In both varieties, LAI was significantly ($p \leq 0.01$) affected at different growth periods except at 8 WAS (Table 1). In case of Samnut 10, highest LAI was recorded at 10 WAS in all the treatments, whereas in MK 373 LAI was higher at 8 WAS (Table 1). During these periods, the weed free check showed significantly higher LAI values of 4.93 and 4.70 in Samnut 10 and MK 373, respectively (Table 2). These were followed by rice straw mulch + one hand weeding at 6 WAS with value of 4.63 in Samnut 10 and 4.65 in MK 373 (Table 1). During these growth periods, lowest LAI was recorded under the weedy check as compared to other weed control treatments. Treatments such as two hand weeding at 3 and 6 WAS and pendimethalin at 1.5 l ha^{-1} + one hand weeding at 6 WAS had LAI values that were statistically similar to those of rice straw mulch + one hand weeding either at 6, 8, 10 and 12 WAS. Sole pendimethalin and sole rice straw mulch showed lower LAI values as compared to other weed control treatments. As observed in this study, treatments with higher weed control efficiency (rice straw mulch + one hand weeding at 6 WAS, two hand weeding at 3 and 6 WAS and pendimethalin at 1.5 l ha^{-1} + one hand weeding at 6 WAS) had higher LAI as compared to those with low weed control efficiency (sole rice straw mulch and sole pendimethalin). Effective weed control has been observed to enhance LAI in groundnut (Kumar 2009). Singh (2003) observed that in Sourashtra region of India, LAI of bunch varieties of groundnut might be 1.7 at 60 days after planting (DAP), and might increased to 4.0 at 90 DAP. McCloud (1974) observed that at 64 DAP, LAI of groundnut might reach 3.0, which at maturity (137 DAP) might reduced to 1.7. In this present study, both varieties, reached highest LAI values of 4.63–4.93, which reduced to 1.93–2.47 at maturity. The reduction in LAI at 12 WAS in Samnut 10 and 10–12 WAS in MK 373 was due to pest attack (Banik et al. 2009).

Table 2. Effect of different weed control methods on above-ground dry weight and relative growth rate at different crop growth stages in Samnut 10 and MK 373 (pooled data of cropping seasons 2010 and 2011)

Variety	Treatment	Week after sowing					
		Above-ground dry weight (g)			Relative growth rate (g g^{-1} day^{-1})		
		8	10	12	6–8	8–10	10–12
Samnut 10	T_1	18.93^d	23.07^d	18.00^d	0.49^a	0.68^a	0.11^c
	T_2	22.50^c	31.77^c	25.27^c	0.51^a	0.83^a	0.28^b
	T_3	20.00^d	23.17^d	19.07^d	0.57^a	0.68^a	0.09^c
	T_4	29.17^b	36.43^b	28.93^a	0.83^a	0.89^a	0.11^c
	T_5	29.53^{ab}	36.17^b	27.33^b	0.72^a	0.81^a	0.13^c
	T_6	31.33^a	38.73^a	29.47^a	0.72^a	0.92^a	0.58^a
	T_7	9.77^e	15.77^c	12.27^e	0.46^a	0.57^a	0.12^c
	Mean	23.03	29.30	22.90	0.62	0.77	0.20
MK 373	T_1	22.27^c	31.33^e	31.47^c	0.67^{ab}	0.15^{ab}	0.01^b
	T_2	32.23^a	35.93^c	32.33^c	0.89^{ab}	0.04^b	0.13^a
	T_3	28.37^b	33.30^d	29.03^d	0.60^{ab}	0.06^{ab}	0.10^{ab}
	T_4	34.47^a	42.07^b	45.13^a	0.89^{ab}	0.10^{ab}	0.04^{ab}
	T_5	33.90^a	46.57^a	40.43^b	0.81^{ab}	0.15^{ab}	0.08^{ab}
	T_6	34.80^a	42.87^b	46.40^a	0.93^a	0.11^{ab}	0.04^{ab}
	T_7	14.53^d	22.40^f	20.47^e	0.49^b	0.18^a	0.01^b
	Mean	28.65	36.35	35.04	0.75	0.11	0.06
	Grand mean	25.84	32.83	28.97	0.69	0.44	0.04
LSD ($p \leq 0.01$)	Variety	5.56	4.30	4.00	0.13	0.06	0.01
	Treatment	1.44	0.93	1.02	0.50	0.06	0.03
	V × T	0.66	0.19	0.31	0.02	0.02	0.03

Within column means followed by the same superscripts are not significantly different at $p \leq 0.01$

T_1: Pendimethalin at 1.5 l ha^{-1}, T_2: Pendimethalin at 1.5 l ha^{-1} + one hand weeding at 6 WAS, T_3: Rice straw mulch, T_4: Rice straw mulch + one hand weeding at 6 WAS, T_5: Two hand weeding at 3 and 6 WAS, T_6: Weed free check (positive control), T_7: Weedy check (negative control)

Dry matter production

Dry matter accumulation at 8, 10 and 12 WAS in the two varieties differed significantly ($p \leq 0.01$) due to different weed control treatments (Table 2). Highest dry matter production was recorded in the weed free check

Among the weed control treatments, rice straw mulch + one hand weeding at 6 recorded highest dry matter accumulation, followed by two hand weeding at 3 and 6 WAS and Pendimethalin at 1.5 l ha^{-1} + one hand weeding at 6 WAS, sole rice straw mulch and pendimethalin at 1.5 l ha^{-1}. The significantly lower dry matter was recorded in weedy check over all other treatments. Variation in dry matter accumulation could be due to differences in the leaf area production and LAI. Similarly, reduction in weed competition for limited resources in treatments that had higher dry matter accumulation could have stimulated stronger carbohydrates sinks via photosynthesis. Highest dry matter accumulation in rice straw mulch + one hand weeding at 6 WAS could be linked to modification of growing environment for enhanced productivity. In this study, LAI showed positive correlation with the dry matter accumulation ($r^2 = 0.690$). Agasimani et al. (2010) observed higher dry matter production in treatments, where integrated weed management was practiced in groundnut-wheat cropping system in Northern Karnataka.

Relative growth rate (RGR)

Relative growth rate was highest between 6 and 8 WAS in MK 373 and 8–10 WAS in Samnut 10 and declined with advent of time (Table 2). All the weed control treatments had higher RGR as compared to the weedy check. This showed that growth of groundnut in weedy check was grossly affected by weed infestation. Better, RGR observed between 6 and 8 and 8–10 WAS in both the varieties in treatments such as weed free check, rice straw mulch + one hand weeding at 6 WAS, two hand weeding at 3 and 6 WAS and pendimethalin at 1.5 l ha^{-1} + one hand weeding at 6 WAS could be attributed to effective weed control and low weed biomass, which enabled plant receiving these treatments higher photosynthetic rate. The lower RGR in sole rice straw mulch and sole pendimethalin, though higher than the weedy check during these periods, could be due to the inability of these treatments to offer season-long weed control. The decline in RGR towards physiological maturity could be due to leaf shedding, shadow of upper leaves over the lower leaves which reduce the photosynthetic capacity of the lower leaves and finally loss of leaves due to pest attack (Banik et al. 2009).

Net assimilation rate (NAR)

Net assimilation rate followed the same trend as RGR (Table 3). At peak periods, between –8 and 8–10 WAS in MK 373 and Samnut 10, respectively, NAR, was highest in weed free plots or plot raised under rice straw mulch + one hand weeding at 6 WAS, followed by two hand weeding at 3 and 6 WAS and pendimethalin at 1.5 l ha^{-1} + one hand weeding at 6 WAS, sole rice straw mulch and pendimethalin at 1.5 l ha^{-1} (Table 3). Lowest NAR was recorded in weedy check. Greater NAR in treatments that showed effective weed control could be due to higher photosynthetic rate on account of reduced weed competition for limited resources such as water, light, nutrient and space. Singh (2004) observed that photosynthetic rate of leaves decrease as the relative water content and water potential decreases in groundnut. This also explains the decrease in NAR in treatments such as sole pendimethalin at 1.5 l ha^{-1}, sole rice straw mulch and the weedy check, where high weed biomass affected the water potential and consequently reduced the photosynthetic efficiency. The decrease in NAR in groundnut towards physiological maturity may be linked to mutual shading, increase in number of old leaves with low photosynthetic efficiency and loss of leaves due to pest attack (Banik et al. 2009).

Table 3. Effect of different weed control methods on net assimilation rate and crop growth rate at different crop growth stages in Samnut 10 and MK 373 (pooled data of cropping seasons 2010 and 2011)

Variety	Treatment	Week after sowing (WAS)					
		Net assimilation rate (g m^{-2} day^{-1})			Crop growth rate (g m^{-2} day^{-1})		
		6–8	8–10	10–12	6–8	8–10	10–12
Samnut 10	T_1	0.006^{a}	0.010^{b}	0.005^{c}	1.73^{d}	3.10^{f}	1.31^{e}
	T_2	0.023^{a}	0.040^{a}	0.008^{c}	5.17^{c}	5.53^{c}	2.47^{d}
	T_3	0.005^{a}	0.014^{b}	0.005^{c}	1.93^{d}	4.87^{d}	1.43^{e}
	T_4	0.025^{a}	0.046^{a}	0.039^{b}	7.40^{ab}	8.83^{a}	6.40^{a}
	T_5	0.025^{a}	0.044^{a}	0.058^{a}	6.50^{b}	6.83^{a}	4.17^{c}
	T_6	0.025^{a}	0.048^{a}	0.010^{c}	8.03^{a}	8.87^{a}	5.79^{a}
	T_7	0.008^{a}	0.008^{b}	0.005^{c}	2.20^{d}	2.33^{f}	1.13^{e}
	Mean	0.017	0.030	0.019	4.71	5.77	3.24
MK 373	T_1	0.019^{ab}	0.010^{b}	0.001^{b}	5.53^{f}	3.47^{de}	2.13^{c}
	T_2	0.031^{abc}	0.019^{a}	0.003^{b}	7.23^{d}	6.43^{b}	4.27^{b}
	T_3	0.017^{bc}	0.009^{a}	0.072^{a}	5.93^{e}	3.22^{e}	2.33^{c}
	T_4	0.042^{ab}	0.008^{b}	0.004^{b}	9.17^{b}	6.17^{b}	1.90^{c}
	T_5	0.041^{ab}	0.009^{b}	0.007^{b}	7.57^{c}	4.27^{c}	4.80^{ab}
	T_6	0.046^{a}	0.008^{b}	0.007^{b}	10.57^{a}	7.00^{a}	5.30^{a}
	T_7	0.012^{c}	0.011^{b}	0.002^{b}	4.37^{g}	3.83^{cd}	2.30^{c}
	Mean	0.030	0.011	0.010	7.20	4.91	3.30
	Grand mean	0.024	0.021	0.015	5.96	5.34	3.27
LSD ($p \leq 0.01$)	Variety	0.0022	0.00016	0.0022	0.14	0.57	0.39
	Treatment	0.020	0.021	0.020	0.27	0.64	0.28
	V × T	0.0001	0.00005	0.0001	0.12	0.01	0.04

Within column means followed by the same superscripts are not significantly different at $p \leq 0.01$

T_1: Pendimethalin at 1.5 l ha^{-1}, T_2: Pendimethalin at 1.5 l ha^{-1} + one hand weeding at 6 WAS, T_3: Rice straw mulch, T_4: Rice straw mulch + one hand weeding at 6 WAS, T_5: Two hand weeding at 3 and 6 WAS, T_6: Weed free check (positive control), T_7: Weedy check (negative control)

Crop growth rate (CGR)

Crop growth rate was significantly ($p \leq 0.01$) affected at different growth periods by weed control treatments (Table 3). Apart from weed free check, from 6–8 to 10–12 WAS, CGR was highest in rice straw mulch + one handing at 6 WAS, followed by two hand weeding at 3 and 6 WAS, pendimethalin at 1.5 l ha^{-1} + one hand weeding, sole rice straw and sole pendimethalin at 1.5 l ha^{-1}. Lowest CGR was recorded in weedy check at all growth periods. The highest CGR in either weed free check or rice straw mulch + one hand weeding at 6 WAS might be due to less weed competition for natural resources such as water, light, nutrients and space for crop growth. Also higher level of LAI and NAR could account for higher CGR recorded under these treatments (Augadi et al. 1990). Varietal differences showed that MK 373 had higher CGR than Samnut 10 and this could be due to more dry matter accumulation in former than latter. Singh and Joshi (1993) have reported Virginia runner cultivar had higher pod number and yield than the erect type due to more dry matter, N-accumulation and higher CGR. The decrease in CGR after 6–8 and 8–10 WAS in MK 373 and Samnut 10, respectively could be due to loss of leaves on account of pest attack and mutual shading of leaves, which was higher in treatments where more leaves were produced.

Yield components and yield

Weed control methods significantly affected yield components and yield in both the varieties (Table 4). In both varieties, highest number of matured pods per plant, seed weight per plant, 100-seed weight, pod yield, seed yield and harvest index were recorded in rice straw mulch + one hand weeding at 6 WAS over all other treatments (Table 4). Lowest yield components and yield were recorded in weedy check. The highest yield recorded in rice straw mulch + one hand weeding at 6 WAS might be due to reduced weed competition for limited resources resulting in increase in 100-seed weight, number of matured pods per plant, seed weight per plants compared to other treatments. Additionally, higher moisture content and the decomposition of the mulch could have also contributed to increase in supply of nutrients for the overall increase in yield in this treatment. The results supported the opinion of Singh and Joshi (1993), where higher pod yield is attributed to better N accumulation, higher dry matter and CGR. Similarly, groundnut yields have been found to be enhanced, where integrated weed control methods were practiced (Kumar 2009; Jat et al. 2011). In the present study, seed yield correlated positively with all the physiological growth parameters ($r^2 = 0.69$–0.95) evaluated.

Table 4. Effect of different weed control methods on yield components and yield in Samnut 10 and MK 37 (pooled data of cropping seasons 2010 and 2011)

Variety	Treatment	Number of matured pod per plant	Seed weight (g plant^{-1})	100-seed weight (g)	Pod yield (g m^{-2})	Seed yield (g m^{-2})	Harvest index
Samnut 10	T_1	24.67^d	10.67^d	37.73^b	397.13^c	280.29^e	0.25^a
	T_2	32.33^c	15.47^c	38.58^b	475.56^b	349.39^a	0.32^a
	T_3	23.17^e	10.53^d	38.10^b	386.76^d	278.36^f	0.26^a
	T_4	45.33^a	22.37^a	40.18^a	542.78^a	384.84^a	0.39^a
	T_5	44.17^b	21.25^b	40.12^a	534.77^a	376.89^b	0.36^a
	T_6	32.67^c	15.78^c	38.48^a	529.63^a	353.19^c	0.30^a
	T_7	13.50^f	6.17^e	35.42^c	264.42^e	172.24^g	0.24^a
	Mean	30.50	14.59	38.43	447.22	313.61	0.30
MK 373	T_1	28.17^b	15.23^b	40.27^c	446.19^d	305.19^e	0.26^a
	T_2	30.00^b	20.60^b	42.63^{ab}	533.57^c	388.51^d	0.35^a
	T_3	21.00^c	12.55^c	40.77^{bc}	416.43^e	292.58^f	0.32^a
	T_4	35.67^a	26.02^a	44.38^a	637.15^a	466.41^a	0.36^a
	T_5	33.33^a	24.08^a	44.12^a	569.86^b	424.85^b	0.34^a
	T_6	33.33^a	21.58^a	42.85^{ab}	531.26^c	391.82^c	0.32^a
	T_7	17.00^d	8.72^d	37.98^d	287.03^f	191.57^g	0.25^a
	Mean	28.36	18.39	41.86	488.78	352.54	0.31
	Grand mean	58.86	16.49	40.15	975.01	693.91	0.31
LSD ($p \le 0.01$)	Variety	0.55	4.24	6.43	31.96	21.04	0.06
	Treatment	0.89	0.67	1.04	25.59	27.56	0.24
	V × T	0.87	0.29	0.13	23.60	17.17	0.02

Within column means followed by the same superscripts are not significantly different at $p \le 0.01$

T_1: Pendimethalin at 1.5 l ha^{-1}, T_2: Pendimethalin at 1.5 l ha^{-1} + one hand weeding at 6 WAS, T_3: Rice straw mulch, T_4: Rice straw mulch + one hand weeding at 6 WAS, T_5: Two hand weeding at 3 and 6 WAS, T_6: Weed free check (positive control), T_7: Weedy check (negative control)

In conclusion, rice straw mulch at 0.1 m depth + one hand weeding at 6 week after sowing increased growth and yield of the two varieties when compared to other weed control methods. The enhancement of crop productivity in this treatment could be attributed to its positive influence on LAI, NAR, RGR and CGR.

OPEN ACCESS

REFERENCES

1. Agasimani, C. A., Shanwad, U. K., Aravndkumar, B. N., Shuvamurth, S. D., Ashork, S., & Jalageri, B. R. (2010). Integrated weed management (IWM): A long time case study in groundnut-wheat cropping system in Northern Karnataka. *Research Journal of Agricultural Sciences, 1*(3), 196–200.
2. Akobundu, I. O. (1987). *Weed science in tropics, principles and practices*. New York: Wiley.
3. Augadi, V. V., Pabl, S. V., Sheelavantar, M. N., & Chittapur, B. M. (1990). Effect of NPK levels and split application of N on growth and yield of bunch groundnut in vertisol soil under irrigation. *Karnataka Journal Of Agricultural Sciences, 3*, 9–14.
4. Banik, C. N., Nath, R., & Chakraborty, P. K. (2009). Effect of dates of sowing on growth and yield of groundnut crop. *Journal of Crop and Weed, 5*(2), 59–62.
5. El-Naim, A. M., & Ahmed, S. E. (2010). Effect of weeding frequencies on growth and yield of two roselle (*Hibiscus sabdariffa* L.) varieties under rain fed. *Australian Journal of Basic and Applied Sciences, 4*(9), 4250–4255.
6. El-Naim, A. M., Eldouma, M. A., Ibrahim, E. A., & Zaied, M. M. B. (2011). Influence of plant spacing and weeds on growth and yield of peanut (*Arachis hypogaea* L.) in rain-fed of Sudan. *Advances in Life Sciences, 1*(2), 45–48
7. Etejere, E. O., Olayinka, B. U., & Wuraola, A. J. (2013). Comparative economic efficacy of different weed control methods in groundnut. *Electronic Journal of Biological Sciences (EJBS), 7*(2), 10–18.
8. Gardner, F. P., Pearce, R. B., & Mithcell, R. L. (1985). *Carbon fixation by crop canopies* (pp. 31–57). Ames: Iowa State University Press.
9. Gill, G. S., & Kumar, V. (1966). Weed index. A new method for reporting weed control trials. *Indian Journal Agronomy, 14*(1), 96–98.
10. Jat, R. S., Meena, H. N., Singh, A. L., Jaya, N. S., & Misra, J. B. (2011). Weed management of groundnut (*Arachis hypogaea* L.) in India—a review. *Agricultural Reviews, 32*(3), 155–171.
11. Kumar, N. S. (2009). Effect of plant density and weed management practices on production potential of groundnut (*Arachis hypogaea* L.). *Indian Journal of Agricultural Research, 43*, 1.
12. McCloud, D. E. (1974). Growth analysis of high yielding peanut. *Proceedings of the Soil and Crop Science Society of Florida, 33*, 24–26.
13. Nakaseko, K., Gotch, K., & Asanuma, K. (1979). Comparative studies on dry matter production, plant type and productivity of soybean, axukibean and kidneybean. I differences in dry matter accumulation patterns under low population density. *Japan Journal Crop Science, 48*, 82–91.
14. Olorunmaiye, P. M., & Olorunmaiye, K. S. (2009). Effect of integrated weed management on weed control and yield components of maize and cassava intercrop in a southern Guinea savanna ecology of Nigeria. *Australian Journal of Crop Science, 3*(3), 129–136.
15. Rahman, M. M., Ahad, A. M., Maninuzzaman, A. F. M., & Khan, K. (1994). Growth analysis of black gram (*Vigna mungo* L.) under varying levels of population densities and its agronomic appraisal. *Bangladesh Journal of Botany, 23*, 155–159.
16. Singh, A. L. (2003). Phenology of groundnut. In A. Hemantranjan (Ed.), *Advances in plant physiology* (Vol. 6, pp. 295–382). Jodhpur: Scientific Publishers (India).
17. Singh, A. L. (2004). Growth and physiology of groundnut. In M. S. Basu & N. B. Singh (Eds.), *Groundnut Research in India* (pp. 178–212). Junagadh: National Research center for groundnut (ICAR).
18. Singh, G. (2009). Effects of wheat straw and farmyard manure mulches on overcoming crust effect, improving emergence, growth and yield of soybean and reducing dry matter of weeds. *International Journal of Agricultural Research, 4*, 418–424.

19. Singh, A. L., & Joshi, Y. C. (1993). Comparative studies on the chlorophyll content, growth, N uptake and yield of groundnut varieties of different habit groups. *Oleagineux, 48*, 27–34.
20. USDA. (2009). Foreign Agricultural Services, Peanut area, yield and production. www.fas.usda.gov.psdaonline.
21. Vijay kumar, B. (1992). Chemical weed control in groundnut. *India Journal of Weed Science,24*, 72–74.

CHAPTER 4

Influence of The Position of Flowers Buds on The Tree on Somatic Embryogenesis of Cocoa (Theobroma Cacao L.)

Rodrigue Pouengue Boutchouang[1,2]*, Olive Flore Zebaze Akitio[2,3], Audrey Germaine Ngouambe Tchouatcheu[2,3] and Nicolas Niemenak[2]

[1]*Department of Biochemistry, Faculty of Sciences, University of Yaounde I, P. O. Box 812, Yaounde-Cameroon.*

[2] *Laboratory of Biochemistry and Plant Physiology, Department of Biological Science, Higher Teachers' Training College, University of Yaounde I, P. O. Box 47, Yaounde- Cameroon.*

[3]*Department of Plant Biology, Faculty of Science, University of Yaounde I, P. O. Box 812 Yaounde, Cameroon.*

ABSTRACT

The recalcitrance of Theobroma cacao L. to somatic embryogenesis, due to non-adapted physiological and metabolical responses to environmental stress, limits its propagation. The present work aims to ameliorate somatic embryogenesis in T. cacao throughout a physiological approach. For this purpose, the influence of the position of flowers buds used as explants was evaluated. Flowers buds were collected from different parts of the tree: orthotropic main stem (OS), primary plagiotropic fan branch (FI) and secondary plagiotropic fan branch (FII). Evolution of some biochemical parameters such as phenolic compounds, soluble sugars, proteins contents and peroxidase activity was followed at different steps of somatic embryogenesis, considering the origin of the explants used. Results obtained show that callogenesis is induced on all explants independently of their origin, with an 80% average frequency. Embryogenesis frequencies were ca 2 fold higher in staminodes-derived calluses from FII and FI than OS. Meanwhile petals of FII do not differentiate embryos. Biochemical analysis shows that the content of phenol is low in calluses during somatic embryo establishment. Explants from FII present the lowest values (after 49th days of culture). Sugars content decrease during callogenesis. When embryos are established the sugars content decrease in explants from OS. During the same period, proteins' and phenols contents increased in staminodes-derived calluses from all origin; while there was decrease in petals from FI and FII. Buds from fan branch are suitable for somatic embryogenesis process and this capacity correlate with peroxidase activity which decrease during embryos dedifferentiation phase.

Keywords: Theobroma cacao L., somatic embryogenesis, proteins, phenols compounds, soluble sugars, peroxidase activity, microclimate.

INTRODUCTION

Theobroma cacao L. (chocolate tree) is grown in the humid tropics and constitutes an important source of incomes for many countries of the West and Central Africa regions. This tree, when grown in wild state, can

reach 12 to 15 m of length (Wood et al., 1985). It is a diploid (2n=20) and preferentially allogamy plant. The root system has a growth dimorphism characterized by an orthotropic development axis, the pivot and plagiotropic lateral ramifications and lateral roots. Cocoa trunk is characterized by a vertical harbor (orthotropic), a phyllotaxis 3/8; long-stalked leaves, axillary orthotropic buds, a defined growth, differentiation of three to five plagiotropic buds under the apex, at the time of the degeneracy of the terminal bud. Cauliflorous tree, inflorescences cymes biparous with reduced inter-node develop at the flowering areas called flower cushions. The flowers are hermaphroditic, with small sizes (diameters ranging from 0.5 to 1 cm), regular and composed of 5 sepals, 5 petals, 5 staminodes, a pistil and an ovary. Pollination is predominantly entomophilous although it can be done manually in the experimental fields. Flowering in cocoa is manifested by the production of a minimum of 50 000 flowers during the term with less than 5% of production pods (Lass, 1999). Cocoa trees demonstrate a high degree of segregation for many traits when propagated by seeds. For this reason, clonal propagation systems such as rooted cuttings and grafting have been applied for multiplication of elite varieties but a vast majority of cocoa plants in production were derived from seeds (Eskes, 2005). The use of in vitro propagation methods for cocoa could potentially contribute to efforts at crop improvement, germplasm conservation, and rapid distribution of new improved varieties.

Somatic embryogenesis is a vegetative propagation method which permits the production of several embryos capable of generating plants similar to the initial one from non-sexual tissues. Somatic embryogenesis has been the most used method these last years for in vitro regeneration of elite genotypes of cocoa (Niemenak et al., 1998; Maximova et al., 2008; Noah et al., 2013). Somatic embryos germinate to yield entire plant with a pivoting root that we cannot obtain by cutting. Temporary immersion culture and suspension culture can enhance embryos multiplication of cocoa (Niemenak et al., 2008). The success of this method is mostly dependent of genotype and culture medium composition. Li et al. (1998) achieved better somatic response from many genotypes using DKW complex salt with floral explants.

Somatic embryogenesis plays an important role in clonal propagation. When integrated with conventional breeding programs and molecular plus cell biological techniques, somatic embryogenesis provides a valuable tool to enhance the pace of genetic improvement of commercial crop species (Stasolla and Yeung, 2003). There are evidences that the main metabolic and developmental processes occurring in the zygotic embryogenesis may be recapitulated in the somatic embryogenesis (Fehér et al., 2003). But the latter are less efficient in converting carbohydrates in lipids and storage proteins during the late developmental stages (Cangahuala et al., 2009). The levels of these substances change along the developmental stages of cells cultures, and their role has been ascribed to the transduction signal cascade or as substrate for cell growth and morphogenesis (Lulsdorf et al., 1992). Storage proteins are the source of amino acids for seed germination (Misra et al., 1993). Proteins could also be involved in the regulation of cell expansion and establishment of biophysical characteristics required for the morphogenesis (Jiménez, 2001). Soluble sugars, such as glucose and sucrose, are involved in the regulation of developmental processes occurring from embryo development to seed maturation (Gibson, 2005).

Generally, when used as source of plant materials for somatic embryogenesis, flowers buttons are collected from the tip of branches on cocoa trees, whereas their production is distributed on the whole plant and is extended all over the year. Due also to the low production and conversion of somatic embryos obtained from different genotypes of T. cacao, leading to important economic lost, a new physiological approach was developed. Assessment of somatic embryogenesis was done using floral explants (staminodes and petals) from different position on the tree: orthotropic main stem (OS), primary plagiotropic fan branch (starting from the first ramification) and secondary plagiotropic fan branch (starting from the second ramification). The present research work objectives were to investigate the mechanism that can help to solve the recalcitrance of T. cacao and to understand the influence of explant position on the tree on biochemical events underlying somatic embryogenesis in this specie.

MATERIALS AND METHODS

Preparation of plant materials

The tissue system used in this investigation was the same as the system previously described by Minyaka et al. (2008). Studies were carried out on the cacao genotype "SCA 6", which is included in the gene-bank of the Institute of Agricultural Research and Development at Nkolbisson (Yaounde, Cameroon). Flowers buds were collected at different positions on the same tree during all the experimentation, including orthotropic main stem (OS), primary plagiotropic fan branch (FI) and secondary plagiotropic fan branch (FII) early in the morning. They were surface sterilized by immersion for 20 min in 3% (w/v) sodium hypochlorite followed by three rinses in sterilized distilled water of 2 min each. Staminodes and petals were excised with scalpels and placed on culture media (in distinct set) into Petri dishes (Figure 1).

Culture medium preparation

All media were defined using Driver and Kuniyuki Walnut (DKW) medium basal salt of Driver and Kuniyuki (1984). The explants were first cultured in primary callus growth medium. Primary callus growth (PCG) medium was supplemented with 250 mgL-1 glutamine, 100 mgL-1 myo-inositol, 1 mLL-1 DKW vitamin stock (100 mgmL-1, 2 mgmL-1 thiamine-HCl, 1 mgmL-1 nicotinic acid and 2 mgmL-1 glycine), 20 gL-1 glucose, 18 μM 2,4-dichloro-phenoxyacetic acid (2,4-D) and 45.4 nM thidiazuron (TDZ). Media were dispensed into sterilized Petri dishes after autoclaving for 20 min at 1 bar pressure and 121°C. Each Petri dish contains either 30 staminodes or 30 petals. Experiments were repeated 5 times with five replicate Petri dishes at each culture initiation. Petri dishes were incubated in dark at 25 ± 1°C for 14 days. After 14 days in PCG medium, explants were

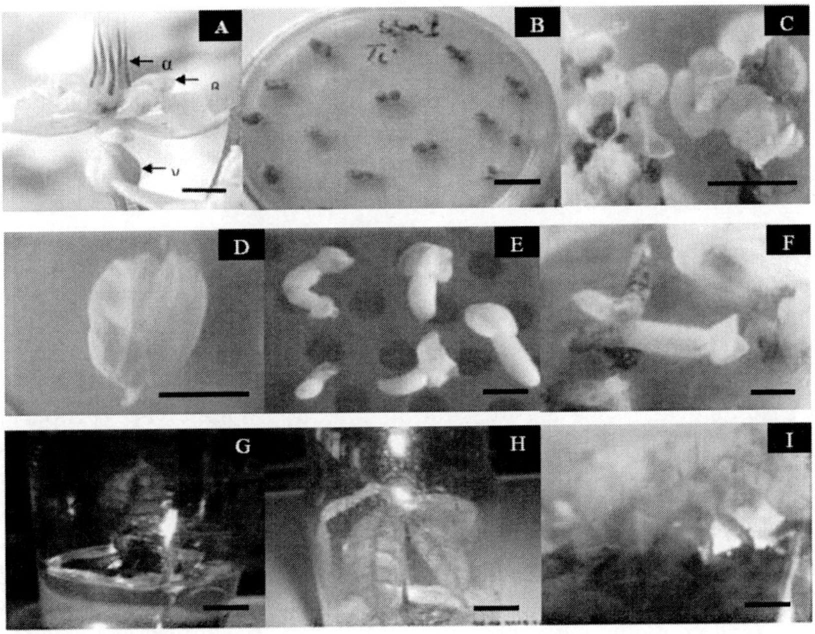

Figure 1. Morphogenic structure of T. cacao during somatic embryogenesis and plantlets conversion. Plant materials (A): staminodes (α), petals (β), and immature floral buds (γ). Explants-derived callus from orthotropic main stem growing in Secondary Callus Growth medium (SCG) (B); Characteristic developmental stages of SE: globular (C), heart shape (D), torpedo (E) and cotyledonnary stages (F). Plantlet originated from somatic embryos matured under high concentration of sucrose, 90 (G) and 120 (H) days after transfer in root development medium. Well-developed plantlet just allocated in the substrate (I). Bar =1 cm.

transferred to secondary callus growth (SCG) medium. SCG medium consisted of DKW basal salts, supplemented with 0.5 mLL-1 DKW vitamin, 20 gL-1 glucose, 9 μM 2,4-D, 250 μgL-1 kinetin and 0.22% (w/v) gelrite. Cultures were also incubated at 25 ± 1°C for 14 days in darkness. Cultures from SCG1 medium were transferred in embryo development (ED) medium. ED medium was made of DKW basal salts supplemented with 6.0 mM MgSO4, 1 mL DKW vitamin, 30 gL-1 sucrose, 1 g.L-1, glucose and 0.22% (w/v) gelrite. Cultures were incubated at 25 ± 1°C in darkness for 21 days. Two others sub-cultures of explants were made every 21 days in ED medium for the development of embryos.

Biochemical analysis

At the end (91st days) of each experience (a given culture), calluses of each characteristic development stages of somatic embryogenesis, that is, calluses aged 14, 28, 49, 70 and 91 days, respectively, were collected from different sources of explant and analyzed independently.

Estimation of phenolic contents

Phenolic compounds were extracted as described by El Hadrami (1995) and Macheix et al. (1990). Embryonic mass (100 mg) were ground in chilled mortars with 2 ml of 80% (v/v) methanol at 4°C. After incubation, tubes were centrifuged thrice at 7000 g for 30 min, supernatant were recuperated each time. Mixture of the three supernatants constituted the crude extract. Total phenols were quantified using the method described by Singleton and Rossi (1965). 15 µl of alcoholic extract were added to Folin-Ciocalteu reagent (250 µl), 2.5 mL of distilled water and 0.5 ml of sodium carbonate (20 %). The mixture was incubated at 40°C for 20 minutes and the blue color was determined at 760 nm. The content of soluble phenolic was expressed in mg-equivalent of gallic acid per fresh weight (FW).

Estimation of soluble sugars contents

Soluble sugars were extracted according to the modified method of Babu et al. (2002). Biological material (400 mg) was ground in mortar with 2 mL of 80% (v/v) ethanol and centrifuge at 6000 g for 20 min. The supernatant was collected and constituted the crude extract. 50 µL of this alcoholic extract was added to 5 mL of Anthron reagent, homogenized and incubated at 80°C for 20 min.

After cooling in melting ice, the absorbance of the green complex formed was determined at 620 nm. Glucose was use as standard.

Estimation of protein contents and POX activity

Calluses (500 mg) were ground in chilled mortar with 2 mL of TAMET buffer (0.5 M Tris, 0.3 M ascorbic acid, 0.2% (v/v) β-mercaptoethanol, 0.01 M EDTA and 0.02% (v/v) Triton X 100, pH 6.7); and 0.125 g polyvinylpyrrolidone was added in the medium. The crude homogenate was centrifuged for 15 min at 20000g and 4°C. The supernatant was removed and used as crude extract for proteins and enzymes assays. Proteins were quantified according to Bradford (1976) method. Peroxidase (POX) activity was measured spectrometrically at 420 nm by the guaiacol-H2O2 method of Erdelsky and Fric (1979). The specific activity of the enzymes was expressed as the change in optical density per mg of fresh weight (FW).

Statistical analysis

Data were subjected to statistical analysis using SPSS software version 16.0. Analysis of variance was performed where applicable and differences between means were determined using LSD and Tukey Test.

RESULTS

The results showed that calluses were distinguishable on floral explants after 4 to 5 days of cultivation in the primary growth medium. Callus development was well established in secondary callus growth medium, and was quite similar in tissues from different origins, according to their morphological aspects (Figure 1). There were no significant differences ($p<0.05$) among the average frequency of callogenesis (after 28 days) when explants were cultivated in DKW medium containing plant regulators. However, callus growth was most influenced by the type and origin of explant used. The average callogenesis frequency was up to 80% in *T. cacao* tissues, with the highest value obtained with staminodes explants collected on FI (93.97±2.43%) (Table 1). By the end of three-week in embryo development medium, some calluses differentiated roots or embryos. The earliest somatic embryos were observed between 49 and 55 day of culture on embryogenic calluses. Petals-derived calluses collected on secondary branches do not differentiate somatic embryos during all the experiment (91 days). The percentage of explants producing somatic embryos is highly influenced by the nature and the origin of explant cultivated in culture media (Figure 2). No statistical differences ($p<0.05$) were found among the rate of petals-derived calluses producing somatic embryos from different origin. Whereas, in staminodes-derived calluses, this rate decrease significantly when moving up from OS to FII. In fact, this rate of explants-derived calluses producing somatic embryos were approximately 3 times higher in staminodes tissues than petals after 91 days of culture, with the highest value obtained with staminodes-derived calluses from FII (13.70±3.21%). As observed, embryogenic and non-embryogenic tissues of *T. cacao* calluses were discernable based on coloration; embryogenic calluses were brown (phenolized) and friable, while non-embryogenic were white and rough. Somatic embryos undergo different developmental stages as shown by the result: globular, heart shaped, torpedo and cotyledonnary stage (Figure 1C, D, E and F). Cotyledonnary stage constituted the last development stage. At this stage, somatic embryos submitted to maturation treatment were able to germinate and generate plantlets (Figure 1G, H, and I).

Table 1. Average frequency of callogenesis in Sca 6 genotype of Theobroma cacao L., 28 days after induction in primary callus growth (PCG) medium.

Callogenic explants	Orthotropic main stem	Primary plagiotropic fan branch	Secondary plagiotropic fan branch
Staminodes-derivezd callus (%)	92.50 ± 4.97[a]	93.97 ± 2.43[a]	81.23 ± 6.40[a]
Petals-derived callus (%)	83.47 ± 9.87[a]	90.17 ± 5.00[a]	93.63 ± 1.78[a]

Each value is mean ± SE of three replicates samples. Values significantly different at the 5% level of significance are indicated with different letters.

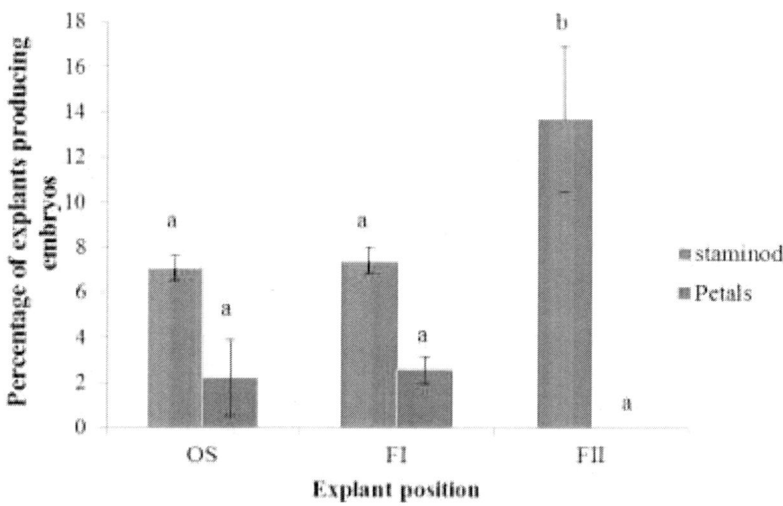

Figure 2. Influence of the position of flowers buds on the percentage of explants of T. cacao producing somatic embryos after 91st days. Data are presented as means of five identical experiments; position of flowers buds: Orthotropic main stem (OS), primary plagiotropic fan branch (FI) and secondary plagiotropic fan branch (FII).Values significantly different at the 5% level of significance are indicated with different letters.

Biochemical analysis showed that the total content of polyphenols compounds in staminodes-derived calluses ranged from 2.16 ± 0.15 to 8.04 ± 0.22 mg g^{-1} of FW in OS, 1.69 ± 0.08 to 6.61± 0.49 mg g^{-1} of FW in FI and from 1.73± 0.23 to 4.9± 0.33 mg g^{-1} of FW in FII (Figure 3). There is no regular evolution of this determinant during somatic embryogenesis process in both types of explants (staminodes or petals) from the same position on the tree. After 28th days (corresponding to induction step), total phenols compounds increased in staminodes-derived calluses (84, 21 and 43%, for OS, FI and FII, respectively). Constitutively, soluble carbohydrates were low in calluses from all origins (Figure 4). Statistical differences ($p<0.05$) were found in the amount of carbohydrates between OS and FII calluses after 28th days. Whereas no statistical difference were found between the amount of sugars in calluses from FI and FII during this period, considering staminodes and petals tissues. Callogenesis in this case was characterized by a decrease in carbohydrates content. Among petals-derived calluses, those from FII displayed the highest amount of carbohydrates (6.37±0.3 mg g^{-1} of FW after 49 days), and OS the lowest (1.6±0.18 mg g^{-1} of FW after 28 days); While decreasing progressively in others stagesbetween 0.8 and 36% (in 70-day-old OS and 90-day-old FI, respectively). Except for calluses from FI and FII aged 14 days, the same observation was made in petals. The protein contents in staminodes-derived calluses varied from 0.537±0.178 mg g^{-1} (FII after 28 days) to 2.181±0.372 mg g^{-1} of FW (OS after 70 days), whereas in petals-derived calluses, it varied from 0.724± 0.03 mg g^{-1} (OS after 70 days) to 1.8495± 0.237 mg g^{-1} of FW (OS after 28 days). Protein content was significantly higher in petals-derived calluses from FII aged 49 days (1.809 ± 0.415 mg g^{-1} of FW) than in OS (0.911 ± 0.178 mg g^{-1} of FW) from the same source. It appears that, cells differentiation is characterized by high protein synthesis in staminodes tissues (Figure 5).

Peroxidases activities presented pattern related to a source of morphogenetic structure. The activity in staminodes derived calluses after 49 days was 2.7 and 2.0 fold higher in OS than in FII and FI, respectively. There was no significant difference ($p<0.05$) among peroxidases activities in staminodes-derived calluses at any developmental stage of this experiment from each source of tissues. However, this value, after 28 days, was significantly higher in petals-derived calluses from BII as compared to OS and FI (Figure 6). Staminodes-derived calluses from OS presented the lowest activity (after 70 days).

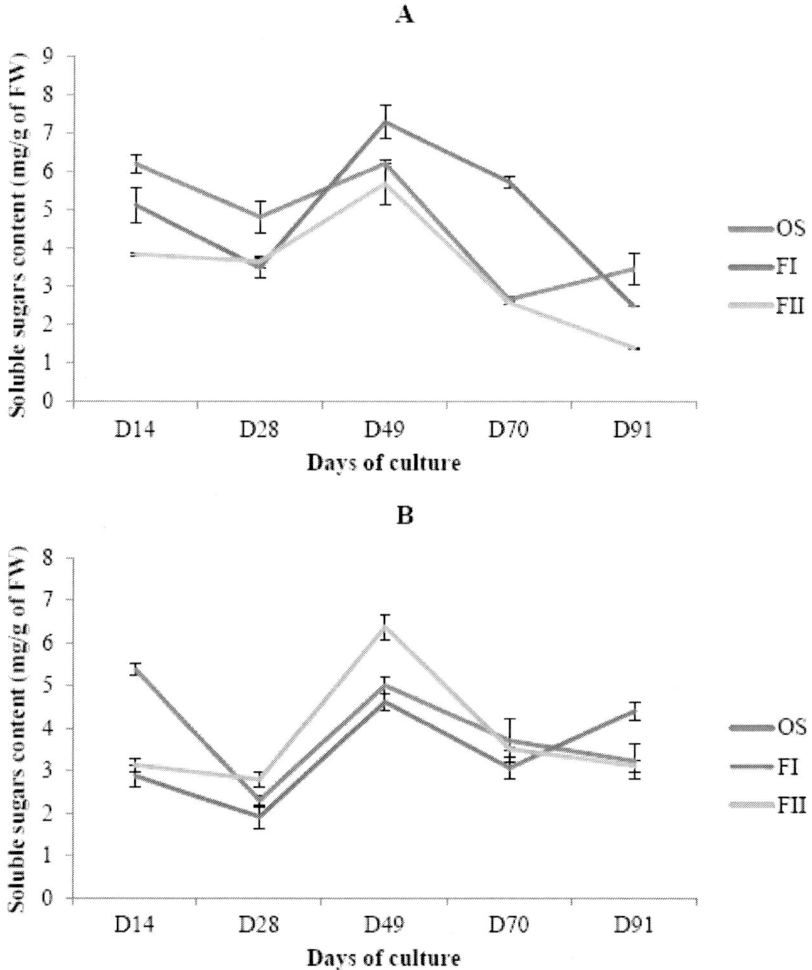

Figure 3. Influence of the position of flowers buds on the tree on total soluble sugars content during somatic embryogenesis of T. cacao: staminodes (A) and petals (B); position of flowers: orthotropic main stem (OS), primary plagiotropic fan branch (FI) and secondary plagiotropic fan branch (FII).Vertical bars represent standard error. Each plot was drawn from means of four identical experiments.

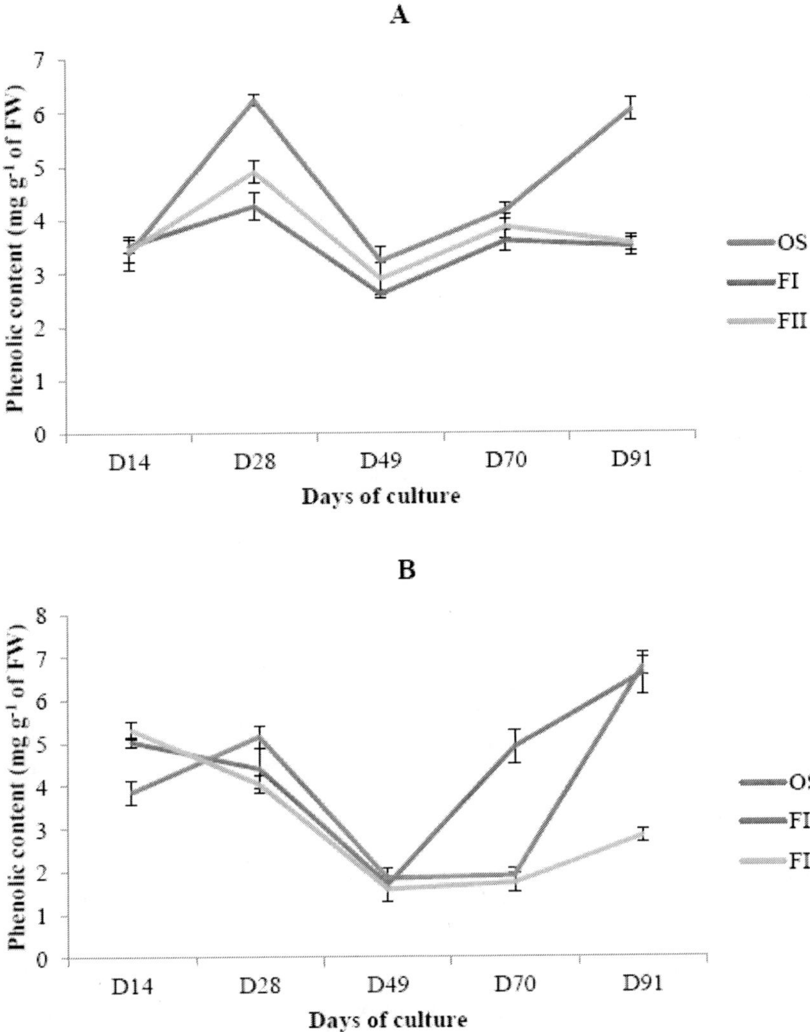

Figure 4. Influence of the position of flowers buds on the tree on the soluble sugars content during somatic embryogenesis. Staminodes (A) and petals (B); position of flowers: orthotropic main stem (OS), primary plagiotropic fan branch (BI) and secondary plagiotropic fan branch (FII). Vertical bars represent standard error. Each plot was drawn from means of four identical experiments.

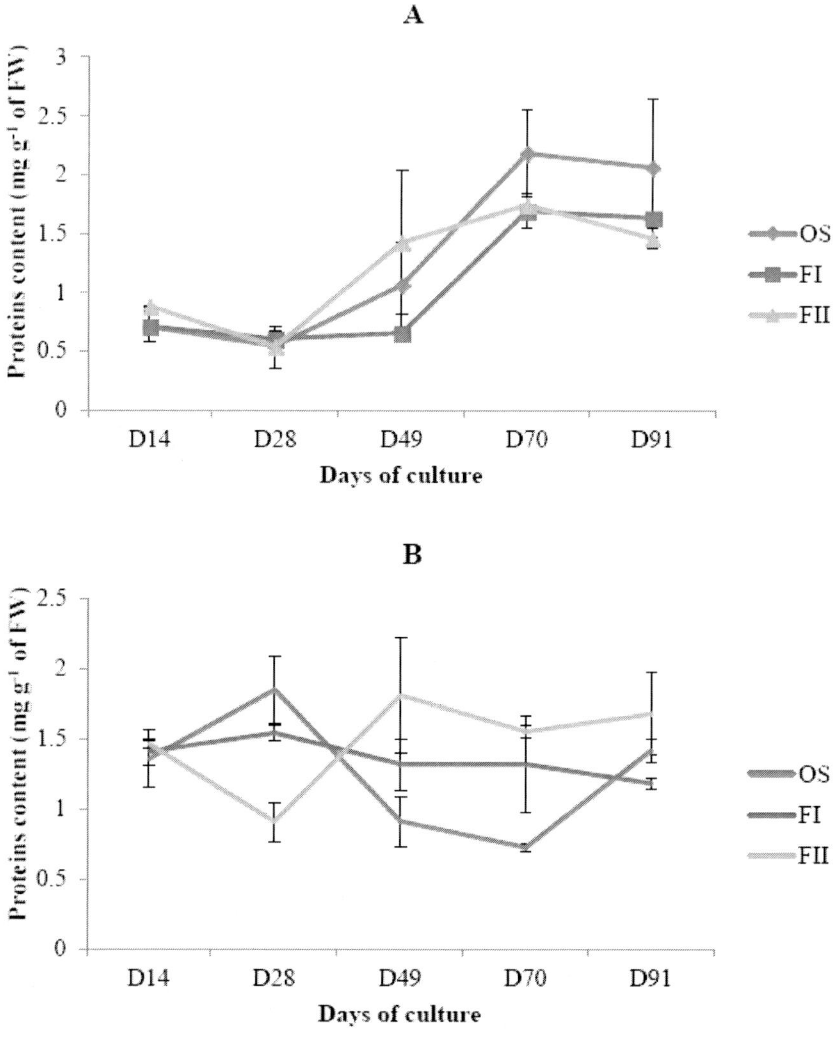

Figure 5. Influence of the position of flowers buds on the tree on proteins content during somatic embryogenesis. Staminodes (A) and petals (B); position of flowers: orthotropic main stem (OS), primary plagiotropic fan branch (FI) and secondary plagiotropic fan branch (FII). Vertical bars represent standard error. Each plot was drawn from means of four identical experiments.

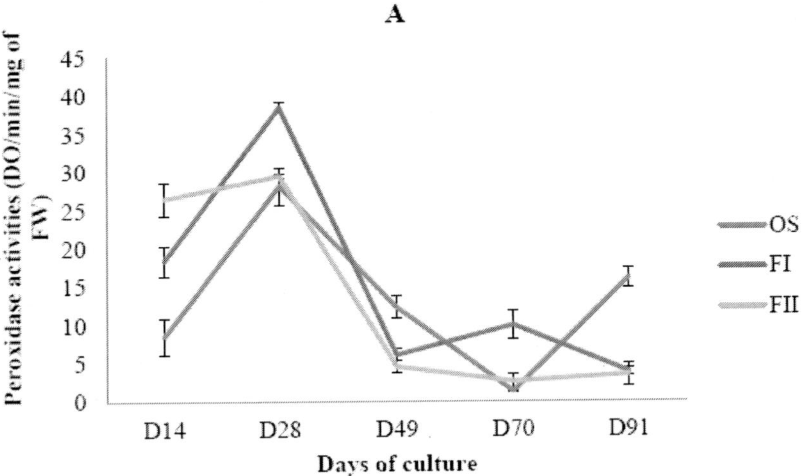

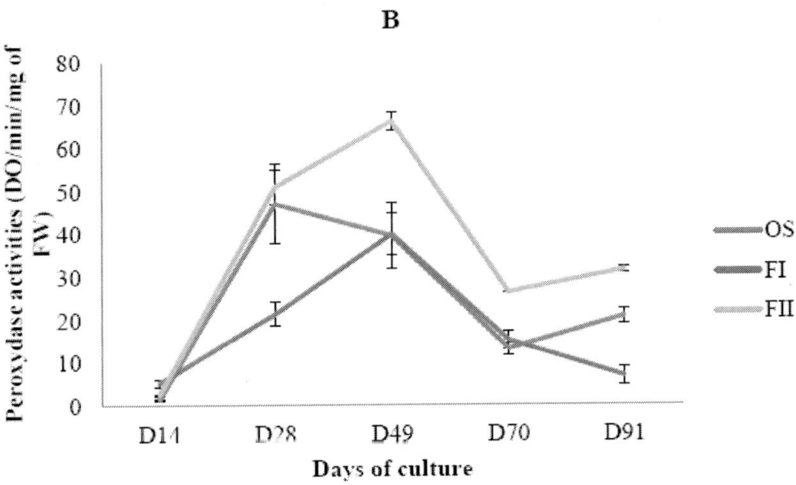

Figure 6. Influence of the position of flowers buttons on the tree on peroxidases activities during somatic embryogenesis. Staminodes (A) and petals (B); position of flowers: orthotropic main stem (OS), primary plagiotropic fan branch (FI) and secondary plagiotropic fan branch (FII). Vertical bars represent standard error. Each plot was drawn from means of four identical experiments.

DISCUSSION

Somatic embryogenesis is the process by which somatic cells, under inductive conditions, generate embryogenic cells, which undergo a series of morphological and biochemical changes resulting in the formation of somatic embryos (Schmidt et al., 1997; Komamine et al., 2005; Businge et al., 2013). Somatic embryogenesis forms the basis of cellular totipotency that is unique to higher plants. Currently, this clonal technique is considered to represent a prominent in vitro regeneration system for cocoa. Somatic embryogenesis allows rapid regeneration of elite genotypes, germplasm conservation and genetic transformation system (Maximova et al., 2002). Unfortunately, the recalcitrance of T. cacao to this technique and numerous factors that control it limit its systematical exploration. In this study, a comparative approach was applied to study physiological differences among T. cacao explants-derived calluses from different origins with the hope to understand the complexity of this phenomenon (recalcitrance). The influence of the position of flowers buds on the tree on somatic embryogenesis of "Sca 6" genotype, known as highly productive in farm was analyzed. The variation of phenols compounds, total soluble sugars, proteins contents and peroxidases activities was also performed in these conditions.

Results showed that both explants types used (staminodes and petals) are favorable to callogenesis with an average frequency above 80% in all callus-derived explants used in this experimentation. These results are in agreement with those previously obtained by Li et al. (1998) and Minyaka et al. (2009). These authors showed that callogenesis was effective in T. cacao using floral explants. The action of 2.4-dichloro-phenoxyacetic acid (2.4-D) and that of thidiazuron (TDZ) are responsible for this callogenesis. A synergic and/or a complementary effect of auxin/cytokinin action in induction of somatic embryogenesis process were reported in many species such as Ricinodendron heudelotti (Fotso et al., 2007) and Vitis vinifera (Olah et al., 2009). This callogenesis success could be due to the high mobilization of soluble sugars during this development step. This might consolidate the fact that reducing carbohydrates are important for calluses formation and cell differentiation (Ana et al., 1997). They regulate osmotic pressure (Blanc et al., 1999) and are major components of cell wall. In cocoa, the use of these metabolites is origin-dependent, as demonstrated by this study. And this could be justified by enzymatic equipment which varies in composition and/or function in cells of flowers buds according to their position on the tree. Study done by Alemano et al. (2003) showed that flowers buds of cocoa contain different types of phenolic compounds, and each type may be expressed qualitatively and quantitatively according to the developmental stage, indicating the key role played by these metabolites in the regulation of cell differentiation events.

Somatic embryogenesis in T. cacao is nature-and origin-dependent as shown by the present study data, but the last factor seems to be the most important. In fact, the average frequency of calluses producing somatic embryos increased when we move up from the orthotropic main stem to the secondary plagiotropic fan branch (with a maximum in staminodes-derived calluses from FII). Except for flowers buds from FII, the same observation is made for calluses of petals, as these last do not differentiate somatic embryos. Thus, somatic embryogenesis is an environmental stress-response Figure 4. Influence of the position of flowers buds on the tree on the soluble sugars content during somatic embryogenesis. Staminodes (A) and petals (B); position of flowers: orthotropic main stem (OS), primary plagiotropic fan branch (BI) and secondary plagiotropic fan branch (FII). Vertical bars represent standard error. Each plot was drawn from means of four identical experiments. Figure 5. Influence of the position of flowers buds on the tree on proteins content during somatic embryogenesis. Staminodes (A) and petals (B); position of flowers: orthotropic main stem (OS), primary plagiotropic fan branch (FI) and secondary plagiotropic fan branch (FII). Vertical bars represent standard error. Each plot was drawn from means of four identical experiments.

process, imposed on culture medium, and this response depends on the expression of genes present in tissues (Johnson et al., 1997). This low rate of somatic embryos produced can be explained by the recalcitrance of this crop to somatic embryogenesis process.

Soluble sugars content was high in calluses from all origins during embryos dedifferentiation step in primary and secondary plagiotropic fan branches. This increase could be due to the important role play by carbohydrates metabolism in organogenesis and morphogenesis. During embryos differentiation steps, this phenomenon is progressively inversed. Moreover during callogenesis, there is a decrease of carbohydrates content in flowers buds suggesting an important utilization of these metabolites. Zygotic and somatic embryos both highly accumulate enzymes of carbohydrate metabolism as demonstrated by several studies (Iraqi et al., 2001; Hendriks et al., 2003; Noah et al., 2013). The explanation for extensive carbohydrate metabolism is the heavy energy demand required for processes that occur during cell division and elongation. One of the factors generally considered as responsible for in vitro recalcitrance of cocoa is the high polyphenol compounds content and their oxidation. In this experiment, a high accumulation of these metabolites was particularly observed during callogenesis and embryos differentiation steps (from 70th to 91st day). Each type of explants used expressed a significant decrease in phenols compounds during embryos dedifferentiation stage (after 49th day) when grown in DKW medium supplemented with sulphate, with the lowest content reported in petals-derived calluses from secondary plagiotropic fan branch. There are several internal and external factors affecting the quality and/or the quantity of polyphenol compounds in plants. The quantitative differences registered within explants-derived calluses from different source could be explained, at least in part, by the interaction of several genetic, physiological, agronomic (position of flower bud on the tree), and environmental factors (microclimate) modifying the final concentration in each flower (Roubelakis-Angelakis and Kliewer, 1986). Besides light and temperature (Wang and Zheng, 2001), the availability of plant nutrients also has a great influence on the accumulation of polyphenols (Francis and Atwood, 1961; Doak and Miller, 1968; Piccaglia et al., 2002). Finally, a high accumulation of polyphenol compounds in floral explants during induction steps has been demonstrated to be not favorable to somatic embryogenesis process. A similar result was obtained in date palm (Phoenix dactylifera L.) by Zouine and El Hadrami (2004). Peroxidase activity was high in calluses during induction steps of somatic embryogenesis and low during dedifferentiation step (after 49th day of subculture). This result might underline the implication of this peroxidase in somatic embryogenesis. In fact, the controversy of the higher peroxidase activity during dedifferentiation step of somatic embryogenesis in calluses of petals could justify their low embryos production capacity as compared to their staminodes counterpart.

In conclusion, the data presented here clearly demonstrate the efficiency of floral explants from different position on the tree to somatic embryogenesis process. The comparison between flowers buds carried by the secondary fan branch, commonly used to regenerate T. cacao plantlets, and those from the main stem and primary fan branch showed many differences in physiological responses. The most pronounced difference among the tree types of explants-derived calluses concern with carbohydrates metabolism: FII and FI biochemical evaluation displayed a low utilization of carbohydrates, while OS explants are characterized by intensive glycolytic activities as documented by the exceptional decreased of sugars content during somatic embryogenesis. The absence of embryos dedifferentiation on petals-derived calluses from FII has been connected, at least in part, to changes in specific enzymes abundances. The results suggest that stress factors (microclimate) and genetic factors affect embryogenic capacity of floral explants in T. cacao and thereby reduce the regeneration frequency.

CONFLICT OF INTERESTS

The authors have not declared any conflict of interests.

REFERENCES

1. Alemano L, Ramos T, Gargadenec A, Andary C, Ferriere N (2003). Localization and identification of phenolic compounds in Theobroma cacao L. somatic embryogenesis. Ann. Bot. 92:613-623.
2. Ana BM, Yolanda C, Hilario G. Piedad G, Oscar H, Luisa M, Ana D, Nieves V (1997). Differences in the contents of total sugars, reducing sugars, starch and sucrose in embryogenic and non-embryogenic calli from Medicago arborea L. Plant. Sci. 154:143-151.
3. Babu S, Shareef M, Shetty P, Shetty T (2002). HPLC method for amino acids profile in biological fluids and inborn metabolic disorders of aminoacidopathies. Indian. J. Clin. Biochem. 17(2):7-26.
4. Blanc G, Michaux- Ferriere N, Teison C, Carron MP (1999). Effect of carbohydrate addition on the induction of somatic embryogenesis in Hevea Brasiliensis. Plant Cell Tissue Organ Cult. 59:103-112.
5. Bradford MM (1976). A rapid and sensitive method for the quantitation of microgram quantities of protein utilizing the principle of protein dye binding. Anal. Biochem. 72:248-254.
6. Businge E, Bygdell J, Wingsle G, Moritz T, Egertsdotter (2013). The effect of carbohydrates and osmoticum on storage reserve accumulation and germination of Norway spruce somatic embryos. Physiol. Planta. 149:273-285.
7. Cangahuala-Innocente GC, Steiner N, Maldonado SB, Guerra MP (2009). Patterns of protein and carbohydrates accumulation during somatic embryogenesis of Acca sellowiana. Pesq. agropec.
8. Doak KD, Miller PR (1968). Influence of mineral nutrition on pigmentation in sorghum. Agro. J. 60:430-432.
9. Driver JA, Kuniyuki AH (1984). In vitro propagation of paradox walnut root stock. Hort. Sci. 19:507-509.
10. El Hadrami I (1995). L'embryogenèse somatique chez Phoenix dactylifera L.. quelques facteurs limitant et marqueurs biochimiques. Thèse de Doctorat, d'Etat Université Cadi Ayyad, Faculté des Sciences-Semlalia, Marrakech, P 227.
11. Erdelsky´ K, and Fricˇ F (1979). Practical and analytical methods in plant physiology. SPN, Bratislava.
12. Fehér A, Pasternak TP, Dudits D (2003). Transition of somatic plant cells to embryogenic state. Plant Cell Tissue Org. Cult. 74:201-228.
13. Fotso, Donfagsiteli TN, Sanonne, Omokolo ND (2007). Effet des phytohormones exogènes sur l'évolution de certains paramètres biochimiques au cours de l'embryogenèse de Ricinodendron heudolotii. Baill. Fruits 62:302-315.
14. Francis FJ, Atwood WM (1961). The effect of fertilizer treatment on the pigment content of cranberries. Proc. Am. Hort. Sci. 77:351-358.
15. Gibson SI (2005). Control of plant development and gene expression by sugar signaling. Curr. Opin. Plant Biol. 8:93-102.
16. Hendriks JHM, Kolbe A, Gibon Y, Stitt M, Geigenberger P (2003). ADP-Glucose pyrophosphorylase is activated by posttranslational redox modification in response

to light and to sugars in leaves of Arabidopsis and other plant species. Plant Physiol. 133:838-849.

17. Iraqi D, Tremblay FM (2001). Analysis of carbohydrate metabolism enzymes and cellular contents of sugar and proteins during spruce somatic embryogenesis suggests a regulatory role of exogenous sucrose in embryo development. J. Exp. Bot. 52:2301-2311.
18. Jiménez VM (2001). Regulation of in vitro somatic embryogenesis with emphasis on the role of endogenous hormones. Rev. bras. Fisiol. Vegetal 13:196-223.
19. Komamine A, Murata N, Nomura K (2005). Mechanisms of somatic embryogenesis in carrot suspension cultures – morphology, physiology, biochemistry, and molecular biology. In vitro Cell. Dev. Biol. Plant 41:6-10.
20. Lass RA (1999). Cacao growing and harvesting practices. In: Knight, I. (Ed.), Chocolate and Cocoa Health and Nutrition. Blackwell Science Ltd., Oxford, London, pp. 11-42.
21. Li Z, Traore A, Maximova S, Guiltinan MJ (1998). Somatic embryogenesis and plant regeneration from floral explant of cacao (Theobroma cacao L.) using thidiazuron. In vitro Cell. Dev. Biol. Plant 34:293-299.
22. Lulsdorf MM, Tautorus TE, Kikcio SI, Dunstan DI (1992). Growth parameter of somatic embryogenic suspension culture of interior spruce (Picea glauca-engelmannii complex) and black spruce (Picea mariana Mill.). Plant Sci. 82:227-234.
23. Macheix JJ, Fleuriet A, Billot J (1990). Fruit Phenolics. Florida: CRC press. Inc. Boca Ranton, P 378.
24. Maximova SN (2008). Field performance of Theobroma cacao L. plants propagated via somatic embryogenesis. In vitro Cell. Dev. Biol. Plant 44:487-493.
25. Maximova SN, Alemano L, Young A, Ferriere N, Traore A, Guiltinan MJ (2002). Efficiency, genotypic variability, and cellular origin of primary and secondary somatic embryogenesis of Theobroma cacao L. In Vitro Cell. Dev. Biol. Plant 38:252-259.
26. Minyaka E (2009). Métabolisme du soufre et embryogenèse somatique chez Theobroma cacao L. (Malvaceae). Thèse de Doctorat/Ph.D, Université de Yaoundé I, Université de Cocody/Abidjan, 132 p.
27. Minyaka E, Niemenak N, Fotso, Sangare A, Omokolo ND (2008). Effect of MgSO4 and K2SO4 on somatic embryo differentiation of Theobroma cacao L. Plant Cell Tissue Organ Cult. 94:149-160. Minyaka E, Niemenak N, Issali EA, Sangare A, Omokolo ND (2010). Sulphur depletion altered somatic embryogenesis in Theobroma cacao L. Biochemical difference related to sulphur metabolism between embryogenic and non-embryogenic calli. Afr. J. Biotechnol. 9(35):5665-5675.
28. Niemenak N (1998). Recherche des marqueurs biochimiques et modification histologique au cours de la callogenèse et de l'embryogenèse somatique in vitro chez Theobroma cacao L. Thèse de doctorat 3ème cycle, Université de Yaoundé, P 138.
29. Niemenak N, Saare-Surminski K, Rohsius C, Omokolo ND, Lieberei R (2008). Regeneration of somatic embryo in Theobroma cacao L. in Temporary immersion bioreactor and analyses of free amino acids in different tissues. Plant Cell Rep. 27:667-676.
30. Noah MA, Niemenak N, Sunderhaus S, Haase C, Omokolo ND, Winkelmann T, Braun HP (2013). Comparative proteomic analysis of early somatic and zygotic embryogenesis in Theobroma cacao L. J. Prot. 78:123-133.
31. Olah R, Zok A, Pedryc S, Howard, Kovacs LG (2009). Somatic embryogenesis in a broad spectrum of grapes genotypes. Sci. Hort. 120:134-137.
32. Piccaglia R, Marotti M, Baldoni G (2002). Factors influencing anthocyanin content in red cabbage (Brassica oleracea var capitata L f rubra (L) Thell). J. Sci. Food Agric. 82(13):1504-1509.
33. Roubelakis-Angelakis KA, Kliewer WM (1986). Effect of exogenous factors on anthocyanins and total phenolics in grape berries. Am. J. Enol Vitic. 37:275-280.

34. Schmidt EDL, Guzzo F, Toonen MAJ, De Vries SC (1997). A leucine-rich repeat containing receptor-like kinase marks somatic plant cells competent to form embryos. Development 124(10):2049-2062.
35. Singleton VL, Rossi JA (1965). Colorimeter of total phenolics with phosphomolybdic-phosphotungstic acid reagents. Am. J. Enol. Viticulture 16:144-158.
36. Wang SY, Zheng W (2001). Effect of plant growth temperature on antioxidant capacity in strawberry. J. Agric. Food Chem. 49:4977- 4982.
37. Wood GAR, Lass RA (1985). Cacao. Longman Scientific and Technical (ed, England), P 620.
38. Zouine L, El Hadrami I (2004). Somatic embryogenesis in Phoenix dactylifera L.: effect of exogenous supply of sucrose on Proteins, Sugars and Peroxidases activities during the Embryogenic Cell Suspension Culture. Biotechnol. 3(2):114-118.

CHAPTER 5

Managing Irrigation of Fruit Trees Using Plant Water Status

Rashid Al-Yahyai

Department of Crop Sciences, College of Agricultural and Marine Sciences, Sultan Qaboos University, Muscat, Sultanate of Oman

ABSTRACT

Optimum growth and production of fruit crops is strongly linked to managing irrigation water. Various method of estimating tree water requirements have been utilized such as direct and indirect soil, water, and climatic measurements. Due to differences in fruit tree anatomical and morphological structures and their adaptation to excess and deficit soil water content, such estimates of irrigation water requirements may be more suitable for herbaceous plants but not as accurate for trees. Studies on temperate and tropical fruit trees, using apple (Malus domestica) and star-fruit (Averrhoa carambola), respectively, showed that tree water potential is highly correlated to soil water status. Irrigation based on climate data (ET) and monitoring of soil water resulted in no significant differences in soil or tree water status of orchard-grown fruit trees under temperate and subtropical climatic conditions. The results indicated the need for better understanding and utilization of tree physiological parameters for management of irrigation water of fruit crops. This will ultimately lead to achieving optimum yield and fruit quality while conserving water resources.

Keywords: apple; star-fruit; water potential; water content; tree physiology; temperate; tropics; irrigation scheduling; tree water status

1. INTRODUCTION

Irrigation is a vital management practice in fruit production regions of the world, particularly in arid and semi-arid climates. Despite the variations in the available water resources among fruit producing countries, the question of frequency of irrigation and amount of water applied to fruit crops is a common concern. Insufficient water supply may result in reduced tree growth, yield and fruit quality due to water stress. Excessive irrigation, on the other hand, may increase nutrient leaching, water-logging problems, incidence of pests and diseases, and the associated cost of frequent operation and maintenance of the irrigation system. In addition, unmanaged irrigation may lead to adverse environmental effects such as agrochemical leaching into groundwater aquifers, reduced water reservoir, and water and soil salinity. Optimizing water applications by scheduling irrigation to fruit orchards may increase water conservation, reduce production costs, and increase tree growth and yield. Irrigation scheduling is especially important in horticultural crops because net returns are normally higher than those of other crops [1].

According to Hillel [2], the main issue with irrigation management is to determine the frequency, quantity, and timing of irrigation to optimize crop growth and productivity. However, irrigation scheduling and management of fruit crops where rain falls in sufficient quantities in temperate, tropical and subtropical regions is not well established. For example, irrigation was not a common practice in fruit orchards in the humid-temperate regions such as New York, USA. However, in recent decades, irrigation has become more common since it was proven to increase growth and promote early production of high-density apple orchards [3]. Availability of weather data allows the use of evapotranspiration (ET) predictions as a basis for irrigation scheduling [4]. Similar irrigation scheduling approaches were reported by Ebel [5, 6]. Irrigation based on climatic data, is the method commonly used to schedule irrigation of apple orchards in North-Eastern United States.

South Florida is the main region in the continental United States where subtropical and tropical fruit crops are grown. In this region, trees are commonly irrigated at rates and frequencies based mainly on grower's experience and observations of crop growth and yield rather than on quantitative scientific information [7-9]. According to a survey by Muñoz-Carpena [10], 64.3% of fruit growers irrigate trees based on rainfall frequency and quantity. Problems concerning variability in irrigation duration and frequency were rated high among tropical fruit growers [11], which highlight the need for a better understanding of irrigation requirements of these crops.

Measurement of soil water content and soil matric potential provide an index of the rate at which water is taken up by the plant or lost from the root zone. Soil water content and potential is therefore most useful in conjunction with information about the soil-plant-atmosphere system [12]. Water extraction from the soil is dependent on plant properties that determine the plant water potential (Ψ) at which a particular plant species can continue to grow and extract water from the soil [13].

Although climateand soil-based methods provide a means for estimating irrigation amount and timing, they do not take into account the variability between fruit tree species and cultivars, growth stage, or the response of trees to soil moisture deficit. The water refill point, which is the lowest possible soil water content with no decrease in yield or fruit quality, varies among different tree varieties, rootstocks, soils and seasons [14,15]. Several physiological variables are used as indicators of tree water status. Among the most frequently used is tree water potential [16,17].

It is often beneficial to use both soil and plant factors for irrigation scheduling. An integrated approach utilizing soil and plant factors suggested by Buss [18] included soil surveying to determine soil properties and available soil moisture content and crop factors that included crop type, canopy size, rooting width and depth, and crop density. Irrigation scheduling was then planned from a combination of available water data, crop water extraction rates, and the irrigation system layout. Physiological processes in fruit trees such as water potential and gas exchange are sensitive to changes in soil water content [17,19]. These physiological variables, growth, and fruit production should be correlated with soil water content prior to determining the appropriate amount of water to apply to an orchard. Little is known about the response of temperate fruit crop, such as apple, and tropical fruit crop, such as star-fruit, to changes in soil water content under field conditions in humid-temperate and subtropical climates. This paper discusses the relationship between soil water content and water potential of apple and star-fruit trees grown in irrigated orchards using climate-based (ET) and soil-based (capacitance probes) irrigation scheduling methods.

2. MATERIALS AND METHODS

2.1. Apple Response to Soil Water Content

Experimental site and treatments. Young apple trees (Malus domestica Borkh.) (cv. "Gala" on M.9 rootstock) were planted on May 28, 1996 at a spacing of 2 m within row and 4 m between rows at the New York State Agricultural Experiment Station at Geneva, NY. The soil at the experimental site is fine Lima silt loam, Glossoboric Hapludalf, which is moderately well-drained [20]. The soil is fine-textured with high water holding capacity and has a hard pan at about 50 cm below the soil surface. A total of 20 treatments consisting of 18 irrigation treatments and two non-irrigated controls were applied during the growing season from Jun. to Sept. of 1997. The irrigation treatments were a factorial combination of two irrigation frequencies (daily and weekly); three irrigation methods (one-line trickle per row, two-trickle lines per row, and microsprinklers); and three irrigation amounts (50% ET, 75% ET, and 100% ET). The two non-irrigated control treatments were rain-excluded and rainfed. The rainfed treatment received only natural rainfall, while plastic rain-shelters were used to exclude rainfall in rain-excluded treatments [21]. A total of six replications were used, within each replication, there were two trees per plot. Each plot was separated from the next by one McIntosh/M.9 apple tree. Blocking was done by location in the field and initial trunk diameter where the total number of test trees was 240. The experiment was bounded by guard rows of Gingergold/M.9 apple trees. After planting, the trees were pruned to five lateral branches located between 53 cm and 95 cm from the rootstock/scion union. All trees were given 20 L of a dilute starter fertilizer solution (31.4 g N, 13.6 g P_2O_5, and 13.6 g K_2O) and then fertilized with 113 grams of $CaNO_3$ per tree two weeks after planting.

Irrigation scheduling. Irrigation was scheduled by calculating ET using pan evaporation measurements adjusted by a crop coefficient. Evaporation data were collected from a US Weather Bureau Class A evaporation pan (Epan) located at about 1.75 km from the site. Epan was adjusted by a crop coefficient (Kcrop) to calculate ET. The crop coefficients used were those developed for mature apple trees [22]. Their values were adjusted for the small canopy size by the method proposed by Ley [22]. Daily water deficits were calculated and then irrigation time in hours was calculated on a daily basis. For the daily treatments, irrigation time was adjusted every day, likewise for the weekly treatments, daily irrigation time was summed and applied once a week.

Soil moisture content measurements. A neutron probe (CPN Model 503 DR Hydroprobe, CPN Corporation, Marines, CA) was used to determine soil moisture content (SMC). Neutron probes provide non-destructive and repetitive soil moisture measurements regardless of the soil condition at several locations in the field [23]. Weekly soil moisture measurements were taken at every 15 cm depth through the soil profile to 75 cm below the soil surface. Neutron probe aluminum access tubes were installed at 30 cm from the trunk of the trees.

Stem water potential measurements. Stem water potential (ψ_s) was measured between 1100 and 1400 HRS using a pressure chamber (Soil Moisture Equipment Corp., Santa Barbara, Calif.). Stem water potential was determined using a fully matured leaf enclosed in a reflective plastic bag for 1 hr to suppress transpiration and allow stem water potential to equilibrate with leaf water potential [24-26]. Three mature leaves were used per tree for stem water potential measurement.

2.2. Star-Fruit Response to Soil Water Content

Experimental site and plant material. An orchard of 8-year-old "Arkin" star-fruit trees grafted onto Golden Star rootstock were used for this study. The trees were planted with 4.5 m within and 6.0 m between rows and surrounded by a windbreak. The orchard soil was Krome very gravelly loam soil, loamy-skeletal, carbonatic, hyperthermic, Lithic Udorthents [27]. Fertilizer application, pest management, and fertilization followed standard practice for South Florida's commercial star-fruit production. Trees were irrigated with 89 $L \cdot h^{-1}$ microsprinklers with a 360° wetting area. In the orchard, trees were irrigated at various times and rates to provide a range of SWD over the course of the experiments. The ranges were determined from a preliminary study (Al-Yahyai, unpublished data) and based on the SWD from soil field capacity (FC) to leaf yellowing, which was the first visual sign of water stress. For orchard trees, irrigation was withheld toward the end of the experiment for 121 d from 1 Aug. to 30 Nov. 2003 to obtain high SWD values.

Soil water measurements. Soil water depletion was determined by continuously measuring soil water content using multisensor capacitance probes (EnviroSCAN, Sentek PTY Ltd., Kent Town, Australia), as recommended for Krome soils in a comparative study with tensiometers and neutron probes [8]. Capacitance probes were installed 60 cm north of tree trunks in the orchard. The sensors recorded soil moisture content every 30 min at depths of 10, 20, 30 and 50 cm. The data were stored in a datalogger and later downloaded to a computer for analysis. Soil water content measurements from sensors placed at the 4 depths were summed and plotted using the EnviroSCAN proprietary system software. Installation of the system in Krome soils was previously described by Al-Yahyai et al. [8]

and Núñez-Elisea et al. [10], and technical multisensor capacitance probe specifications were described by Paltineanu and Starr [28].

Irrigation scheduling. Irrigation scheduling was based on soil water depletion from continuously monitoring soil water content with the EnviroScan. EnviroSCAN software was used for plotting soil water depletion (SWD) over time and determining the "full point" (field capacity), the "refill point" (time to irrigate), and the theoretical "onset of water stress", based on the rate of SWD [28]. Irrigation of orchard trees was initiated when SWD reached one of the following levels (where 0% SWD = FC): 0% - 8% SWD, 9% - 11% SWD, 12% - 14% SWD, or 15% - 17% SWD.

Stem water potential measurements. The procedure for measuring stem water potential (ψ_s) of star-fruit was similar to that of apple trees mentioned above. Stem water potential was determined periodically on four mature leaves per tree from the orchard that were randomly selected for water potential measurements.

2.3. Statistical Analysis

Linear regression analysis was performed with SigmaPlot (SYSTAT Software Inc., Richmond, Cal.) and Inflection point was determined using NONLIN Procedure for SAS (SAS Institute, Inc., Cary, North Carolina, USA).

3. RESULTS AND DISCUSSION

3.1. Apple Response to Soil Water Content

Soil water content. Neutron probe measurements taken at 15, 30, 45, 60 and 75 cm indicated that soil moisture content was lowest in the top 15 cm of the soil profile. Rain-excluded and rainfed treatments showed significantly lower soil moisture content throughout the season in the upper 45 cm of the soil profile than irrigation treatments. However, there were no significant differences in soil water content among treatments of irrigation frequencies and systems and amounts. Moreover, no significant differences in soil water content were observed among irrigation amounts of 50% ET, 75% ET and 100% ET or between rainfed and rain-excluded treatments throughout the growing season of 1997. During the growing season, calculated water deficit as determined from ET was greater than rainfall for most of the season (**Figure 1**). In fruit orchards, soil moisture varies according to soil type and characteristics, cultivar and rootstock, and climatic conditions [19,29,30]. Lack of significant differences in soil water content in this study may be attributed to one or all of these factors.

Stem water potential. Stem water potential of rainfed and rain-excluded treatments was significantly lower than that of irrigated trees. However, no significant differences were found among trees irrigated at 50% ET, 75% ET, and 100% ET. No significant correlation between soil water depletion (SWD) and stem water potential was observed (**Figure 2**). This indicated that soil moisture content was not the only factor that influenced the reduction of stem water potential in non-irrigated treatments when compared to irrigated treatments. According to Goode and Higgs [31], if soil moisture is high and evaporative demand is low, plant water potential will remain high and irrigation will have a negligible effect on plant water status. Although plant stem water potential and soil water content were increased by irrigation, trees have not reached a critical water stress level and did not respond to various amounts of water applied. Ebel et al. [32] reported that stem water potential of fruit trees change little over a range of soil water depletion that can reach 80%. Thus corresponding reduction in growth and yield response can only be detected when trees are severely stressed at soil water depletion (SWD) levels below 25% [30].

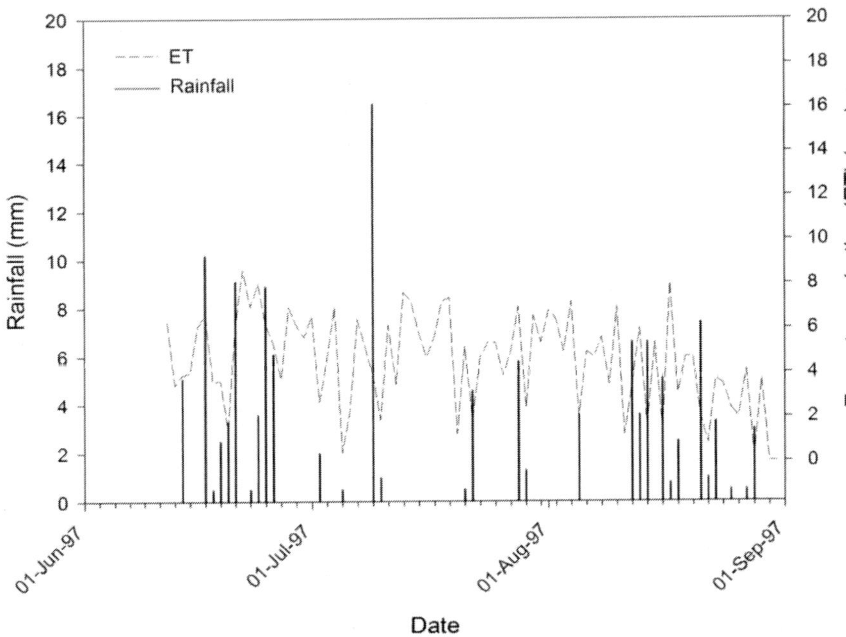

Figure 1. Rainfall and evapotranspiration (ET) during apple growing season of 1997 in Geneva, New York.

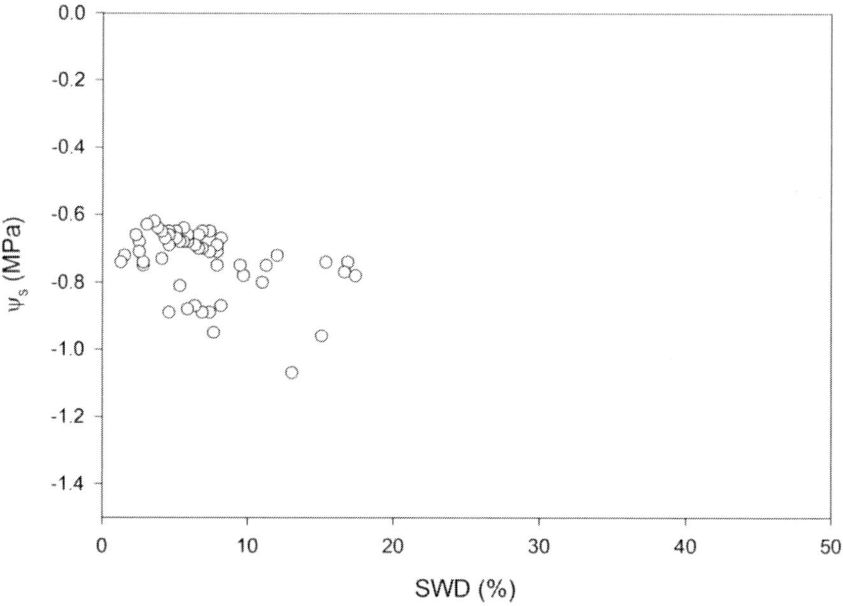

Figure 2. Soil water depletion (SWD) and stem water potential (Ψ_S) of "Gala" apple on M.9 rootstock grown in Geneva, New York.

3.2. Star-Fruit Response to Soil Water Content

Soil water content. In southern Florida, where tropical and subtropical fruit crops are grown, rainfall patterns and evapotranspiration (ET) followed historical records. Of the 1611 mm of total rainfall in 2003, 79.5% occurred between May and Oct. (**Figure 3**). Total ET was estimated as 1129 mm with 67% occurred during the same period (**Figure 3**). Nonetheless, the distribution within the year, however, varied greatly among and within months and daily rainfall ranged from 0 to 93 mm and ET from 0.99 mm to 5.33 mm. During the winter months there was 20% higher ET than rainfall. Irrigation of star-fruit orchard was scheduled using changes in soil moisture content over time as determined by multisensor capacitance probes. Continuous measurement of soil water content in tropical fruit orchards in Krome very gravelly loam soils was determined using multisensor capacitance probes [7,8,33,34]. Irrigation was applied to the "full point" (field capacity) to maintain soil water content above the theoretical "onset of water stress", the point at which the ET rate is significantly reduced [7] and [34]. In the present study, soil water content declined in a typical stepwise pattern over the course of the experiment and never reached the "onset of stress" levels in the orchard.

Stem water potential. The soil water depletion (SWD) of star-fruit orchard remained below 30% throughout the year 2003. Within the range of SWD from 0% to 30%, stem water potential remained (Ψs) above −1.0 MPa and did not significantly correlate with SWD (**Figure 4**). Therefore, within a 0% to 30% SWD range, there was no significant effect of Ψs on leaf gas exchange, presumably due to sufficient soil water content in the orchard from frequent rainfall and capillary rise to the root zone from the water table which was 1 - 2 m below the soil surface [35]. Similarly, in a study of mango trees in Krome soil, predawn water potential ranged from −0.3 to −0.5 MPa and there were no significant differences in tree physiological processes were observed between irrigated and non-irrigated trees [36]. Ismail et al. [37,38] observed that withholding water for 1, 2, or 4 weeks resulted in midday Ψ_L of −1.3 to −1.5 MPa, −2.0 to −2.5 MPa, or −2.9 MPa of container-grown star-fruit trees, respectively, in comparison to trees irrigated at field capacity that had a Ψ_L of −0.9 MPa.

Al-Yahyai et al. [17] reported that stem water potential of star-fruit trees in containers in the greenhouse was above −1.0 MPa when SWD increased from 0% to 60% above which there was a sharp decrease in stem water potential. A nonlinear reanalysis of the data showed that the reflection point at which stem water potential was reduced linearly with a decline in SWD was 58% (**Figure 5**). A similar trend was observed in container-grown trees with a sharp decrease in stem water potential when SWD reached 50% under field conditions [17]. The ability of star-fruit trees to regulate their water potential was attributed to osmotic adjustment through increased levels of proline [37] (Ismail et al., 1994) and a reduction in stomatal conductance [37-39].

Stem water potential is a useful indicator of water stress and thus can be used for irrigation management [26,40,41]. Al-Yahyai et al. [17] found that physiological processes such as leaf gas exchange of container-grown star-fruit trees in the field and in a greenhouse was more closely correlated with stem water potential than with SWD. In this study, Ψs was not influenced by SWD of up

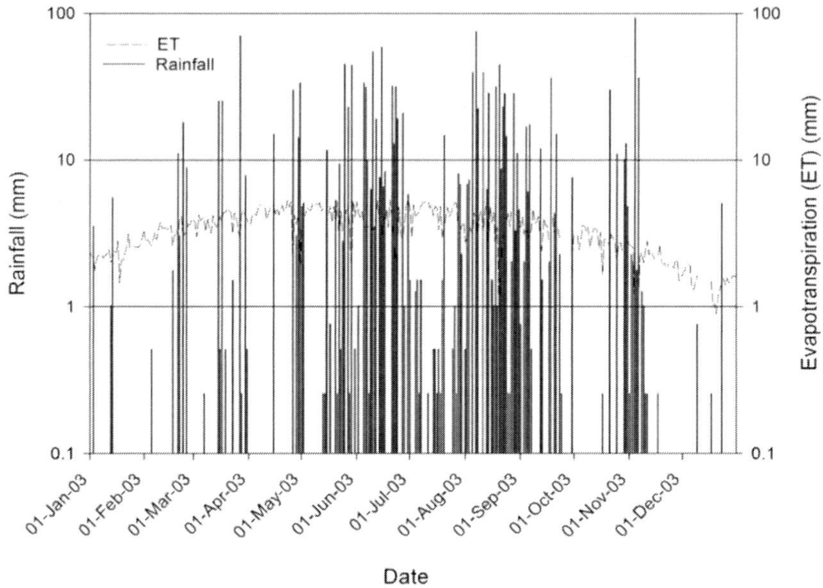

Figure 3. Rainfall and evapotranspiration (ET) during star-fruit growing season in 2003 in Homestead, Florida (Y axes on log scale).

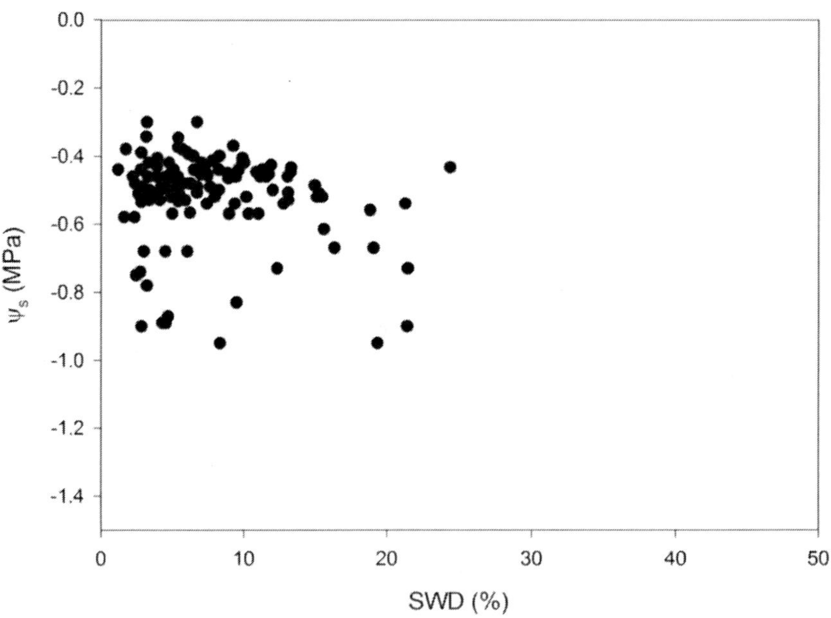

Figure 4. Soil water depletion (SWD) and stem water potential (Ψ_S) of "Arkin" star-fruit on Goldenstar rootstock grown in Homestead, Florida.

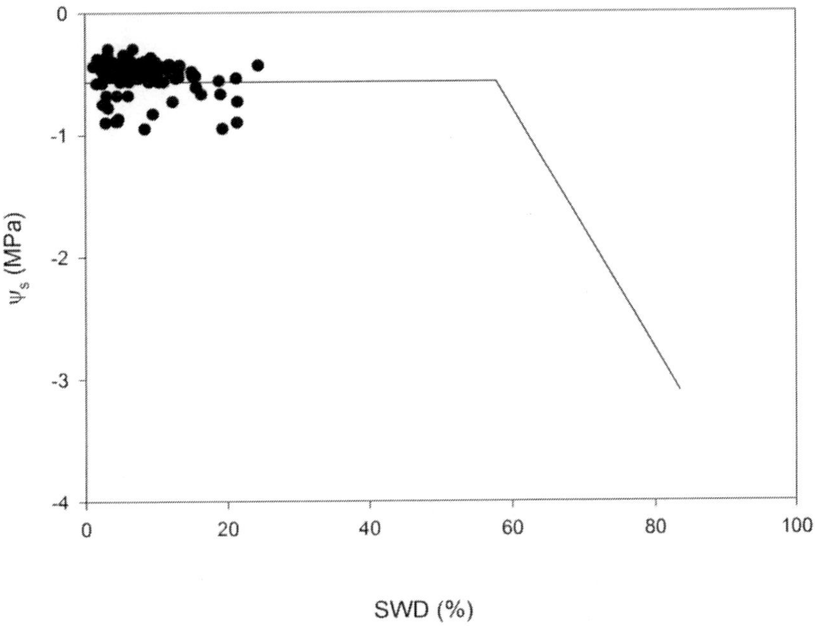

Figure 5. Soil water depletion (SWD) and stem water potential (Ψs) of "Arkin" star-fruit on Goldenstar rootstock grown in an orchard in Homestead, Florida. The data are plotted against nonlinear model of greenhouse-grown star-fruit subjected to continuous soil water depletion (model data from Al-Yahyai et al. [17]).

to 30% under field conditions (Figures 4 and 5). This is despite continuous monitoring of water status which indicated that irrigation was required. Measurements of Ψs and net CO_2 assimilation (A) chosen randomly at a point beyond the theoretical "onset of water stress" point as measured with multisensor capacitance probes indicated that growers frequently applied more water than needed by trees (**Figure 6**). This also suggests that using soil water monitoring devices to schedule irrigation without correlation to tree water status may lead to over irrigation. This was confirmed with results from Al-Yahyai et al. [9] who applied from 623 to 890 liters per tree per year during 22 hours of irrigation. This is considerably less than 0.1% of the amount of water applied by the star-fruit growers who irrigate from 548 to 2190 hrs of irrigation yearly as much as to 21,700 to 115,739 liters per tree per year.

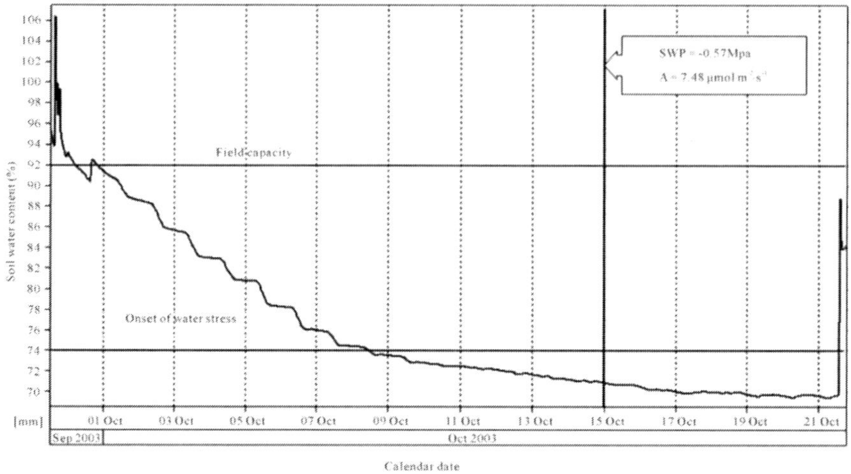

Figure 6. Soil water depletion (SWD) as measured with capacitance probes (EnviroSCAN) in an 8-year-old starfruit orchard in Krome soils. Points of "field capacity" and "onset of water stress" were determined based on the pattern of soil water depletion. Measurements of stem water potential (SWP) and net CO_2 assimilation (A) indicated by the vertical double-line on the x-axis were chosen randomly at a point beyond the theoretical "onset of water stress" point.

Stem water potential relationship to leaf water potential. It has been suggested that leaf water potential (Ψ_L) is a less reliable measure when making irrigation scheduling decisions for fruit trees than stem water potential (Ψ_S) due to its inherent variability under field conditions [25,40,42]. Enclosing leaves in a reflective, dark plastic bag for 1 - 2 h prior to water potential measurements will reduce transpiration until an equilibrium is achieved between the Ψ_L and the stem water potential [24], thus providing a more precise method of determining tree water status [25,26,40,42]. In this study, Ψ_L and Ψ_S from the same branch were measured within 2 min of each other to determine their relationship. There was a strong relationship ($r^2 = 0.85$) between Ψ_L and Ψ_S in orchard trees (**Figure 7**), despite the narrow water potential range of −0.3 to −1.1 MPa. Jones [13] suggested that either stem water potential or Ψ_L may be used to assess the effect of water stress on a particular tree response that is assumed being affected by tree water status. Our results with star-fruit trees appeared to confirm this.

4. CONCLUSIONS

Results from this study showed that irrigation scheduling of apple and star-fruit based on climatic parameters (ET) and measurements of soil water content, respectively, did not affect stem water potential under field conditions. Soil water content and stem water potential (Ψs) of apple trees irrigated at 50% ET, 75% ET, and 100% ET did not significantly differ and no correlation between soil and tree water status was found. Similarly, in a subtropical climate of south Florida, stem water potential of star-fruit trees was not influenced by irrigation based on continuous measurement of soil water depletion below 30%. In addition to tree factors, lack of correlation between soil water depletion and stem water potential in apple may have been due to frequent rainfall and high water holding capacity of the soil. In, Star-fruit high rainfall and capillary rise to the root zone from the water table, located at 1 to 2 m below the soil surface, was sufficient to maintain adequate soil water even after several weeks of no irrigation.

Measurements of soil water content, plant water status or climatic data are all practical and feasible methods of irrigation scheduling of fruit crops. Irrigation scheduling based on soil water content has been widely adopted due to the availability and ease of use of various devices that monitor soil water status either discretely (e.g. tensiometers, neutron probes) or continuously (e.g. multisensor capacitance probes). In recent decades, management of irrigation based on tree water status has gained interest. Several studies have determined that tree stem water potential, gas exchange measurements, sap flow, and stem daily shrinkage, among other tree measurements, can be utilized for making decisions about irrigation of fruit crops. For several fruit crops, yield is more highly correlated with stem water potential than with soil water potential, thus validating the use of irrigation decisions based on plant water status as well as soil water content.

Despite the lack of differences among treatments, substantial savings of water can be gained from irrigation scheduling using climatic and soil water measurement devices of fruit crops. However, irrigation based on such methods should be correlated to tree water status as indicated by physiological processes such as stem water potential. This is especially important in regions where water is scarce such as the Arabian Peninsula.

5. ACKNOWLEDGEMENTS

Rashid Al-Yahyai The author acknowledges the support of Terence Robinson and Alan Lakso, Cornell University, in conducting the apple study.

The author is also grateful to Bruce Schaffer and Frederick S. Davies, University of Florida, for their support during the study of star-fruit. Results of this work was presented during the International Conference on Economic Incentives & Water Demand Management, 18-22 March 2006, Sultan Qaboos University, Oman. Sultan Qaboos University is acknowledged for partially funding this research work.

REFERENCES

1. Fereres, E. (1997) Irrigation scheduling of horticultural crops. Acta Horticulturae, 449, 253-258.
2. Hillel, D. (1998) Environmental soil physics. Academic Press, New York.
3. Robinson, T. and Stiles, W. (1995) Maximizing the performance of young apple trees. New York Fruit Quarterly, 3, 10-16.
4. Hutmacher, R.B., Nightingale, H.I., Rolston, D.E., Biggar, J.W., Dale, F., Vail, S.S. and Peters, D. (1994) Growth and yield responses of almond (Prunus amygdalus) to trickle irrigation. Irrigation Science, 14, 117-126.
5. Ebel, R.C., Proebsting, E.L. and Patterson, M.E. (1993) Regulated deficit irrigation may alter apple maturity, quality, and storage life. HortScience, 28, 141-143.
6. Ebel, R.C., Proebsting, E.L. and Evans, R.G. (1995) Deficit irrigation to control growth in apple and monitoring fruit growth to schedule irrigation. HortScience, 30, 1229-1232.
7. Schaffer, B. (1998) Flooding response and water-use efficiency of subtropical and tropical trees in an environmentally-sensitive wetland. Annals of Botany, 81, 475-481.
8. Al-Yahyai, R., Schaffer, B. and Davies, F.S. (2003) Monitoring soil water content for irrigation scheduling in a carambola orchard in a gravelly limestone soil. Proceedings of the Florida State Horticultural Society, 116, 37-41.
9. Al-Yahyai, R., Davies, F.S., Schaffer, B. and Crane, J.H. (2005) Effect of soil water depletion on growth, yield, and fruit quality of carambola in gravelly loam soil. Proceedings of the Florida State Horticultural Society, 118, 237-241.
10. Muñoz-Carpena, R., Crane, J., Israel, G. and Yurgalevitch, C. (2003) Water conservation survey of Miami-Dade County agricultural and golf course commercial water users. Proceedings of the Florida State Horticultural Society, 116, 15-21.
11. Li, Y., Crane, J., Boman, B. and Balerdi, C. (2000) Irrigation management survey for tropical fruit crops in South Florida. Proceedings of the Florida State Horticultural Society, 113, 40-42.
12. Campbell, G.S. and Campbell, M.D. (1982) Irrigation scheduling using soil moisture measurements: Theory and practice. Advances in Irrigation, 1, 25-42.
13. Jones, H.G. (1990) Physiological aspects of the control of water status in horticultural crops. HortScience, 25, 19-26.
14. Braun, P., Aspinall, D. and Lenz, F. (1989) The use of physiological indicators of water status in irrigation scheduling in citrus. Acta Horticulturae, 240, 267-270.
15. Jones, H.G. (2004) Irrigation scheduling: Advantages and pitfalls of plant-based methods. Journal of Experimental Botany, 55, 2427-2436.
16. Hsiao, T. (1990) Measurements of plant water status. In: Steward, B. and Nielsen, D., Eds., Irrigation of Agricultural Crops, Agronomy Monographs No. 30, ASA, Madison, Wis, 243-279.
17. Al-Yahyai, R., Schaffer, B. and Davies, F.S. (2005) Physiological response of carambola to soil water depletion. HortScience, 40, 2145-2150. [Citation Time(s):4]

18. Buss, P. (1989) Irrigation scheduling for horticulture: An integrated approach. Acta Horticulturae, 240, 261-270.
19. Naor, A. and Cohen, S. (2003) Sensitivity and variability of maximum trunk shrinkage, midday stem water potential, and transpiration rate in response to withholding irrigation from field-grown apple trees. HortScience, 38, 547-551.
20. USDA (United States Department of Agriculture) (1958) Soil survey. Ontario and Yates Counties, New York, Series 1949, No. 5, 56-57.
21. Jones, H.G., Luton, M.T., Higgs, K.H. and Hamer, P.J. (1983) Experimental control of water status in an apple orchard. Journal of Horticultural Science, 58, 301-316.
22. Ley, T.W. (1994). Using PAWS and AgriMet for irrigation scheduling. In: Williams, K.M. and Ley, T.E. Eds., Tree Fruit Irrigation, Good Fruit Grower, Washington DC, 71-77.
23. Chanasyk, D.S. and Naeth, M.A. (1996) Field measurement of soil moisture using neutron probes. Canadian Journal of Soil Science, 76, 317-323.
24. Begg, J.E. and Turner, N.C. (1970) Water potential gradients in field tobacco. Plant Physiology, 46, 343-346.
25. McCutchan, H. and Shackel, K. (1992) Stem-water potential as a sensitive indicator of water stress in prune trees (Prunus domestica L. cv. French). Journal of the American Society for Horticultural Science, 117, 607- 611.
26. Shackel, K.A., Ahmadi, H., Biasi, W., Buchner, R., Goldhamer, D., Gurusinghe, S., Hasey, J., Kester, D., Krueger, B., Lampinen, B., McGourty, G., Micke, W., Mitcham, E., Olson, B., Pelletrau, K., Philips, H. Ramos, D., Schwankl, L., Sibbett, S., Snyder, R., Southwick, S., Stevenson, M., Thorpe, M., Weinbaum, S. and Yeager, J. (1997) Plant water status as an index of irrigation need in deciduous fruit trees. HortTechnology, 7, 23-29.
27. Noble, C.V., Drew, R.W. and Slabaugh, V. (1996) Soil survey of Dade County area, Florida. U.S. Department of Agriculture, Natural Resources Conservation Service, Washington DC.
28. Paltineanu, I.C. and Starr, J.L. (1997) Real-time soil water dynamics using multisensor capacitance probes: Laboratory calibration. Soil Science Society of America Journal, 61, 1576-1585.
29. Lord, W.J., Michelson, L.F. and Filed, D.L. (1963) Response of irrigation and moisture use by McIntosh apple tree in Massachusetts. Experiment Station Bulletin, 537, 1-23.
30. Forshey, G.G. and Dominick, B.A. (1965) Irrigation of apples in the Hudson Valley. New York State Agricultural Experiment Station Bulletin, 809.
31. Goode, A.J. and Higgs, K.H. (1973) Water, osmotic and pressure potential relationships in apple leaves. Journal of Horticultural Science, 48, 203-215.
32. Ebel, R.C., Proebsting, E.L. Evans, R.G. and Ley, T.W. (1988) "Delicious" tree and fruit response to duration of drought. HortScience, 23, 723.
33. Núñez-Elisea, R., Schaffer, B., Zekri, M., O'Hair, S.K. and Crane, J.H. (2001) In situ soil-water characteristic curves for tropical fruit orchards in trenched calcareous soil. HortTechnology, 11, 65-69.
34. Zekri, M., Núñez-Elisea, R., Schaffer, B., O'Hair, S.K. and Crane, J.H. (1999) Multisensor capacitance probes for monitoring soil water dynamics in the oolitic limestone soil of South Florida. Proceedings of the Florida State Horticultural Society, 112, 178-181.
35. Al Al-Yahyai, R., Schaffer, B., Davies, F.S. and Crane, J.H. (2005) Four levels of soil water depletion minimally affect carambola phenological cycles. HortTechnology, 15, 623-630.
36. Larson, K., Schaffer, B. and Davies, F.S. (1989) Effect of irrigation on leaf water potential growth and yield of mango trees. Proceedings of the Florida State Horticultural Society, 102, 226-228.
37. Ismail, M.R., Burrage, S.W., Tarmizi, H. and Aziz, M.A. (1994) Growth, plant water relations, photosynthesis rate, and accumulation of proline in young carambola plants in relation to water stress. Scientia Horticulturae, 60, 101- 114.

38. Ismail, M.R., Yusaf, M.K. and Masturi. A. (1996) Growth and flowering of water stressed star-fruit plants and response to ameliorated water stress. Proceedings of the International Conference on Tropical Fruits, Kuala Lampur, Malaysia, 2, 97-106.
39. Razi, M.I., Awang, M. and Razlan, S. (1992) Effect of water stress on growth and physiological processes of Averrhoa carambola. Acta Horticulturae, 321, 505-509.
40. Naor, A. (2000) Midday stem water potential as a plant water stress indicator for irrigation scheduling in fruit trees. Acta Horticulturae, 537, 447-454.
41. Stern, R.A., Meron, M., Naor, A. Wallach, R., Bravdo, B. and Gazit, S. (1998) Effect of fall irrigation level in "Mauritius" and "Floridian" lychee on soil and plant water status, flowering maturity, and yield. Journal of the American Society for Horticultural Science, 123, 150- 155.
42. Garnier, E. and Berger, A. (1985) Testing water potential in peach trees as an indicator of water stress. Journal of Horticultural Science, 60, 47-56.

CHAPTER 6

Research Progress in Physiological and Molecular Biology Mechanism of Drought Resistance in Rice*

Qiang Chen[1,2], Siyuan Tao[1], Xiaohua Bi[2], Xin Xu[1], Lanlan Wang[3], Xuemei Li[3#]

[1]*Experimental Center, Shenyang Normal University, Shenyang, China*

[2]*Department of Horticulture, Shenyang Agricultural University, Shenyang, China*

[3]*College of Chemistry and Life Science, Shenyang Normal University, Shenyang, China*

ABSTRACT

Rice is one of the most important crops, providing staple food for about half population of the world. Drought stress affects plant growth and development seriously. This article reviewed the research progress of the physiological and molecular biology mechanism including osmotic adjustment, scavenging oxidative radicals, endogenous hormones, drought-resistance genes and epigenetic modification, it may be afford interrelated reference for increasing rice drought resistance and breeding drought resistance rice varieties.

Keywords: Rice; Drought Resistance; Physiology and Biochemistry; Methylation; Epigenetics

1. INTRODUCTION

Rice is one of the most important crops, providing staple food for about half population of the world [1]. Rice production must be increased 60% so as to meet for the contention by the year 2025 [2]. A growing population, urbanization, industrialization, pollution, and drought have compounded the shortage of water resources [3]. Drought is one of the most important constraints in crop resulting in large yield losses and limiting the average yield increase. In fact, since the 1990s China's average annual drought-affected area was up to 26.67 million hectares decreasing food production by 70 - 80 billion kg [4]. Therefore, the development and production of drought-resistant rice varieties is be of great significance in ensuring food security, shortage of water resources, protecting environment and increasing income. The article reviewed the research progresses of the Physiological and molecular biology mechanism, which could supply interrelated reference for increasing rice drought resistance and breeding drought resistance rice varieties.

2. PHYSIOLOGICAL MECHANISM DROUGHT RESISTANCE IN RICE

2.1. Osmoregulation Substance

Study has shown that cellular holds turgidity through inducing solute accumulation and decrease of osmotic potential [5]. Osmoregulation substance could be classified two categories according to mechanism of action: One is inorganicions adjusting osmotic potential of vacuole, such as K^+, Na^+etc., the other is organic matter adjusting osmotic potential of cytoplasm, such as proline, betaine etc. [6]. Cai et al. [7] argued that ability of leaf osmoregulation is greater than the root after subjecting to water stress at different growth stages; In addition, K^+ gave the greatest contribution to whether roots or leaves, secondly Ca^{2+}. The sequence ordered based on content is K^+ > Ca^{2+} > soluble sugar > Mg^{2+} > free amino acid > proline, however, the sequence is proline > free amino acid > soluble sugar > K^+ > Mg^{2+} > Ca^{2+} according to extent of increase. (Cai et al., 2008). Cabuslay et al. [8] revealed that ability of leaf osmoregulation is also greater than the root after subjecting to water stress. The results of experiment in wheat also showed that K^+ gave the greatest contribution in drought stress, yet the sequence (K^+ > soluble sugar > free amino acid > Ca^{2+} > Mg^{2+} > Pro) is different from that in rice [9]. The difference of Osmoregulation in contribution rate could be concerned with plant varieties, growth stages, intensity, environment, and time in stress.

In the last years, studies have showed small organic molecules play an important role such as proline, glycine, betaine, trehalose, mannitol, fructosan, etc. [10,11]. Studies indicated that accumulation of proline and betaine could decrease extent of damage [12]. Since 1954 accumulation of free Proline was discovered, the relation of accumulation of free proline and water stress was studied widely [13]. Content of free praline would had a multiplication in crops such as sorghum, sorghum, rice etc. [14-16]. Proline was synthesized through P5CS(Δ^1- pyrrolin-5-carboxylate synthetase), but it was synthetized by ornithine in normal conditions (Delauney et al. [17] and. Sheela et al. [18] indicated that there was close relationship between free proline and drought resistance because of the strong hydratability to hold on water. Wu et al.'s [19] study showed that the gene for Na^+/H^+ of reverse transport protein was expressed efficiently in rice and then activated biosynthesis of proline. Moreover, content of Na^+ and proline in transgenic were above them in contrast. Storey et al. [20] indicated that salinity and water result in accumulation of Glycine-beyaine in barley.

2.2. The ROS-Eliminating System

Production and elimination of ROS (reactive oxygen species) was in balance [21,22]. The ROS could pose a hazard to plant cells in adversity. The plant eliminated the ROS by antioxidase system and antioxidant [23-25]. Antioxidase system includes superoxide dismutase (SOD), peroxidase (POD) catalase (CAT) and ascorbate peroxidase (APX-POD). Generally, high and low of antioxidase activity is connected to stress resistance in plant. The stronger stress resistance, the higher antioxidase activity [26,27]. There are diversities of antioxidase activity in different plant. SOD is in centre of antioxidase system and Widespread in the plant body [28,29]. In addition, the POD and CAT are also major members of antioxidase system, playing a very important in stress drought. The research showed the protective enzymes of seedlings were increased significantly by pre-treating seeds with water and 10% - 15% PEG and that effect of appropriate pre-treating with PEG is super to effect of pre-treating with water [30]. Wu et al. [31] proved that drought stress could result in increasing of significantly. In further, the study showed Antioxidase activities in plant leaf were higher than them in plant root significantly. In the ROS-eliminating system, antioxidant also played a significant role such as ascorbic acid (ASA), glutathione (GSH), carotenoid, etc. These substance quenched reactive oxygen species by several ways directly or indirectly [32,33].

2.3. Signal Molecule—Hormone

Plant hormone the generic terms of growth regulation with trace by synthesizing in plant. Although simpleness in chemical structure, it has the complicated physiological effect [34]. The growth and development of every stages was regulated by the hormone [35]. During the reaction process to water stress, the plant hormone perform a key role in the signal transduction. At present, the research on this field is one of the most pop content in life science [36-40]. ABA (abscisic acid) was mostly researched in plant hormone [41]. Most research showed that ABA was the major signal substance, especially for root. In the early 1960s, Wright and Hiron [42] had proved that osmosis stress could induce synthesis of ABA in cellular and synthesis of ABA had great relationship with stress drought. At the same time, ABA could be used for evaluation index in authenticating stress drought. Research showed that ABA could adjust stomatal behavior of plant to enhance stress drought[41]. Root is earlier than leaf in the synthesis of ABA [43]. Chen et al. [44] proved that content of cytokine in the rice organs (overground part) change unconspicuously, yet ABA changed remarkably in content. Studies showed that hormonal balance was effected in water stress. Content of cytokinin and ethylene decreased, yet content of ABA increased significantly [45,46]. In addition, Yang et al. [47] argued that antagonism of ABA and ethylene regulated the development of spikelet, furthermore Ratio of ABA and ethylene is a physiological feature coping with water stress.

In recent years, researches about SA (salicylic acid), BR (brassinolide) [48,49], JA (jasmonic acid) [50-54,] were studied in water stress. In addition, some secondary metabolites such as peptide, NO, SL (strigol-actones), etc, play a part like hormone in growth and development [55]. Wu et al. [56] argued that a new regulatory factor, protein (14-3-3), was in signal transduction pathway of brassinolide. In addition, they also reveal a new regulating mechanism for protein OsBZR1. This Would supply a new way and means for plant in stress drought.

3. RESEARCH ADVANC OF DROUGHT RESISTANCE INMOLECULAR BIOLOGY

During adapting to the drought stress, the plant has some reactions in molecular level. The reactions were regulated by several genes forming response systems to the drought [57]. Drought genes were divided into two classes according to the way of action [58]. The first is function genes with protecting directly in drought resisting; the second is regulation genes, productions which encold could regulate function genes in signal transduction and gene expression.

3.1. Function Gene with Drought-Resistance

Research showed that there were several significant differences between transgenic rice with gene for GST and CAT1 and normal rice in growth, photosynthesis, reduction extent of RWC (relative water content), accumulation of H_2O_2 and MDA (malon dialdehyde) and so on. These results implied that the GST and CAT1 transgene mitigated oxidative damage from water stress [59]. Excessive expression of the coli trehalose synthesis A (otsA) and coli trehalose synthesis B (otsB) transgene rice, accumulation of trehalose increased and oxidative damage caused by photooxidation decreased [60]. Jang et al. [61] revealed that the bifunctional fusion of the TPS and TPP (TPSP) transgene enhanced resistance to drought, salty and cold. By activating gene GH3.13 of rice and adapting the content of IAA in leaf, stem and tuber, adaptability to the drought was enhanced [62].

3.2. Regulator Gene with Drought Resistance

Hou et al. [63] separated a gene named osSKIPa. Studies implied that the gene osSKIPa, an upstream gene for regulating, would mobilize the other gene of drought resistance, promote the cell vitality, increased viability in water stress. The mechanism similar to chain reaction was never founded before. Huang et al. [64] acquired the rice mutant named drought and salt tolerance (DST) through screening from mutant library large-scaly. Further; they cloned the gene DST with stress-resistance. They argued that DST down-regulated expression of the gene related to metabolization of H_2O_2, and the ability with eliminating H_2O_2 declined. This resulted in accumulation of H_2O_2, and shut of stoma in guard cells, and then improved the ability of drought-resistance by decreasing evaporation of water.

3.3. Epigenetics Mechanism

Epigenetics refers to the heritable changes in gene expression without any alteration in DNA sequence, including DNA methylation, histone modification and chromatin conformation [65,66]. DNA methylation is one of the important modifications in eukaryotic genome. It regulates genetic information from the epigenetic level including regulation of gene expression, growth and development, genomic imprinting and so on. In the recent years, researches have showed that DNA methylation would be effected in stress such as heat, drought, etc. [67], and the genes with DNA methylation were connected with stress [68]. That is to say that DNA methylation involved the expression and regulation progress of gene in stress. Pan et al. [69] argued that the level of DNA methylation increased significantly in water stress, the level and state of DNA methylation under drought was temporal-spatial specific and variety specific, and the change of DNA methylation was related to drought responsiveness.

4. CONCLUSION

It is important issue of water resource shortage which is harm to agriculture development of the world. Efficient use of water resources is an urgent problem around the world [70]. However, drought-resistance is a complicated problem and is regulated by several pathways involved with many genes [71]. Thus, there are several problems to resolve in the signal perceiving, the relations between several factors in signal transduction, epigenetics mechanism, transgene silence etc. Therefore, these researches would help us understand molecular mechanism in drought-resistance in depth. In addition, we should further study by integrating molecular level, cell, organ, and the whole etc. The studies would improve resistance of crops and water utilization efficiency, and establish basement for cultivating the varieties of resistance.

5. ACKNOWLEDGEMENTS

Qiang Chen, Siyuan Tao, Xiaohua Bi, Xin Xu, Lanlan Wang, Xuemei Li# (30870205) and Shenyang Normal University Experimental center director foundation (SY201104).

REFERENCES

1. Fageria, N.K. (2003) Plant tissue test for determination of optimum concentration and uptake of nitrogen at different growth stages in low land rice. Communications in Soil Science and Plant Analysis, 34, 259-270.
2. Fageria, N.K. (2007) Yield physiology of rice. Plant Nutrition, 30, 843-879.
3. Belder, P., Spiertz, J.H.J., Bouman, B.A.M., Lu, G. and Tuong, T.P. (2005) Nitrogen economy and water productivity of lowland rice under water-saving irrigation. Field Crops Research, 93, 169-185.
4. Jing, R.L. (2007) Advances of research on drought resistance and water use efficiency in crop plants. Review of China Agricultural Science and Technology, 9, 1-5.

5. Bais, H.P., Vepachedu, R. and Gilroy, S. (2003) Allelopathy and etotic plant invasion: From molecules and genes to species interaction. Science, 301, 1377-1380.
6. Rania, B.S. and Nabil, Z. (2010) Improved drought and salt stress tolerance in transgenic tobacco overexpressing a novel A20/AN1 zinc-finger "AlSAP" gene isolated from the halophyte grass Aeluropus littoralis. Plant Molecular Biology, 72, 171-190.
7. Cai, K.Z., Wu, X.Z. and Luo, S.M. (2008) Effects of water stress on osmolytes at different growth stages in rice leaves and roots. Journal of Plant Ecology, 32, 491-500.
8. Cabuslay, G.S., Ito, O. and Alejar, A.A. (2002) Physiological evaluation of responses of rice (Oryza sativa L.) to water deficit. Plant Science, 163, 815-827.
9. Li, D.Q., Zou, Q. and Cheng, B.S. (1992) Osmotic adjustment and osmotica of wheat cu|tivars with different drought resistance under soil drought. Acta Phytophysiologica Sinica, 18, 37-44.
10. Bianchi, G., Limiroli, R. and Pozzi, N. (1993) The unusual sugar composition in leaves of the resurrection plant Myrothamnus flabellifolia. Physiologia Plantarum, 87, 223-226.
11. Crowe, J.H. and Crowe, L.M. (1984) Preservation of membrances in anhydrobiotic organisms: The role of thehhslose. Science, 223, 701-703.
12. Zonglie, H., Karuna, L., Zhang, Z.M. and Desh, P.S. (2000) Removal of Feedback inhibition of delta 1-pyrroline-5-carboxylate synthetase results in increased proline accumulation and protection of plants from osmotic stress. Plant Physiology, 122, 1129-1136.
13. Kemple, A.R. and Macpherson, H.T. (1954) Liberation of amino acid in perennial rye grass during wilting. Biochemistry, 58, 46-49.
14. Zaifnejad, M., Clark, R.B. and Sullivan, C.Y. (1997) Aluminum and water stress effects on growth and proline of Sorghum. Journal of Plant Physiology, 153, 338-344.
15. Carceller, M., Prystupa, P. and Lemcoff, J.H. (1999) Remobilization of proline and other nitrogen compounds from senescing leaves of maize under water stress. Journal of Agronomy and Crop Science, 183, 61-66.
16. Alkaraki, G.N., Clark, R.B. and Sullivan, C.Y. (1996) Phosphorus nutrition and water stress effects on proline accumulation in sorghum and bean. Journal of Plant Physiology, 148, 745-751.
17. Delauney, A.J., Hu, C.A.A. and Kavi Kishor, P.B. (1993) "Cloning of ornithine delta-aminotransferase CDNA from vigna-aconitifolia by transcomplementation in Escherichia coli and regulation of proline biosynthesis. Journal of Biological Chemistry, 268, 18673-18678.
18. Sheela, K.R and Alexallder, V.T. (1995) Physiological response of riee varieties as influenced by soil moisture and seed hardening India. Plant Physiology, 38, 269-271.
19. Wu, L.Q., Fan, Z.M and Guo, L. (2005) Over-expression of the bacterial nhaa gene in rice enhances salt and drought tolerance. Plant Science, 168, 297-302.
20. Storey, R., Ahmad, N. and Wyn Jones, R.G. (1977) "Taxonomic and ecological aspects of the distribution of glycine-betaine and related compounds in plants. Oecologia (Berl), 27, 319-332.
21. Foyer, C.H. and Noctor, G. (2000) Oxygen processing in photosynthesis: Regulation and signaling. New Phytologist, 146, 359-388.
22. Porcel, R., Barea, J.M. and Ruiz-Lozano, J.M. (2003) Antioxidant activities in mycorrhizal soybean plants under drought stress and their possible relationship to the process of nodule senescence. New Phytologist, 135-143.
23. Foyer, C.H., Descourvieres, P. and Kunert, K.J. (1994) Protection against oxygen radicals: An important defence mechanism studied in transgenic plants. 17, Plant Cell and Environment, 507-523.
24. Smirnoff, N. (1998) Plant resistance to environmental stress. Current Opinion in Biotechnology, 9, 214-219.

25. Reddya, A.R., Chaitanyaa, K.V. and Vivekanandanb, M. (2004) Drought-induced responses of photosynthesis and antioxidant metabolism in higher plants. Journal of Plant Physiology, 161, 1189-1202.
26. Prasad, T.K. (1993) Role of active oxygen in the response of plants to water deficit and dessication. New Phytologist, 125, 27-58.
27. Shao, H.B., Liang, Z.S., Shao, M.A. and Wang, B.C. (2005) Changes of antioxidative enzymes and membrane peroxidation for soil water deficits among 10 wheat genotypes at seedling stage. Colloids and Surfaces B: Biointerfaces, 42, 19-25.
28. Seel, W.E. and Hendry, G.A. (1992) Effect of desiccations on some activated oxygen processing enzymes and anti-oxidants in mosses. Journal of Experimental Botany, 43, 1031-1037.
29. Scandaliol, J.G. (1993) Oxygen stress and super oxide dimutases. Plant Physiology, 101, 7-12.
30. Sun, Y.Y., Sun, Y.J., Wang, M.T., Li, X.Y., Guo, X., Hu, R. and Ma, J. (2010) Effects of seed priming on germination and seedling growth of rice under water stress. Acta Agronomica Sinica, 36, 1931-1940.
31. Wu, X.Z., Cai, Z.K. and Luo, S.M. (2008) Effects of soil drying at heading stage on physiological characteristics in root and leaf of rice. Chinese Agricultural Science Bulletin, 24, 202-207.
32. Yin, Y.Q., Hu, J.B. and Deng, M.J. (2007) Latest development of antioxidant system and responses to stress in plant leaves. Chinese Agricultural Science Bulletin, 23, 105-110.
33. Shao, H.B., Chu, L.Y., Shao, M.A., Jaleel, C.A. and Mi, H.M. (2008) Higher plant antioxidants and redox signaling under environmental stresses. Comptes Rendus Biologies, 331, 433-441.
34. Santner, A., Calderon-Villalobos, L.I.A. and Estelle, M. (2009) Plant hormones are versatile chemical regulators of plant growth. Nature Chemical Biology, 5, 301-307.
35. Wolters, H. and Gurgens, J. (2009) Survival of the flexible: Hormonal growth control and adaptation in plant development. Nature Reviews Genetics, 10, 305-317.
36. Yin, Z.M., Hennig, J. and Szwacka, M. (2004) Tobacco PR-2D promoter is induced in transgenic cucumber in response to biotic and abiotic stimuli. Journal of Plant Physiology, 161, 621-629.
37. Hieng, B., Ugrinovic, K. and Sustar-Vozlic, J. (2004) Different classes of proteases are involved in the response to drought of Phaseolus vulgaris L. cultivars differing in sensitivity. Journal of Plant Physiology, 161, 519-530.
38. Wu, G.H., Robertson, A. and Liu, X.J. (2004) A lipid transfer protein gene BG-14 is differentially regulated by abiotic stress, ABA, anisomycin, and sphingosine in bromegrass (Bromus inermis). Journal of Plant Physiology, 161, 449-458.
39. Serna, L. (2005) A network of interacting factors triggering different cell fates. Plant Cell, 16, 2258-2263.
40. Larkindale, J. and Huang, B.R. (2004) Thermotolerance and antioxidant systems in Agrostis stolonifera: Involvement of salicylic acid, abscisic acid, calcium, hydrogen peroxide, and ethylene. Journal of Plant Physiology, 161, 405-413.
41. Davies, W.J. and Zhang, J. (1991) Root signals and the regulation of growth and development of plants in drying soil. Annual Review of Plant Physiology and Plant Molecular Biology, 42, 55-76.
42. Wrigh, S.T.C. and Hiron, R.W.P (1969) (+)-Abscisic acid, the growth inhibitor induced in detached wheat leaves by a period of wilting. Natrue, 224, 719-720.
43. Jackson, M.B. (1993) A replant homones in volved in the root to shoot communication. Advances in Botanical Research, 85, 183-187.
44. Chen, X.H., Liu, K., Xi, L.L., Wang, Z.Q. and Yang, J.C. (2005) Effects of soil moisture and nitrogen on abscisic acid and cytokinin contents in the above ground plant parts of rice. Acta Agronomica Sinica, 31, 1406-1414.
45. Chen, Y.M. and Yu, S.W. (1988) Effects of ethylene release and content of ACC and MACC in drought stressses of different modes to wheat. Plant Physiology Journal, 14, 281-288.

46. Chen, J., Wan, J., Gao, X.L., Jiang, H. and Wang, P.R. (2006) Study on drought resistance physiology and drought resistance genes of rice. Chinese Agricultural Science Bulletin, 22, 56-60.
47. Yang, J.C., Liu, K., Zhang, S.F., Wang, X.M., Wang, Z.Q. and Liu, L.J. (2008) Hormones in rice spikelets in responses to water stress during meiosis. Acta Agronomica Sinica, 34, 111-118.
48. Singh, J., Nakamum S. and Ota, Y. (1993) Effect of epi-brassi-nolide on gram (Cicer arietinum) plant grown under water stress in juvenilstage. Indian Journal of a cultural Science, 1993, 63, 395-401
49. Kairam, S.R. (1994) Effect of homobrassinilide application on plant metalolism and grain yield under irrigated and moisture stress conditions of two wheat varieties. Plant Growth Regulation, 14, 173-179.
50. Moons, A., Prinsen, E. and Bauw, G. (1997) Antagonistic effects of abscisic acid and jasmonates on salt stress-inducible transcripts in rice roots. The Plant Cell, 9, 2243-2259.
51. Gao, X.P., Wang, X.F. and Lu, Y.F. (2004) Jasmonic acid is involved in the water-stress-induced betaine accumulation in pear leaves. Plant Cell and Environment, 27, 497- 507.
52. Gonzaliez, A., Tiznado, H.M. and Zavaleta, G.R. (2004) Methyl jasmonate treatments reduce chilling injury and activate the defense response of guava fruits. Biochemical and Biophysical Research Communications, 313, 694- 701.
53. Thaler, J.S., Owen, B. and Higgns, V.J. (2004) The role of the jasmonate response in plant susceptibility to diverse pathogens with a range of lifestyles. Plant Physiology, 135, 530-538.
54. Creelm, R.A. and Mulent, J.E. (1995) Jasmonic acid distribution and action in plants: Regulation during development and response to biotic and abiotic stress. Proceedings of the National Academy of Sciences of the USA, 92, 4114-4119.
55. Muhammad, F., Abdul, W. and Lee, D.J. (2009) Exogenously applied polyamines increase drought tolerance of rice by improving leaf water status, photosynthesis and membrane properties. Acta Physiologiae Plantarum, 31, 937-945.
56. Wu, G., Wang, X.L., Li, X.B., Kamiya, Y., Otegui, M.S. and Joanne, C. (2011) Methylation of a phosphatase specifies dephosphorylation and degradation of activated brassinosteroid receptors. Science, 4, 29.
57. Shinozaki, K. and Yamaguchi-Shinozaki K. (2000) Molecular responses to dehydration and low temperature: Differences and cross-talk between two stress signaling pathways. Current Opinion in Plant Biology, 3, 217-223.
58. Holmberg, N. (1998) Improving stress tolerance in plants by gene transfer. Trends in Plant Science, 3, 1361-1366.
59. Zhao, F.Y. and Xu, Z.J. (2009) Physiological changes of transgenic rice under drought and heat stress. Acta Botanica Boreali-Occidentalia Sinica, 29, 240-248.
60. Garg, A.K., Kim, J.K. and Owens, T.G. (2002) Trehalose accumulation in riceplants confershightolerance levelsto different abiotic stresses. Proceedings of National Academy of Science of USA, 99, 15898-15903.
61. Jang, I.C., Oh, S.J. and Seo, J.S. (2003) Expression of a bifunctional fusion of the Escherichia coli genes for trehalose-6-phosphate synthase and trehalose-6-phosphate phosphatase in transgenic rice plants increases trehalose accumulation and abiotic stress tolerance without stunting growth. Plant Physiology, 131, 516-524.
62. Zhang, S.-W., Li, C.-H., Cao. J., et al. (2009) Altered architecture and enhanced drought tolerance in rice via the down-regulation of indole-3-acetic acid by TLD1/ OsGH3.13 activation. Plant Physiology, 151, 1889-1901.
63. Hou, X., Xie, K.B., Yao, J.L., Qi, Z.Y. and Xiong, L.Z. (2009) A homolog of human ski-interacting protein in rice positively regulates cell viability and stress tolerance. Proceedings of National Academy of Science of USA, 106, 6410-6415.

64. Huang, X.Y., Chao, D.Y., Gao, J.P., Zhu, M.Z., Shi, M. and Lin, H.X. (2009) A previously unknown zinc finger protein, DST, regulates drought and salt tolerance in rice via stomatal aperture control. Genes & Development, 23, 1805-1817.
65. Wu, C.T. and Morris, J.R. (2001) Genes, genetics and epigenetics: A correspondence. Science, 293, 1103-1105.
66. Bird, A.P. (1986) CpG-rich islands and the function of DNA methylation. Nature, 321, 209-212.
67. Finnegan, E.J., Bretell, R.I. and Dennis, E.S. (1993) The role of DNA methylation in the regulation of plant gene expression. In: Jost, J.P. and Saluz, H.P., Eds., DNA Methylation: Molecular Biology and Biological Significance, Birkhauser Verlag, Basel, 218-261.
68. Wada, Y., Miyamoto, K., Kusano, T. and Sano, H. (2004) Association between up-regulation of stress-responsive genes and hypomethylation ofgenomic DNA intobaccoplants. Molecular Genetics and Genomics, 271, 658- 666.
69. Pan, Y.J., Fu, B.Y., Wang, D., Zhu, L.H. and Li, Z.K. (2009) Spatial and temporal profiling of DNA methylation induced by drought stress in rice. Scientia Agricultura Sinica, 42, 3009-3018.
70. Luo, L.J. (2010) Breeding for water-saving and droughtresistance rice (WDR) in China. Journal of Experimental Botany, 61, 3509-3517.
71. Elizabet, P. (2008) The blue revolution, drop by drop, gene by gene. Science, 320, 171-173.

NOTES

*National Natural Science Foundation of China (30870205); Shenyang Normal University Experimental Center director funds (Sy201104).

CHAPTER 7

FLC Expression is Down-Regulated by cold Treatment in Diplotaxis Tenuifolia (Wild Rocket), but Flowering time is Unaffected

Jemma L. Taylor[a], Andrea Massiah[a], Sue Kennedy[b], Yiguo Hong[a,c], Stephen D. Jackson[a]

[a] *School of Life Sciences, University of Warwick, Coventry CV4 7AL, UK*

[b] *Elsoms Seeds Ltd, Pinchbeck Road, Spalding PE11 1QG, UK*

[c] *Research Centre for Plant RNA Signaling, College of Life and Environmental Sciences, Hangzhou Normal University, Hangzhou 310036, China*

ABSTRACT

Wild rocket (*Diplotaxis tenuifolia*) has become a very popular salad leaf due to its peppery taste. It is part of the *Brassicaceae* family and thus has a high level of homology at the DNA level to other Brassica species including *Arabidopsis thaliana*. The vernalization and photoperiodic requirements of wild rocket have not been reported to date. Photoperiodic experiments described here demonstrate that rocket is a facultative long day plant. To investigate the vernalization requirement, both seed and young plants were given vernalization treatments at 4 °C for different lengths of time. A rocket homologue of *FLOWERING LOCUS C* (*DtFLC*) was isolated and shown to functionally complement the *Arabidopsis* FRI^{+}*flc3* null mutant. Whilst the expression of *DtFLC* was significantly reduced after just one week of cold treatment, cold treatments of two to eight weeks had no significant effect on bolting time of wild rocket indicating that rocket does not have a vernalization requirement. These findings illustrate that important fundamental differences can exist between model and crop plant species, such as in this case where down-regulation of *DtFLC* expression does not enable earlier flowering in wild rocket as it does in *Arabidopsis* and many other Brassica species.

Keywords: Bolting time; ***Diplotaxis tenuifolia***; Photoperiod; Vernalization; Wild rocket

1. INTRODUCTION

Wild rocket (*Diplotaxis tenuifolia*) has increased in popularity over the last 20 years in the leafy salads market (Chun et al., 2013), in the UK alone over 80 tons of rocket is consumed per week (Gill, 2008). The genus *Diplotaxis* is found in the *Brassicaceae* family in the *Oleracea* clade (Arias and Pires, 2012) and is therefore closely related to species *Brassica rapa*, *Brassica juncea*, *Brassica napus* and *Brassica oleracea,* as well as *Arabidopsis thaliana* (Arabidopsis). Flowering is undesirable in commercial rocket production as pre-harvest flowering can lead to the crop being unsaleable (Figure 1a), despite this very little work has been reported on the control of flowering in wild rocket. Here we investigate the vernalization and photoperiodic requirements of this relatively new salad crop species.

Regulation of flowering time involves a complex gene network which allows a plant to respond to both internal and external signals in order to control the timing of the transition from the vegetative phase to the reproductive phase (Srikanth and Schmid, 2011). The timing of this transition is crucial to enable reproduction to occur when the plant is at its fittest and the environmental conditions are most favorable (Thomas et al., 2006). Most research into flowering time has been conducted in the model plant Arabidopsis through the generation and testing of many mutant and transgenic overexpression lines to elucidate gene function (Corbesier and Coupland, 2005). Using the knowledge gained from Arabidopsis and applying it to crop species is helping our understanding of the control of flowering in crop plants. Currently there are six key pathways that regulate flowering time: the photoperiodic, autonomous, vernalization, gibberellic acid, age dependent and ambient temperature pathways (Brambilla and Fornara, 2013; Fornara *et al.*, 2010; Jarillo and Pineiro, 2011 ; Srikanth and Schmid, 2011). All of these pathways converge on the floral pathway integrator genes such as *FLOWERING LOCUS T* and *SUPPRESSOR OF*

OVEREXPRESSION OF CONSTANS 1 which activate the downstream floral meristem identity genes. When flowering is induced the shoot apical meristem changes from forming vegetative tissues such as leaves to form flowers.

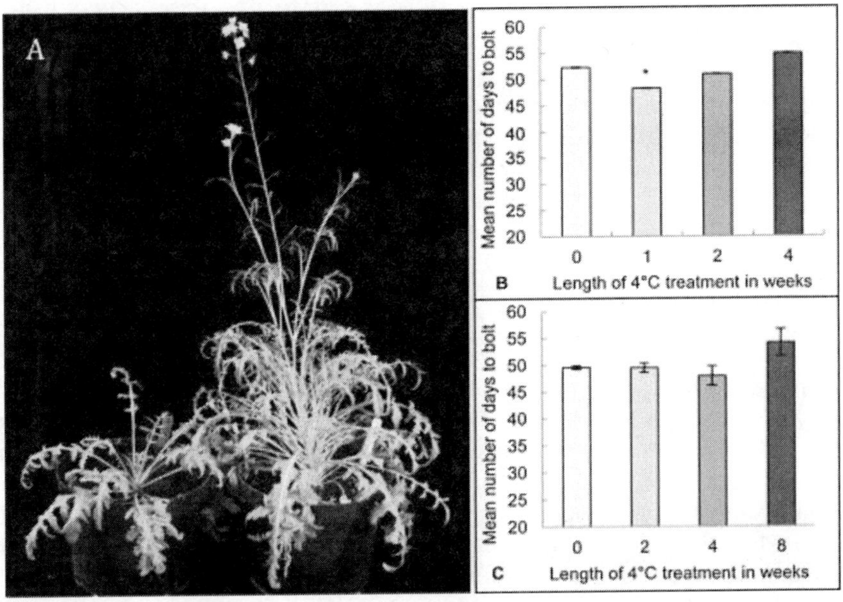

Figure 1. Effect of vernalization at 4 °C on D.tenuifolia. A) Photo of a non-bolting rocket plant (left) and one that has bolted and flowered (right).B) The mean number of days to bolt of plants where the seed was subjected to cold treatment at 4 °C for one, two or four weeks. Bars show mean ± SE (n = 14 (0 weeks), n = 15 (1 week and 2 weeks), n = 10 (4 weeks)). Student's *t*-test was used to compare the number of days to bolt of each vernalization treatment (one, two and four weeks) against the ambient conditions (0 weeks). *Statistical significance of $p < 0.05$ C) Mean number of days to bolt for 4 week old plants subjected to cold treatment at 4 °C for two, four and eight weeks. The length of the vernalization treatment was subtracted from the number of days to bolt. Bars show mean ± SE (n = 14 (0 weeks), n = 15 (2 weeks and 4 weeks), n = 13 (8 weeks)). Student's *t*-test was used to compare the number of days to bolt of each vernalization treatment (2, 4 and 8 weeks) against the ambient conditions (0 weeks). There was no significant difference in number of days to bolt between vernalization treatments and ambient conditions ($p < 0.05$)

Vernalization is the attainment or acceleration of floral competence as a consequence of prolonged exposure to cold temperature (Chouard, 1960). In vernalization-responsive plant species, vernalization conveys the ability of the plant to flower and can greatly reduce the time it takes to flower (Guo *et al.*, 2004; Hackett and Hartmann, 1967; Jung and Mueller, 2009 ; Strange *et al.*, 2011). The cold treatment causes epigenetic changes that reduce levels of a floral repressor gene *FLOWERING LOCUS C* (*FLC*). When the plant is returned to a higher temperature these epigenetic changes are stable enough to maintain the repression of *FLC* expression and the reduced levels of this repressor thus enables flowering to be induced (Boss et al., 2004). Studies of Arabidopsis ecotypes show that there are differences in the requirement for vernalization. Rapid cycling ecotypes will flower without the need for a vernalization treatment, whereas winter annual ecotypes will flower very late unless exposed to a period of vernalization (Nordborg and Bergelson, 1999 ; Wang, 2014). It was found that these vernalization response phenotypes all relate to the levels of expression of *FLC* (Song et al., 2012). Plants which have a requirement for vernalization tend to be those from seasonal climates where adaptation to extended periods of cold can be advantageous. The stage in the plant's lifecycle which is responsive to vernalization varies between species as does the length of treatment needed in order to fully satisfy the vernalization requirement. Arabidopsis is able to respond to vernalization as both seed and young plants (Nordborg and Bergelson, 1999 ; Strange *et al.*, 2011), whereas many *Brassica* species only respond to a vernalization treatment as young plants (Jung and Mueller, 2009). Rocket is closely related to both Arabidopsis and *Brassica* and it is not currently known at which stage of its lifecycle it is able to respond, what the optimum vernalization period might be, or even if it has a vernalization response at all.

The ability of a plant to measure and respond to the length of day, or photoperiod, is often critical to its growth and development, particularly in the timing of the floral transition. Most plants fall into one of three main photoperiodic categories; long day (LD) plants which flower when the photoperiod is longer than a critical length, short day (SD) plants which flower when the photoperiod is shorter than a critical length, and day neutral plants which flower regardless of the photoperiod (Matsoukas et al., 2012). LD plants flower quite quickly in LD, but if subjected to SD conditions they will flower much more slowly (facultative response) or not at all (obligate response), the opposite being true for SD plants.

Understanding the vernalization and photoperiod requirements of commercial crops, and the underlying genes that control these responses in these crops, is of huge benefit to crop breeding programs aiming to create varieties that can be grown successfully in a range of latitudes and climatic conditions. In this paper, we investigate the photoperiodic and vernalization responses of *D.tenuifolia*. We show that wild rocket is a facultative long day plant and that it does not respond to vernalization treatments of several weeks duration even though the expression of the rocket *FLC* orthologue is strongly reduced by these treatments.

2. MATERIALS AND METHODS

2.1. Plant material

Elsoms Seeds Ltd provided seed of *Diplotaxis tenuifolia* bred through four generations of single seed descent for uniformity of leaf shape, color, and bolting time. *Arabidopsis thaliana* mutant FRI$^+$*flc3* null (Col0 background) seed was originally sourced from NASC (http://arabidopsis.info/) and used in the complementation experiments.

2.2. DNA and RNA extraction

Rocket genomic DNA was extracted using CTAB extraction method adapted from Stewart and Via (1993) on frozen leaf material, which was ground to a powder using a Dremel Drill with a 1.5 ml Eppendorf tube drill bit. 300 µl CTAB B buffer (100 mM Tris/Cl pH 8.0, 1.4 M NaCl, 20 mM EDTA, 2% hexadecyltrimethyl ammoniumbromide) was added and homogenized. Samples were incubated at 65 °C for 30 min and centrifuged. The supernatant was removed into a new tube and twice extracted using 300 µl Chloroform:Isoamyl alcohol (24:1) and centrifugation. The aqueous top phase was transferred each time. 300 µl CTAB C buffer (1% hexadecyltrimethyl ammoniumbromide, 10 mM EDTA, 50 mM Tris/Cl pH 8.0) was added and left overnight at room temperature. On day two, the tubes were centrifuged and the pellet dissolved in 400 µl 1 M CsCl. DNA is then precipitated using 100% Ethanol and centrifuged. The pellet was then washed twice using 70% Ethanol before drying and resuspending in 25 µl TE buffer (pH 8) with RNase A (Invitrogen) (20 µg/ml).

RNA extraction was done using the Z6 extraction buffer (containing 8 M guanidine hydrochloride) method (Logemann et al., 1987). Frozen leaf, or imbibed seed, samples were ground to a powder using liquid N_2 and a pestle and mortar. The powder was transferred to a 1.5 ml Eppendorf tube and Z6 buffer plus 2-mecaptoethanol (50 mM final concentration) added. This was homogenized using a Dremel Drill with a 1.5 ml Eppendorf tube drill bit before the rest of the published method was followed. 5 µg RNA was DNase treated using Ambion® Turbo DNA-*free*™ DNase and then resuspended in 12 µl DEPC treated H_2O. 1 µg of DNase treated RNA was synthesized into cDNA using an Invitrogen Thermoscript cDNA synthesis kit or BioRad iScript cDNA synthesis kit. Arabidopsis RNA extraction was performed using the same method, but with a reduction in the volume of reagents used due to the smaller amount of starting leaf material.

2.3. Vernalization experiments

D. tenuifolia seed was sown onto damp paper towels and covered with aluminum foil and placed at 4 °C for one, two or four weeks. At the end of the vernalization period, seeds were removed from the paper and put onto F2S soil (Levington) in P40 trays at one seed per cell and placed in the glasshouse at 20 °C with a 16 h photoperiod. The germination date was recorded. Control seed to stay at ambient temperature was sown in the same way but was placed into the glasshouse at 20 °C for one week by which time the seeds had germinated. These were then transferred to F2S soil in P40 trays at one seed per cell. At four weeks post germination, plants were transplanted into 5 inch pots of M2 soil (Levington) to continue growing. In the plant experiment, seed was sown directly to F2S soil (Levington) in P40 cells and put in a controlled environment cabinet (MLR-352, Panasonic Co. Ltd) at 20 °C to aid germination with 16 h photoperiod for five days (light level 104 $\mu mol\ m^{-2}\ s^{-1}$). The germination date was recorded and plants grown until three weeks old in a glasshouse at 20 °C with 16 h photoperiod. These were then transplanted to 5 inch pots of M2 soil (Levington). After one week (four weeks after germination), the plants were moved to 4 °C with 16 h photoperiod to vernalize for two, four or eight weeks. The plants were then returned to glasshouse conditions. Control plants were kept at 20 °C with 16 h photoperiod (ambient conditions) for the entirety of the experiment. Plant material was harvested at the end of each week at 9 h after lights on. The number of days from germination to bolting was recorded, bolting was taken as the point when the height of the bolt was about 10 cm and is the measurement normally used by growers and industry to tell whether a plant has bolted. As the plants were grown in controlled environment most plants in a given treatment bolted around the same time. It is known that in Arabidopsis *FLC* is expressed in seed, shoot and root tissue (Chiang *et al.*, 2009 ; Michaels and Amasino, 1999) so for analysis of FLC expression all material above soil level was sampled from up to six plants for the seed vernaliszation experiment and for weeks 1–3 of the plant experiment,. From week 4 onwards of the plant vernalization experiment when the plants were larger newly expanded leaves were collected from five plants and pooled together.

2.4. Photoperiod experiments

Rocket seeds were sown onto F2S soil (Levington) in p24 trays and topped with vermiculite. These were placed into a controlled environment cabinet (MLR-352, Panasonic Co. Ltd) at 22 °C with 16 h light, 8 h dark. The germination date was recorded and at 10 days after sowing, plants were moved into the different photoperiods. These were short day conditions with 8 h light and 16 h dark, intermediate day conditions with 12 h light and 12 h dark, and long day conditions with 16 h light and 8 h dark. Plants were grown in the controlled environment cabinets under fluorescent light (light levels 124 μmol m^{-2} s^{-1}). Plants were transplanted at three weeks from sowing into 5 inch pots of M2 soil (Levington) and returned to the correct photoperiod and time to bolting was recorded for each plant. A second experiment was carried out using incandescent bulbs (Bell Striplite, 30 W, light levels 33 μmol m^{-2} s^{-1}) to provide the day extension lighting. The experimental design was as described above but with short day (8:16) conditions with 8 h fluorescent bulb light and 16 h dark, intermediate day conditions (12:12) with 8 h fluorescent bulb light + 4 h incandescent bulb day extension and 12 h dark, and long day conditions (16:8) with 8 h fluorescent bulb light +8 h incandescent bulb day extension and 8 h dark.

2.5. Isolation of the *DtFLC* gene from *D. tenuifolia*

Degenerate primers for *FLC* were designed using aligned sequences for *Arabidopsis* and *Brassica* *FLC* genes and used to PCR amplify *DtFLC* using *D. tenuifolia* cDNA as the template. Specific *DtFLC* primers (Supplementary data Table 1A) were then designed to isolate the remaining sequence using RLM RACE (Invitrogen). 5′ and 3′ UTR primers specific to *DtFLC* (see supplementary data for sequences) were used to amplify the gene from rocket genomic DNA and cDNA. The gene sequence data has been submitted to the GenBank database under accession number KX148480.

2.6. Functional complementation in *Arabidopsis*

The *DtFLC* coding sequence was isolated from cDNA using primers which incorporated the start and stop codons. This was cloned into pGEM-T easy and amplified out using GATEWAY™ (Invitrogen) adapter primers specific to *DtFLC* (see supplementary data for sequences). Using BP clonase (Invitrogen), this was cloned into pDONR 207 and then pB2GW7 using LR clonase (Invitrogen). The sequence was verified before *DtFLC* containing pB2GW7 plasmid DNA was transformed into *Agrobacterium tumefaciens* strain c58pGV3101. 500 ml LB (10:5:5) plus 25 μg/ml

Gentamycin, 100 μg/ml Spectinomycin and 12.5 μg/ml Rifampicin was inoculated using 5 ml cell culture of *A. tumefaciens* strain c58pGV3101 containing the *DtFLC*-pB2GW7 plasmid. This was incubated at 28 °C for 16 h. The culture was centrifuged and the supernatant removed. 500 ml 5% (w/v) sucrose solution was used to resuspend the cells and 100 μl silwet L-77 added before dipping the inflorescences of the Arabidopsis FRI$^+$*flc3* null plants (Clough and Bent, 1998). The plants were sealed in a bag for 24 h before putting at 22 °C with 16 h light, 8 h dark in the glasshouse facility. T_1 seed was harvested and sown onto soil (Levington F2S:sand:vermiculite fine grade 6:1:1). BASTA (Ammonium glyfosinate (150 g/L)) soil soaking was used at 1:1000 as the selection method. The first treatment was given and the trays were covered and placed at 4 °C for three days. These were removed and put under a propagator lid in a 16 h photoperiod at 22 °C. Four further treatments were done before transplanting. Transformed plants were transplanted into individual pots and leaf number and number of days to bolt were recorded when the primary bolt was 1 cm. Seed was collected from plants showing the expected complementation phenotype and sown onto soil. The pots were stratified at 4 °C for 3 days before being placed at 22 °C with 16 h photoperiod in controlled environment cabinets (MLR-352, Panasonic Co. Ltd). The number of days to flower and rosette leaf number at 1 cm bolt were recorded. Leaf material was collected and RNA extracted as outlined above.

2.7. Real time RT-PCR

Real time RT-PCR was performed on samples from plants during the rocket vernalization experiments and also on the Arabidopsis *DtFLC* complementation T_2 plants. To measure the expression of *DtFLC* in rocket from the vernalization experiments, a mixture of RNA from samples where *DtFLC* expression was likely to be highest, were combined to synthesize cDNA to make the standard for the real time reactions. The standard was diluted 10-fold from 10^0–10^{-4} to make the standard curve. All real time RT-PCR carried out for the vernalization samples used *DtTIP41*, *DtCACS* and *Dtα-tubulin* as housekeeping genes for normalization (Supplementary data Table 2A). To measure the expression of the *DtFLC* gene in Arabidopsis FRI + *flc3* null plants, *AtActin*, *AtTIP41* and *Atβ-tubulin* were used as housekeeping genes for normalization. The standard was made from PCR products amplified from cDNA using real time primer pairs for the housekeeping genes, and *DtFLC* (see supplementary data for sequences). These were combined and purified using a PCR purification kit (Qiagen) before diluting from 10^0 to 10^{-10} for the standard curve. cDNA synthesized for each sample was used with the reaction mix iTaq™Universal SYBR® Green Supermix (Bio-Rad Laboratories Ltd., UK). All real time PCR experiments were run on a

CFX384 TouchTM Real-time PCR machine (Bio-Rad Laboratories Ltd., UK) using Bio-Rad CFX manager 3.0 software. Results were analyzed using Biogazelle qBase Plus software version 2.5 (http://www.biogazelle.com/qbaseplus).

3. RESULTS

3.1. The effect of vernalization on flowering of *D.tenuifolia*

To investigate the vernalization response of wild rocket, both seed and young plants were tested for their response to different lengths of vernalization treatment at 4 °C. *D.tenuifolia* seed were subjected to a vernalization treatment of one, two or four weeks at 4 °C in the dark before being grown under ambient conditions (20 °C, 16 h photoperiod) in the glasshouse until they initiated bolting. The number of days to bolting was recorded starting from when the seed were taken out of the vernalization treatment (i.e. the time the seed spent in the vernalization treatment was not included). Control seed were not vernalized and were grown in ambient conditions for the duration of the experiment. There was no great effect on bolting time of the different lengths of vernalization treatment at 4 °C, except for a slight reduction of 4 days after one week of treatment which was not observed after two or four weeks of treatment (Figure 1b). The effect of a 4 °C vernalization treatment on young four week old plants was also tested. The plants were subjected to two, four or eight weeks of cold treatment whilst a subset of control plants remained at constant ambient temperature throughout the experiment. There was no significant reduction in bolting time between the non-vernalized plants kept at ambient temperatures to those subjected to the different vernalization treatments (Figure 1c). The plants which had been subjected to eight weeks of vernalization in fact bolted slightly later (approximately five days) than the control plants and the other vernalization treatments.

As wild rocket is a Mediterranean plant it may have a higher optimum vernalization temperature. To test this, a higher vernalization temperature of 10 °C was used for the vernalization treatment of both seed and young four week old plants. Overall the results mirror those subjected to a 4 °C vernalization as there was no great difference in bolting time between those kept under ambient conditions and the different vernalization treatments, although a small reduction in bolting time was observed following the six week treatment of seed (Supplementary data Figure S1a). Vernalization of young plants at 10 °C did not reduce bolting time compared to non-vernalized plants kept at ambient temperature; indeed this treatment seemed to cause a delay in bolting (Supplementary data Figure S1b).

3.2. Isolation of a rocket *FLC* gene

We isolated an *FLC* homologue from wild rocket to investigate whether expression of this gene changed in response to the vernalization treatment. PCR using primers designed to Arabidopsis and *Brassica FLC* genes gave fragments that had a sequence of an *FLC-like* gene, so specific RLM RACE (Invitrogen) primers were designed. This allowed the isolation of the full genomic sequence of the *DtFLC* gene revealing that it contained 7 exons and 6 introns (Figure 2a). The intron sizes differ to that of Arabidopsis *FLC* but the exon sizes and placement are very similar (Michaels and Amasino, 1999).

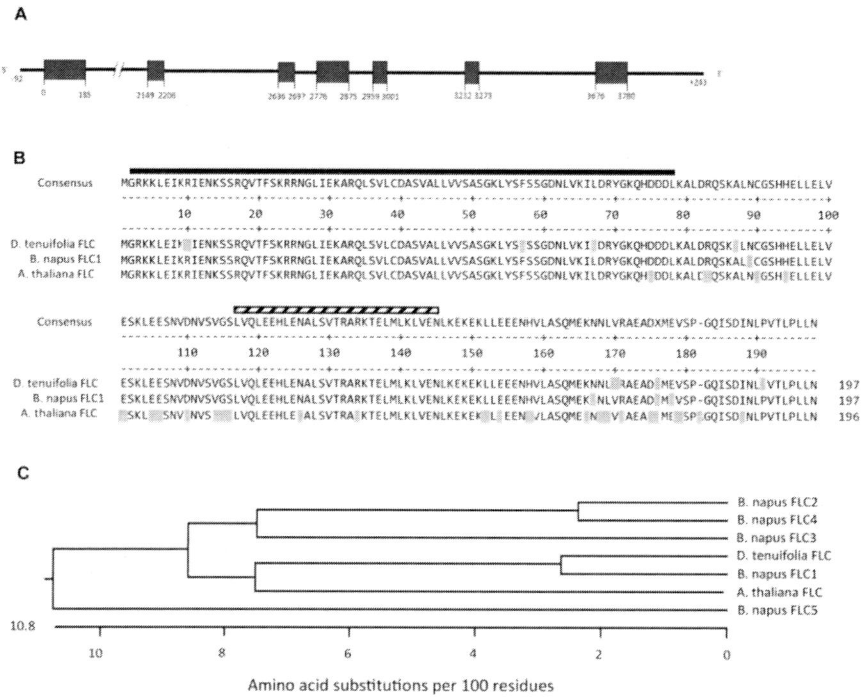

Figure 2. DtFLC gene and protein. A) Structure of DtFLC gene from D.tenuifolia. Boxes represent exons and lines represent introns and UTR. B) DtFLC protein alignment with B.napus FLC1 and A.thaliana FLC proteins. Grey shading identifies the residues that differ to the consensus sequence shown above. The black line denotes the MADS box domain and the hatched bar denotes the K-box motif. C) Phylogenetic tree showing the relationship between DtFLC, AtFLC and the five B.napus FLC proteins (1–5). DtFLC is most closely related to B.napus FLC1.

The full length coding sequence for *DtFLC* has 94% identity to the *B.napus FLC1* and *Sinapis alba FLC* genes and 86% identity to *Arabidopsis thaliana FLC*. The translated protein sequence has 95% identity to *B.napus* FLC1 and *S.alba* FLC, and 84% identity to *A.thaliana* FLC, and contains a MADS box at the N-terminus and a K-box domain in the middle of the sequence (Figure 2b). Phylogenetic comparison of DtFLC protein sequence with that of AtFLC and the five known FLC-like proteins of B.napus (FLC1-5) shows that DtFLC is most closely related to B.napus FLC1 (Figure 2c).

3.3. Functional complementation of the Arabidopsis FRIGIDA (FRI)$^+$*flc3* null mutant with *DtFLC*

The coding sequence for *DtFLC* was cloned into the vector pB2GW7 containing a 35S promoter and *Bar* gene which was transformed into Arabidopsis FRI$^+$*flc3* null mutant plants by the floral dip method. The transformed seed were selected using BASTA and the T_1 seed were grown and scored for bolting. The expression of the FLC floral repressor should result in delayed flowering/bolting and so T_1 plants which were late bolting compared to the untransformed Arabidopsis *Col0* and FRI$^+$*flc3* null mutant plants were harvested for seed. The T_2 seed were sown out for the FRI$^+$*flc3:DtFLC 5* and FRI$^+$*flc3:DtFLC 23* lines and grown in a 16 h photoperiod to screen for the flowering phenotype. The T_2 population for each line showed the expected segregation of 3:1 late:early flowering phenotype associated with the insertion of one copy of the transgene (Figure 3a & b). A Chi-squared test was performed on the data and shows no significant difference to the 3:1 ratio. Overall the data suggest that *DtFLC* is able to complement the lack of function of *AtFLC* in FRI$^+$*flc3* null mutant plants. RNA was extracted from a few T_2plants of the FRI$^+$*flc3:DtFLC 5* and FRI$^+$*flc3:DtFLC 23* lines to investigate the levels of expression of the transgene present in the transformed lines. Figure 3c shows the number of leaves at flowering for the plants tested from both the FRI$^+$*flc3:DtFLC 5* and FRI$^+$*flc3:DtFLC 23* lines compared to the non-transformed *AtCol0* wild type and FRI$^+$*flc3* mutant plants, and Figure 3d shows the expression level of the *DtFLC* transgene in each of these plants. As expected there is no transgene expression detectable in the non-transformed *AtCol0* wild type and FRI$^+$*flc3* mutant plants.

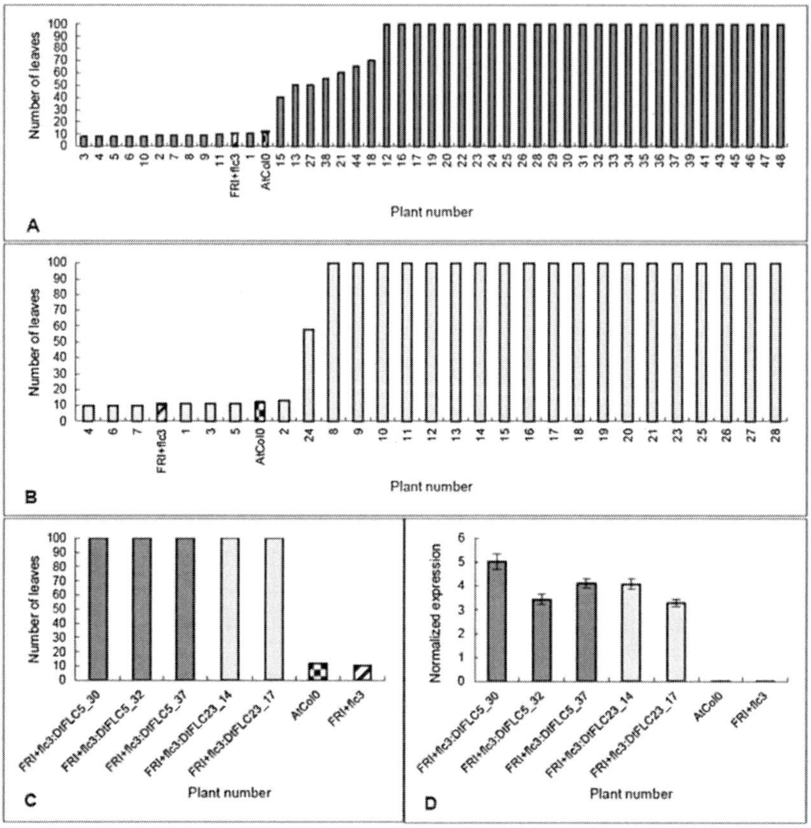

Figure 3. Flowering time and DtFLC expression data of FRI+flc3:DtFLC T_2 lines. A) Flowering time data for line FRI+flc3:DtFLC 5. Rosette leaf number was recorded when bolt reached 1 cm. B) Flowering time data for line FRI+*flc3:DtFLC 23.* Checked bar shows mean number of rosette leaves at flowering (1 cm bolt) for *AtCol0* wild type (mean 11.8 ± 1.5 $n = 24$). Hatched bar shows mean number of leaves at flowering for FRI+*flc3* mutant (mean 10.7 ± 1.2 $n = 21$). Black bars at 100 leaves show plants that did not bolt before the end of the experiment. Error bars denote the standard error of the mean. Chi-squared tests assessing goodness of fit for data to an expected 3:1 late:early bolting phenotype show that lines FRI+*flc3:*DtFLC 5 and FRI+*flc3:DtFLC 23* were not significantly different to the 3:1 ratio ($p < 0.05$). C) Number of rosette leaves at 1 cm bolt of T_2 plants from lines *FRI + flc3:DtFLC 5* (dark grey bars) and *FRI + flc3:DtFLC 23* (light grey bars) alongside WT *AtCol0* (checked bar) and *FRI + flc3* null mutant (hatched bar). Bars with 100 leaves did not bolt before the end of the experiment. Error bars denote standard error of the mean. D) Transgene expression analysis using real time PCR for T_2 plants of lines *FRI + flc3:DtFLC 5* (dark grey bars) and *FRI + flc3:DtFLC 23* (light grey bars). Expression was normalized to the combined average expression of *Atβ-tubulin, AtActin2* and *AtTIP41*. Error bars denote standard error of three technical replicates.

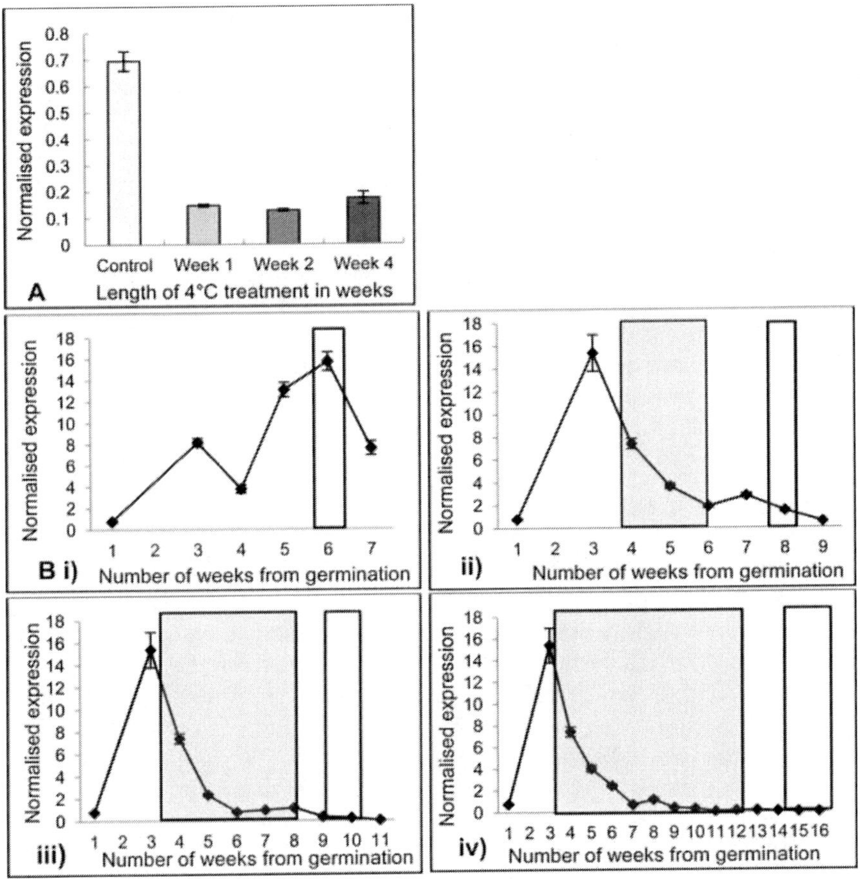

Figure 4. Effect of vernalization at 4 °C on DtFLC expression in D.tenuifolia. A) Expression of *DtFLC* where seed was subjected to vernalization at 4 °C for one, two or four weeks. Error bars denote the standard error of three biological replicates. B) Expression of *DtFLC* over development where 4 week old plants were subjected to vernalization at 4 °C for two, four or eight weeks. Change in levels of *DtFLC* expression over the course of the experiment under i) ambient conditions, ii) two weeks vernalization, iii) four weeks vernalization, iv) eight weeks vernalization. Error bars denote the standard error of three biological replicates. Range of bolting time of plants shown by white boxes and the respective 4 °C vernalization treatments shown by the grey boxes.

3.4. Expression of *DtFLC* during cold treatment of *D.tenuifolia*

The isolation of *DtFLC* allowed the expression of the gene to be investigated in wild rocket following vernalization. In the 4 °C seed vernalization experiments, *DtFLC* expression was analyzed in the seed or seedlings collected at the end of each vernalization treatment (Figure 4a) and the results revealed that those subjected to one week of ambient temperature conditions (control) had a much higher level of expression than in seed subjected to the vernalization treatments. This demonstrates that one week of cold treatment is sufficient to dramatically reduce the levels *DtFLC* expression. There is little change in the level of *DtFLC* expression during subsequent vernalization treatments indicating that once the expression level has been reduced in the first week the length of the vernalization treatment thereafter has no further effect on *DtFLC* expression.

The expression of *DtFLC* throughout plant development was investigated for each of the 4 °C vernalization treatments of four week old plants, and the non-vernalized control. In non-vernalized plants maintained in ambient conditions *DtFLC* expression increased up until three weeks after germination, fell slightly at four weeks, then increased again up until bolting and fell after the transition to flowering had occurred (Figure 4bi). In plants that had been given the vernalization treatment, the pattern of *DtFLC* expression was the same as under the ambient conditions up to the point the vernalization treatment started (at four weeks after germination). When the vernalization treatment began, the levels of *DtFLC* fell to a low level and the level remained low, even when plants were returned to ambient conditions (Figure 4bii, iii, iv). This shows that all of the cold treatments tested were effective in reducing the expression of *DtFLC* and maintaining it at a low level after vernalization.

3.5. Photoperiodic requirements of *D. tenuifolia*

Three different photoperiods were used to investigate the effect of photoperiod on the number of days to initiate bolting in wild rocket. The experiment was carried out using fluorescent bulbs to provide photoperiods of 8 h, 12 h or 16 h (light levels 124 $\mu mol.m^{-2}.s^{-1}$). The data show that the plants respond differently to each photoperiod, with shorter photoperiods delaying bolting (Figure 5a). There was a significant difference in the number of days to bolting between plants in each photoperiod ($p < 0.05$) (Figure 5a). The experiment was repeated using fluorescent bulbs to provide a standard SD photoperiod of 8 h, with low levels of incandescent light (33 $\mu mol\ m^{-2}\ s^{-1}$) used to provide the additional 4 and 8 h day extensions for the 12 h and 16 h photoperiods respectively.

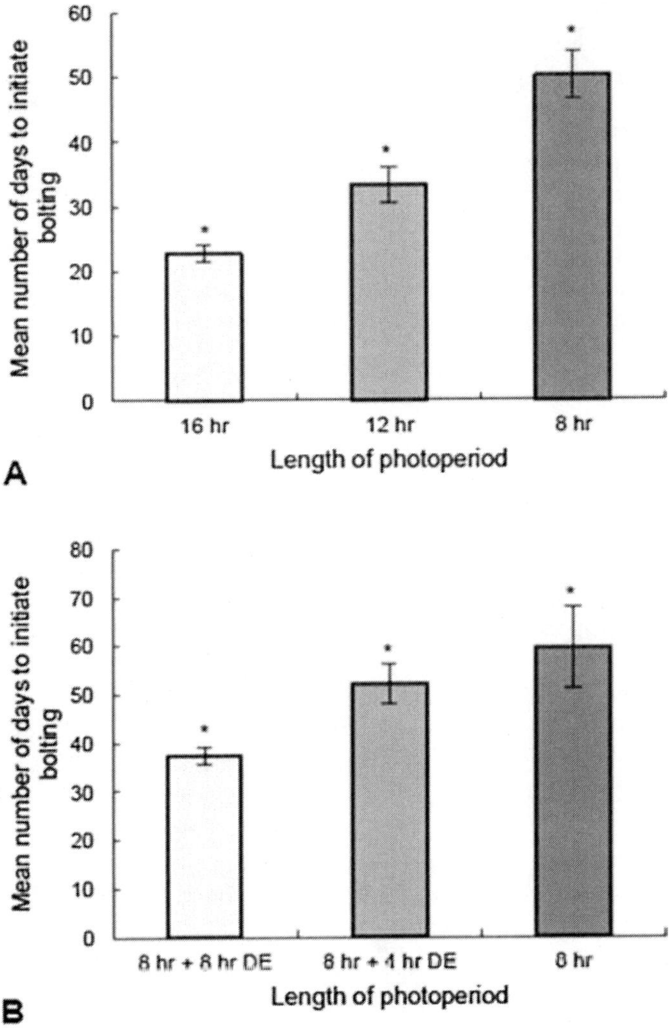

Figure 5. Effect on bolting time of varying photoperiod lengths in Diplotaxis tenuifolia plants. Mean number of days to initiate bolting for plants grown in 8 h, 12 h and 16 h photoperiods. A) Photoperiod experiment with light provided by fluorescent bulbs. Bars show mean ± SE (n = 9 (16 h + 12 h), n = 6 (8 h)). B) Photoperiod experiment with 8 h light provided by fluorescent bulbs and the 4 h or 8 h day extensions by incandescent bulbs. Bars show mean ± SE (n = 5 (8 h + 8 h DE, 8 h + 4 h DE), n = 6 (8 h)). Error bars denote the standard error of the mean. Ordinary one-way ANOVA showed a significant difference (*) between all photoperiod lengths on the initiation of bolting time ($p < 0.05$).

Low light levels were used for the day extension treatments to reduce photosynthesis levels therefore reducing any effects on flowering time that may be caused by photosynthesis/growth due to the additional 4 or 8 h light in the 12 h and 16 h photoperiodic treatments respectively. Incandescent bulbs rather than the fluorescent bulbs were used for the day extension treatments as these contain higher amounts of far-red light and are much more effective than fluorescent light for photoperiodic responses (**Johnson et al., 1994**). This experiment gave the same result as the first experiment in that shorter photoperiods caused a delay in bolting with there being a significant difference in the number of days to bolt between each photoperiod length ($p < 0.05$) (**Figure 5**b).

4. DISCUSSION

The vernalization and photoperiodic pathways are two key pathways involved in the perception of, and response to, environmental conditions. Different responses to vernalization and photoperiod are observed in different ecotypes of Arabidopsis (Strange et al., 2011) and in different varieties of crop plants such as wheat and oilseed rape (Bluemel *et al.*, 2015 ; Srikanth and Schmid, 2011). Understanding the vernalization and photoperiodic requirements of wild rocket is important for commercial production in different parts of the world, and for breeding of new varieties with different flowering behaviours.

Prior to this set of experiments it was not known whether wild rocket has a vernalization requirement, and if so at which developmental stage and temperature the response is greatest. Our data shows that wild rocket does not respond to vernalization neither as seed nor as four week old plants, at either 4 °C or 10 °C. Vernalization of vernalization-responsive Arabidopsis accessions results in a large difference in flowering time between non-vernalized plants and vernalized plants, which flower much earlier (Sanchez-Bermejo et al., 2012), and if wild rocket has a vernalization requirement, we would expect to see similar large differences in bolting time. Although there is a small decrease in the mean number of days to bolt after vernalizing wild rocket seed for one week at 4 °C, it clearly doesn't satisfy a putative vernalization requirement as longer periods of vernalization of two or four weeks at 4 °C resulted in a later bolting time (Figure 1b). Similarly, whilst a small reduction in the number of days to bolt was observed following vernalization of seed at 10 °C for six weeks (Supplementary data Figure S1a), a much greater reduction in bolting time would be expected if wild rocket possessed a vernalization response. Experiments with young four week old plants showed that there is no decrease in the mean number of days to bolt in response to vernalization treatments at either 4 °C or 10 °C (Figs. 1 c & S1b). In some cases, the vernalized plants take slightly longer to bolt than the non-vernalized controls grown under constant ambient conditions, which may reflect the need for a recovery period from the cold temperature which is likely to have slowed metabolic and physiological processes in the treated seed/young plants.

A homologue of the Arabidopsis *FLC* (*AtFLC*) gene was isolated from *D.tenuifolia* (*DtFLC*), the predicted protein sequence was found to have a higher identity to *B.napus* FLC1 than to AtFLC and other Brassica FLC proteins. *AtFLC* functions as a repressor of flowering and Kim et al. (2007) investigated how candidate *FLC* genes from Chinese cabbage (*B.rapa* L. ssp. Pekinensis) behaved when used to complement Arabidopsis *flc* mutants. They found that all *BrFLC* genes tested caused a late flowering phenotype showing that these genes were able to complement the *flc* mutant and were repressors of flowering. A similar approach was taken to investigate the function of *DtFLC*. The Arabidopsis *FRI*$^{+}$*flc3* null mutant was transformed with the *DtFLC* coding sequence and late flowering lines were obtained in the T_1 and T_2 generations showing that *DtFLC* was able to complement the lack of functional *AtFLC* in the mutant *flc* lines. Analysis of transgene expression levels confirmed that the complemented lines were expressing the transgene (Figure 3d). These results showed that *DtFLC* is a functional orthologue of *AtFLC* and enabled us to investigate the pattern of *DtFLC* gene expression during the various vernalization treatments used in our experiments.

When expression levels of *DtFLC* were examined in seed vernalized at 4 °C the results showed a large drop in expression during the first week of cold treatment and that levels remained low during additional weeks of vernalization (Figure 4a), this reflects results obtained in Arabidopsis (Sheldon et al., 2000). The expression of *DtFLC* observed in the four week old plants subjected to vernalization at 4 °C showed a similar pattern in that one week of cold treatment was sufficient to cause a large reduction in expression levels. The levels of *DtFLC* were shown to remain low during longer periods of 4 °C cold, and were still low after the plants were returned to ambient conditions for the remainder of the experiment (Figure 4b). Together, these results show that just one week of 4 °C cold treatment is effective in reducing the levels of *DtFLC* expression but that this does not have an effect on reducing the time to bolting.

The *DtFLC* expression data showed that the vernalization-induced reduction in expression levels of *DtFLC* is maintained even when plants are returned to ambient conditions (Figure 4b). This suggests that wild rocket may have genes homologous to the Arabidopsis *VERNALIZATION INSENSITIVE 3*, *VERNALIZATION 1* and *VERNALIZATION 2* genes which reduce the expression of *FLC* and maintain the low levels after vernalization (Gendall *et al*., 2001; Levy *et al*., 2002 ; Sung and Amasino, 2004). However, as the reduction of *DtFLC* expression during and post-vernalization treatment does not affect bolting time in rocket, this suggests that the relationship between *FLC* expression and flowering in wild rocket may not be the same as in Arabidopsis. This may be due to some disruption of the flowering time pathway between *DtFLC* and the floral integrator genes meaning that the latter are unaffected by changes in *DtFLC* expression. It could also possibly be that in non-vernalized wild rocket *DtFLC* is expressed at levels which are insufficient to actively repress flowering and so the observed reduction in expression of *DtFLC* during vernalization would not have any significant effect on flowering. This would be similar to Arabidopsis ecotypes *Ler, Da* and *Shakhdara* which have a low expression of *AtFLC* but flower rapidly and are classified as summer annuals. It was found that the *AtFLC* gene in *Ler, Da* and *Shakhdara* did not contain any mutations in the coding sequence compared to *Col0*, but the low expression was caused by mutations in intron 1 (*Ler*) or the 5′ UTR (*Da* and *Shakhdara*) sequences (Michaels et al., 2003). Overall the results demonstrate that rocket does not have a vernalization requirement as the cold treatments (either 4 °C or 10 °C) did not reduce the time the plants took to bolt compared to non-vernalized plants. This could be due to its origins in the Mediterranean where a vernalization response would be a disadvantage as has been found in *B.napus* varieties grown in Mediterranean climates where the lack of a vernalization requirement is important for its early flowering phenotype (Dahanayake and Galwey, 1999). Other studies in *B.rapa* have also shown that vernalization isn't a requirement for plants bred in warmer climates (Franks et al., 2015).

Studies in Arabidopsis have explored the pattern of ecotype distribution and flowering time response particularly in relation to photoperiod and vernalization responses (Caicedo *et al*., 2004; Lempe *et al*., 2005 ; Stinchcombe *et al*., 2004). They linked the different degrees of response to differences in the activity of *FRI* (which promotes FLC expression) and *FLC*. Further investigation, for example the identification of a *FRI* homologue in rocket, may provide a greater understanding of vernalization responses in Mediterranean crops.

The photoperiodic requirement of wild rocket has not been properly defined and so experiments were set up to test the response of *D.tenuifoila* to different photoperiods. Plants were able to flower in LD (16 h), intermediate (12 h) and SD (8 h) photoperiods but flowered much more rapidly in longer photoperiods (Figure 5), this shows that wild rocket is a facultative LD plant. Results of photoperiodic experiments using day extensions of low levels of incandescent light day to minimise the effects of additional photosynthesis on the control of bolting in the longer photoperiods were consistent with the first experiment, confirming the conclusion that rocket is a facultative LD plant.

5. CONCLUSIONS

A rocket homologue of *FLOWERING LOCUS C* (*DtFLC*) has been isolated and shown to functionally complement the *Arabidopsis* FRI^{+}*flc3* null mutant. The expression of *DtFLC* is significantly reduced by one week of cold, however vernalization treatments of up to eight weeks had no significant effect on bolting time of rocket demonstrating that rocket does not have a vernalization requirement. The flowering response of rocket to different photoperiods show that flowering is promoted by LD and that rocket is a facultative long day plant.

AUTHOR CONTRIBUTIONS

J.T., A.M. and S.J. designed the experiments, J.T. performed the experiments, J.T. and S.J wrote the manuscript, A.M., S.J., S.K. and Y.H. supervised the study and revised the manuscript.

CONFLICT OF INTEREST

The authors declare that there are no conflicts of interest.

ACKNOWLEDGEMENTS

We acknowledge the technical assistance of Mrs. Anushree Choudhary on the 10 °C vernalization experiments and advice from Professor Brian Thomas for the photoperiod experiments. We thank Elsoms Seeds Ltd for the *D.tenuifolia* seed stocks and Nottingham Arabidopsis Stock Centre for the Arabidopsis *Col0* and FRI^{+}*flc3* mutant seeds.

This work was supported by a BBSRC CASE award (BB/I016120/1) in partnership with Elsoms Seeds Ltd., and BBSRC grant no. BB/G007330/1.

REFERENCES

1. Arias, T., Pires, J.C., 2012. A fully resolved chloroplast phylogeny of the brassica crops and wild relatives (Brassicaceae: Brassiceae): novel clades and potential taxonomic implications. Taxon 61, 980–988.
2. Bluemel, M., Dally, N., Jung, C., 2015. Flowering time regulation in crops – what did we learn from Arabidopsis? Curr. Opin. Biotechnol. 32, 121–129.
3. Boss, P.K., Bastow, R.M., Mylne, J.S., Dean, C., 2004. Multiple pathways in the decision to flower: enabling, promoting, and resetting. Plant Cell 16, S18–S31.
4. Brambilla, V., Fornara, F., 2013. Molecular control of flowering in response to day length in rice. J. Integr. Plant Biol. 55, 410–418.
5. Caicedo, A.L., Stinchcombe, J.R., Olsen, K.M., Schmitt, J., Purugganan, M.D., 2004. Epistatic interaction between Arabidopsis FRI and FLC flowering time genes generates a latitudinal cline in a life history trait. Proc. Natl. Acad. Sci. U. S. A. 101, 15670–15675.
6. Chiang, G.C., Barua, D., Kramer, E.M., Amasino, R.M., Donohue, K., 2009. Major flowering time gene, FLOWERING LOCUS C, regulates seed germination in Arabidopsis thaliana. Proc. Natl. Acad. Sci. U. S. A. 106, 11661–11666.
7. Chouard, P., 1960. Vernalization and its relation to dormancy. Annu. Rev. Plant Physiol. 11, 191–238.
8. Chun, J.-H., Arasu, M.V., Lim, Y.-P., Kim, S.-J., 2013. Variation of major glucosinolates in different varieties and lines of rocket salad. Hortic. Environ. Biotechnol. 54, 206–213.
9. Clough, S.J., Bent, A.F., 1998. Floral dip: a simplified method for Agrobacterium-mediated transformation of Arabidopsis thaliana. Plant J. 16, 735–743.
10. Corbesier, L., Coupland, G., 2005. Photoperiodic flowering of Arabidopsis: integrating genetic and physiological approaches to characterization of the floral stimulus. Plant Cell Environ. 28, 54–66.
11. Dahanayake, S.R., Galwey, N.W., 1999. Diallel analysis of vernalisation responses in spring rape (Brassica napus L.): a basis for adaptation to a Mediterranean environment. Aust. J. Agr. Res. 50, 1417–1423.
12. Fornara, F., de Montaigu, A., Coupland, G., 2010. SnapShot: control of flowering in arabidopsis. Cell 141, 50–550.
13. Franks, S.J., Perez-Sweeney, B., Strahl, M., Nowogrodzki, A., Weber, J.J., Lalchan, R., Jordan, K.P., Litt, A., 2015. Variation in the flowering time orthologs BrFLC and BrSOC1 in a natural population of Brassica rapa. PEER J. 3, e1339.
14. Gendall, A.R., Levy, Y.Y., Wilson, A., Dean, C., 2001. The VERNALIZATION 2 gene mediates the epigenetic regulation of vernalization in Arabidopsis. Cell 107, 525–535.
15. Gill, M., 2008. Wild rocket sales on the up. The Vegetable Farmer. ACT Publishingpp. 27–28.
16. Guo, D.P., Shah, G.A., Zeng, G.W., Zheng, S.J., 2004. The interaction of plant growth regulators and vernalization on the growth and flowering of cauliflower (Brassica oleracea var. botrytis). Plant Growth Reg. 43, 163–171.
17. Hackett, W.P., Hartmann, H.T., 1967. The influence of temperature on floral initiation in the olive. Physiol. Plant 20, 430–436.
18. Jarillo, J.A., Pineiro, M., 2011. Timing is everything in plant development. The central role of floral repressors. Plant Sci. 181, 364–378.
19. Johnson, E., Bradley, M., Harberd, N.P., Whitelam, G.C., 1994. Photoresponses of lightgrown phyA mutants of Arabidopsis (phytochrome A is required for the perception of daylength extensions). Plant Physiol. 105, 141–149.
20. Jung, C., Mueller, A.E., 2009. Flowering time control and applications in plant breeding. Trends Plant Sci. 14, 563–573.

21. Kim, S.-Y., Park, B.-S., Kwon, S.-J., Kim, J., Lim, M.-H., Park, Y.-D., Kim, D.Y., Suh, S.-C., Jin, Y.-M., Ahn, J.H., Lee, Y.-H., 2007. Delayed flowering time in Arabidopsis and Brassica rapa by the overexpression of FLOWERING LOCUS C (FLC) homologs isolated from Chinese cabbage (Brassica rapa L. ssp pekinensis). Plant Cell Rep. 26, 327–336.
22. Lempe, J., Balasubramanian, S., Sureshkumar, S., Singh, A., Schmid, M., Weigel, D., 2005. Diversity of flowering responses in wild Arabidopsis thaliana strains. PLoS Genet. 1, 109–118.
23. Levy, Y.Y., Mesnage, S., Mylne, J.S., Gendall, A.R., Dean, C., 2002. Multiple roles of Arabidopsis VRN1 in vernalization and flowering time control. Science 297, 243–246.
24. Logemann, J., Schell, J., Willmitzer, L., 1987. Improved method for the isolation of RNA from plant-tissues. Anal. Biochem. 163, 16–20.
25. Matsoukas, I.G., Massiah, A.J., Thomas, B., 2012. Florigenic and antiflorigenic signaling in plants. Plant Cell Physiol. 53, 1827–1842.
26. Michaels, S.D., Amasino, R.M., 1999. FLOWERING LOCUS C encodes a novel MADS domain protein that acts as a repressor of flowering. The Plant Cell 11 (5), 949–956.
27. Michaels, S.D., He, Y.H., Scortecci, K.C., Amasino, R.M., 2003. Attenuation of FLOWERING LOCUS C activity as a mechanism for the evolution of summer-annual flowering behavior in Arabidopsis. Proc. Natl. Acad. Sci. U. S. A. 100, 10102–10107.
28. Nordborg, M., Bergelson, J., 1999. The effect of seed and rosette cold treatment on germination and flowering time in some Arabidopsis thaliana (Brassicaceae) ecotypes. Am. J. Bot. 86, 470–475.
29. Sanchez-Bermejo, E., Mendez-Vigo, B., Xavier Pico, F., Martinez-Zapater, J.M., AlonsoBlanco, C., 2012. Novel natural alleles at FLC and LVR loci account for enhanced vernalization responses in Arabidopsis thaliana. Plant Cell Environ. 35, 1672–1684.
30. Sheldon, C.C., Rouse, D.T., Finnegan, E.J., Peacock, W.J., Dennis, E.S., 2000. The molecular basis of vernalization: the central role of FLOWERING LOCUS C (FLC). Proc. Natl. Acad. Sci. U. S. A. 97, 3753–3758.
31. Song, J., Angel, A., Howard, M., Dean, C., 2012. Vernalization – a cold-induced epigenetic switch. J. Cell Sci. 125, 3723–3731.
32. Srikanth, A., Schmid, M., 2011. Regulation of flowering time: all roads lead to Rome. Cell Mol. Life Sci. 68, 2013–2037.
33. Stewart, C.N., Via, L.E., 1993. A rapid CTAB DNA isolation technique useful for rapid fingerprinting and other PCR applications. Biotechniques 14, 748–751.
34. Stinchcombe, J.R., Weinig, C., Ungerer, M., Olsen, K.M., Mays, C., Halldorsdottir, S.S., Purugganan, M.D., Schmitt, J., 2004. A latitudinal cline in flowering time in Arabidopsis thaliana modulated by the flowering time gene FRIGIDA. Proc. Natl. Acad. Sci. U. S. A. 101, 4712–4717.
35. Strange, A., Li, P., Lister, C., Anderson, J., Warthmann, N., Shindo, C., Irwin, J., Nordborg, M., Dean, C., 2011. Major-effect alleles at relatively few loci underlie distinct vernalization and flowering variation in Arabidopsis accessions. PLoS One 6, 1–11.
36. Sung, S.B., Amasino, R.M., 2004. Vernalization in Arabidopsis thaliana is mediated by the PHD finger protein VIN3. Nature 427, 159–164.
37. Thomas, B., Carre, I., Jackson, S., 2006. Photoperiodism and flowering. Molecular Biology and Biotechnology of Flowering, 2nd edition. pp. 3–25.
38. Wang, J.W., 2014. Regulation of flowering time by the miR156-mediated age pathway. J. Exp. Bot. 65, 4723–4730.

CHAPTER 8

Senescence, Stress, and Reactive Oxygen Species

Ivan Jajic [1], Tadeusz Sarna [2] and Kazimierz Strzalka [1,3,*]

[1]*Department of Plant Physiology and Biochemistry, Faculty of Biochemistry, Biophysics and Biotechnology, Jagiellonian University in Krakow, Gronostajowa 7, Krakow 30-387, Poland*

[2]*Department of Biophysics, Faculty of Biochemistry, Biophysics and Biotechnology, Jagiellonian University in Krakow, Gronostajowa 7, Krakow 30-387, Poland*

[3]*Malopolska Centre of Biotechnology, Jagiellonian University in Krakow, Gronostajowa 7, Krakow 30-387, Poland*

ABSTRACT

Generation of reactive oxygen species (ROS) is one of the earliest responses of plant cells to various biotic and abiotic stresses. ROS are capable of inducing cellular damage by oxidation of proteins, inactivation of enzymes, alterations in the gene expression, and decomposition of biomembranes. On the other hand, they also have a signaling role and changes in production of ROS can act as signals that change the transcription of genes that favor the acclimation of plants to abiotic stresses. Among the ROS, it is believed that H_2O_2 causes the largest changes in the levels of gene expression in plants. A wide range of plant responses has been found to be triggered by H_2O_2 such as acclimation to drought, photooxidative stress, and induction of senescence. Our knowledge on signaling roles of singlet oxygen ($^1O_2$) has been limited by its short lifetime, but recent experiments with a *flu* mutant demonstrated that singlet oxygen does not act primarily as a toxin but rather as a signal that activates several stress-response pathways. In this review we summarize the latest progress on the signaling roles of ROS during senescence and abiotic stresses and we give a short overview of the methods that can be used for their assessment.

Keywords: leaf senescence; abiotic stresses; reactive oxygen species; hydrogen peroxide; singlet oxygen; superoxide anion

1. INTRODUCTION

Senescence in plants is a complex deterioration process that can lead to the death of whole organisms or a single organ. It is regulated by autonomous (internal) factors (age, reproductive development, and phytohormone levels) and by environmental signals, including photoperiod, stresses such as drought, ozone, nutrient deficiency, wounding, and shading [1]. The generation of reactive oxygen species (ROS) is one of the earliest responses of plant cells under abiotic stresses and senescence [2,3]. In plants, ROS are formed as byproducts of aerobic energy metabolism and of plants being exposed to various biotic and abiotic stresses [4,5,6]. Under normal conditions, the production of ROS in cells is maintained at low levels by antioxidant enzymes. This balance can be disrupted by a depletion of antioxidants or the excess accumulation of ROS, leading to oxidative stress, and consequently to damage to cellular macromolecules and membranes and an increase in lipid peroxidation [7,8]. ROS-induced oxidative stress limits agricultural yields worldwide [9]; in the United States alone, it is estimated that the deleterious effects of abiotic stresses on agricultural production are responsible for losses amounting to billions of dollars annually [10]. Plants have evolved different mechanisms to protect themselves from adverse environmental conditions, such as the process of acclimation, which involves less ROS production coupled with an efficient antioxidant defense [11,12] and the activation of different signaling pathways [13,14].

Initially, ROS were exclusively considered toxic metabolic products that can damage cellular components, but now it is clear that ROS play a dual role in plants both as toxic compounds and as key regulators of many biological processes [15,16,17]. The important role of ROS in signaling has been demonstrated in many studies. It has been shown that ROS modulate the activity of key signaling compounds such as MAP kinases [18], provide protection against pathogen invasion [19,20], stimulate abiotic stress tolerance [21], and have an important role during early responses to wounding [22]. Despite the importance of ROS, our knowledge of the mechanism of their action is still limited. In this review we will try to summarize the latest progress on the roles of reactive oxygen species during senescence and abiotic stresses.

2. ROS DETECTION

The main problem with an accurate determination of the role of ROS in senescence and abiotic stresses is the simultaneous generation of ROS [23,24] and the limited number of non-invasive and specific methods that can be used for their determination. For example, in plants suffering from moderate light stress singlet oxygen ($^1O_2$), superoxide anion ($O^{\bullet}{}_2{}^-$) and hydrogen peroxide (H_2O_2) are released simultaneously [23], making it difficult to establish their individual roles. This is further complicated by significant differences in the production of ROS when senescence is induced artificially and when the plant ages naturally [25]. The effects of H_2O_2 on gene expression have also been reported to be different when the H_2O_2 was applied exogenously and when it was induced in response to high light [26]. Finally, several studies revealed that multiple stressors, as usually encountered by plants in nature, could substantially change the expression patterns of genes determined in a single factor analysis [27,28,29,30]. There is a need for an experimental model that can take into account all the factors that can influence the outcome of research into the role of ROS.

There are a great number of user-friendly assays that can be used for the measurement of ROS production in plant tissues, including the fluorometric determination of H_2O_2 with Amplex Red [31] and the colorimetric determination of $O^{\bullet}{}_2{}^-$ with XTT [32]. Although easy to use, these methods lack specificity. This is because Amplex Red is somewhat unstable—it can be autooxidized and produce $O^{\bullet}{}_2{}^-$ and H_2O_2 [33], and it can react not only with H_2O_2 but also with other redox active compounds [34], while XTT can be reduced by short-chain sugars [35]. Higher specificity can be achieved by using the spin trapping technique by electron paramagnetic resonance (EPR) spectroscopy. Spin trapping involves the reaction between a nitrone or nitroso compound and a free radical to form a stable spin adduct [36]. The spin adduct usually yields a distinctive EPR spectrum characteristic of the particular free radical that is trapped. Under normal conditions, the flux of ROS generated in cells is maintained at low levels by the action of antioxidant enzymes and low molecular weight antioxidants, making the detection of ROS difficult. In order to successfully detect ROS, an imbalance between the production of

ROS and their decay needs to be created. This can be achieved by the illumination of samples with visible light in the presence of an appropriate spin trap or with the addition of the spin trap immediately after illumination [37,38,39,40]. There are numerous publications with detailed descriptions of how to use EPR spectroscopy for the detection of hydrogen peroxide [39], the superoxide anion [40], the hydroxyl radical [41], and singlet oxygen [38] in plant tissues. One of the limitations of spin trapping is that the method does not provide information on the specific sites of ROS production in tissues due to solvent incompatibility with living tissue and high concentrations of spin traps needed [42], Also, it typically underestimates actual *in situ* ROS concentrations because only a small fraction of the radicals produced is usually trapped [43]. Non-invasive, *in vivo* measurement of ROS can be achieved using fluorescent probes in combination with confocal laser scanning microscopy (CLSM) [44] or fluorescence microscopy [45,46,47]. One advantage of CLSM methodology is the possibility of studying the intracellular location of ROS using simultaneously specific fluorescent probes for different organelles. 2′,7′-Dichlorofluorescein diacetate (DCF-DA) can be used for monitoring H_2O_2 in plant tissues [48]; however, it lacks specificity since it was demonstrated that it can also react with other peroxides [49]. Dihydroethidium (DHE) is a more specific probe that can be used for detection of $O^{\bullet}{}_2^-$ in different plant tissues [50,51]. The determination of singlet oxygen with a singlet oxygen sensor green (SOSG) reagent showed high specificity to $^1O_2$ without the interference of hydroxyl radicals or superoxide [46]. Unfortunately, even in the absence of external $^1O_2$ photosensitisers, the indicator can be converted to its green fluorescent form upon exposure to either UV or visible radiation. This could obviously lead to a wrong estimation of $^1O_2$ levels [52]. Real-time monitoring of reactive oxygen species in living cells and tissues can be achieved with a genetically encoded redox probe such as HyPer and reduction-oxidation sensitive green fluorescent protein (roGFPs) [53]. These probes offer high specificity and can be used for determination of ROS in various subcellular compartments. HyPer is highly sensitive to hydrogen peroxide, is insensitive to other oxidants, and it does not cause artifactual ROS generation, thus having great potential in plant biology [54], while roGFP can be used for determination of H_2O_2, but also for determination of thiol redox state depending on its form [53]. Disadvantages of these probes include the necessity of pH control and possible antioxidant activity, which need to be taken into consideration [53]. In Table 1 we summarize the methods that can be used for measurement of ROS production in plant samples.

Table 1. Overview of the methods for determination of ROS in plan samples.

Method/Probe	Advantage	Disadvantage	Used for
Amplex Red	Easy to use and fast	Can be autooxidized, reacts with other redox reactive compounds	H_2O_2
XTT	Easy to use and fast	Non-specific, can be reduced by short-chain sugars	$O^{\bullet}{}_2^-$
EPR spin trapping	Highly specific, can be used for determination of numerous ROS	Expensive, laborious, method does not provide information on the specific sites of ROS production in tissues	H_2O_2, $O^{\bullet}{}_2^-$, $^1O_2$, OH^-
DCF – DA	*In vivo* determination of intracellular ROS production	Non-specific, reacts with other peroxides, expensive equipment	H_2O_2
DHE	*In vivo* determination of intracellular ROS production	Expensive equipment	$O^{\bullet}{}_2^-$
SOSG	*In vivo* determination of intracellular ROS production	Wrong estimation of $^1O_2$ upon exposure to visible or UV light	$^1O_2$
HyPer and roGFP	Real time monitoring of ROS in living cells and tissues	Necessity for pH control and possible antioxidant activity	H_2O_2

3. SUPEROXIDE ANION

In plants $O^{\bullet}{}_2^-$ is generated in different cell compartments, including chloroplasts, peroxisomes, apoplast, the mitochondrial electron transport chain, and the plasma membrane [55,56,57]. The primary source of the superoxide anion in chloroplasts are Mehler reactions, during which O_2 is reduced by electrons from the photosynthetic electron transport chain [58]. Generated $O^{\bullet}{}_2^-$ is then converted to hydrogen peroxide (H_2O_2), mostly by the action of CuZn superoxide dismutase (SOD) [59]. Thus, SOD determines the lifetime of $O^{\bullet}{}_2^-$ in cells and the probability of its involvement in biochemical processes. Furthermore, $O^{\bullet}{}_2^-$ is a moderately reactive, short-lived ROS with a half-life of approximately 2–4 μs [17], and it cannot cross the chloroplast membrane [60]. For the reasons listed above, a signaling role of $O^{\bullet}{}_2^-$ during senescence and abiotic stresses seems unconvincing. However, chloroplasts are not the only sites of $O^{\bullet}{}_2^-$ production. In peroxisomes, $O^{\bullet}{}_2^-$ is being generated by two different sources: in peroxisomal matrix via action of enzyme xanthine oxidase [61,62] and by electron transport chain (ETC) in peroxisomal membrane [63]. Peroxisomes can be considered as an important source of signaling molecules since they have capacity to rapidly produce and scavenge H_2O_2 and $O^{\bullet}{}_2^-$ due to the presence of many antioxidants in these organelles. Another important source of $O^{\bullet}{}_2{}^-$ in plant cells are NADPH oxidases (NOX), in plants commonly known as respiratory burst oxidase homologs (Rbohs), which catalyze the production of $O^{\bullet}{}_2{}^-$ [64,65]. Plant Rbohs have been intensively studied recently since they play key roles in many physiological processes, such as ROS signaling and stress responses [66,67]. Finally, $O^{\bullet}{}_2{}^-$ is also produced in cytosol by action of xanthine dehydrogenase and the aldehyde oxidase [68,69]. Numerous studies have reported an increase in the production of $O^{\bullet}{}_2{}^-$during natural and artificially induced senescence [70,71]; however, attributing a specific signaling role to this increase is extremely difficult since the increase in most cases is accompanied by the production of other ROS and the quick conversion of $O^{\bullet}{}_2{}^-$ to H_2O_2. High production of ROS is damaging to the cell due to oxidative modifications of key cellular components and may ultimately lead to plant cell death [72]. In a recent study [73] it was shown that under the high temperature treatments, large amounts of $O^{\bullet}{}_2{}^-$ and H_2O_2 were generated and accumulated in cucumber leaves, leading to premature senescence, which is indicated by the changes in protein, lipid peroxidation (LPO), and chlorophyll content. Nevertheless, a signaling role of $O^{\bullet}{}_2{}^-$ was demonstrated in *Arabidopsis thaliana* plants exposed to methyl viologen, a superoxide anion propagator, under light. The generation of $O^{\bullet}{}_2{}^-$ in the absence of H_2O_2 accumulation revealed a subset of nuclear encoded genes that are likely to be specific for an $O^{\bullet}{}_2{}^-$-mediated signaling pathway [74]. Data analysis identified a strong upregulation of genes belonging to categories functioning in abiotic stress responses, among them WRKY6, which has previously been reported to play a role during senescence and in defense-related processes [75]. Recently, we investigated the production of $O^{\bullet}{}_2{}^-$ during the development and senescence of secondary barley leaves by using EPR––a spin trapping method with DMPO as a spin trap [76]. It was shown that the production of $O^{\bullet}{}_2{}^-$ increases during the development of barley, reaching its highest level right after the onset of senescence. Thereafter, the production of $O^{\bullet}{}_2{}^-$started to decline till the end of the senescence process. This was accompanied with an increase in membrane fluidity during the same period [77], which could be a factor facilitating the increase in the generation of ROS. An increase in $O^{\bullet}{}_2{}^-$was also observed in the interveinal area of senescing tobacco leaves, as well as in the minor veins of mature and senescent leaves, while it was absent in the major veins [78]. It is hypothesized that spatial differences in the superoxide anion are important for the non-uniform downregulation of photosynthesis-associated genes. A further role of $O^{\bullet}{}_2{}^-$ as a signaling molecule was demonstrated during the early wound response in an experiment with *Medicago* leaves where ROS production was inhibited with diphenyleneiodonium (DPI). The rapid (≤3 min) DPI inhibition of phase I $O^{\bullet}{}_2{}^-$ production suppressed the differential regulation of 7 out of 19 wound responsive proteins, showing that early, wound-related

$O^{\bullet}{}_2^-$production (phase I) provides an essential signal for wound-related changes in the leaf apoplast proteome [22]. Increased production of $O^{\bullet}{}_2^-$ was observed in plant responses to cadmium stress in pea (*Pisum sativum* L.) [48]. Exposure to Cd leads to an oxidative stress as a result of disturbance in antioxidant defense and a decrease in NO level. It was demonstrated that NO can mitigate the deleterious effect of Cd on lupine roots [79] and that it has a possible antioxidant effect in its ability to react with $O^{\bullet}{}_2^-$ to prevent oxidative damage [80]. In this way, $O^{\bullet}{}_2^-$ could contribute to plant responses to abiotic stresses.

4. HYDROGEN PEROXIDE

Hydrogen peroxide plays an important role in plants under stress conditions as a signaling molecule that mediates between different physiological processes [81]. It is involved in the regulation of the senescence process [82], protection against pathogen attack [83], the reduction of stress intensity at low light [84], and the alleviation of drought stress [85], and it can influence the expression of hundreds of genes [86]. Hydrogen peroxide is produced in plants via two possible pathways: dismutation of $O^{\bullet}{}_2^-$ with the involvement of SOD [59], and via oxidases such as amino and oxalate oxidases [87]. The level of H_2O_2 is kept under control by a fine-tuned network of enzymatic and low-molecular-weight antioxidants that prevent the excess accumulation of H_2O_2 [88]. Production and scavenging of H_2O_2 in plant cells has been summarized in Figure 1. The balance between SODs and the different H_2O_2-scavenging enzymes in cells is considered to be crucial in determining the steady-state level of H_2O_2 [89]. In comparison with other ROS, H_2O_2 is the most stable and least reactive ROS, and it can easily cross the membrane [81,90], which makes it a good signaling molecule. H_2O_2 plays a versatile role in plants; as a signaling molecule it is involved in the regulation of various abiotic and biotic stresses [81] and, at high concentrations, it has an important role in cell death and during the final stages of senescence, when it contributes to cell degradation [16,17]. The dual role of H_2O_2 was confirmed in a recent study, in which treatment with 600 μM H_2O_2 caused an increase in the vase life of a cut Oriental × Trumpet hybrid lily "Manissa," while concentrations of 800 and 1200 μM resulted in negative effects [91]. Further evidence that the effects of H_2O_2 are dose dependent comes from a study in which wax apple trees were spray-treated with different concentrations of H_2O_2 under field conditions [92]. Spraying wax apple fruits with 5 and 20 mM of H_2O_2 once a week produced better fruit growth and maximized the yield and quality in comparison with the control and with a higher dose of 50 mM of H_2O_2.

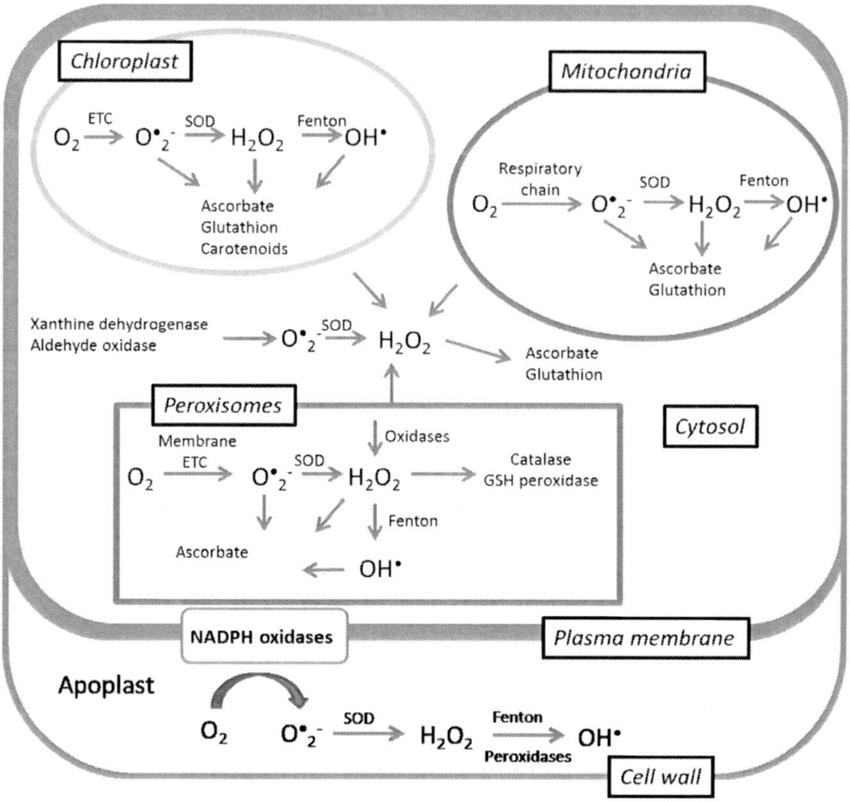

Figure 1. Production and scavenging of ROS in a plant cell. Figure legend: Superoxide dismutase (SOD); Fenton (decomposition of hydrogen peroxide to highly reactive hydroxyl radical in presence of iron); ETC (electron transport chain).

H_2O_2 plays an important role during the senescence process, where it was shown that it could be used as a signal to promote senescence in different plant species, and to be part of a complex regulatory network [93]. It was shown that H_2O_2increases at the point when the plants start to bolt and flower, which is supported by a decrease in ascorbate peroxidase 1 activity at the same time [16]. This probably induces the expression of transcription factors and senescence-associated genes. Indeed, expression analysis showed that H_2O_2 treatment induced 14 out of 36 salt-triggered senescence-associated genes and 15 senescence-associated NAC genes [94], indicating that salt-triggered senescence at least in part involves H_2O_2-mediated signaling through NAC transcription factors. This is supported by studies in which it was demonstrated that

H_2O_2 treatment induces the expression of NAC transcription factors ORS1 [95] JUB1 [96] and ATAF1 [97], which have a significant impact on progression of senescence. Overexpression of ORS1 triggers the expression of senescence-associated genes and accelerates senescence in transgenic plants, whereas its inhibition delays it. Contrary to ORS1, the overexpression of JUB1 strongly delays senescence, dampens intracellular H_2O_2 levels, and enhances tolerance to various abiotic stresses. *ATAF1* overexpression stimulates the progression of senescence by activating senescence promoting transcription factor *ORE1* and repressing chloroplast maintenance transcription factor *GLK1*. The generation of H_2O_2during the development and senescence of barley was investigated in our recent study [76]. The results showed an increase in the production of H_2O_2 with the development of barley, with the highest levels observed right after the induction of senescence and at the very end of the senescence process, which is similar to the pattern observed in the study discussed above [16]. Our research provides further evidence of the important role of H_2O_2 during leaf senescence in two different aspects: as a signal molecule for the induction of senescence, and in the degradation of molecules at later stages of senescence.

Several studies have indicated that H_2O_2 can interplay with other signal molecules that are important for plant development and during senescence such as abscisic acid (ABA) and ethylene [12,98]. It was shown that H_2O_2 could be involved in the signaling of plant growth regulators such as ethephon [98]. The application of ethephon results in an elevation in H_2O_2 levels, which is accompanied by the increased expression of sweet potato catalase. The elimination of H_2O_2 influence by exogenous-reduced glutathione alleviates ethephon-mediated effects. Recently, the interaction between ABA, H_2O_2, and ascorbic acid in Mediterranean shrubs during summer drought was investigated. It was shown that the drought stress–ABA–H_2O_2 interaction can induce an increase in ascorbic acid, maintaining and even decreasing the ascorbate oxidative status under summer drought conditions, thereby protecting plants from oxidative damage.

There are numerous studies demonstrating the important role of H_2O_2 in the acquisition of tolerance to different abiotic and biotic stresses [84,85,99,100]. It was recently shown that pre-treatment with H_2O_2 provides protection against heat stress and low light induced oxidative stress by modulating the activity of antioxidant enzymes. The exogenous application of H_2O_2 can induce tolerance to heat stress in seedlings of *Cucumis sativus* cv Lvfeng no. 6 [100]. The pre-treatment of cucumber leaves with H_2O_2 and heat increased antioxidant enzyme activities, decreased lipid peroxidation, and thus protected the ultrastructure of chloroplasts under heat stress. Similarly, it has been shown that exogenous H_2O_2 can have a beneficial effect on low light induced oxidative stress [84]. Low light induces an oxidative stress [101], which increases ROS and causes lipid peroxidation. H_2O_2 pre-treatment of cucumber leaves resulted in decreased levels of $O^{\bullet}{}_2{}^-$, endogenous H_2O_2, and malonaldehyde by moderating the activities of antioxidant enzymes, thus reducing lipid peroxidation and stress intensity at low light. Pre-treatment with H_2O_2 can also increase drought stress tolerance in soybean leaves by promoting the expression of stress-response genes [85]. Exogenous application of H_2O_2 caused an increase in the mRNA levels of key enzymes for the biosynthesis of oligosaccharides, which are known to help plants tolerate drought stress. This enabled the soybean plant to avoid drought stress through the maintenance of leaf water content and thus to delay foliar wilting. Finally, hydrogen peroxide contributes to defense responses against pathogens. It was demonstrated that H_2O_2 is important for the greater tolerance of kumquat leaves infected with *Xanthomonas axonopodis* than that of grapefruit [19]. Infected kumquat leaves have a high accumulation of H_2O_2, which is promoted by the suppression of ascorbate peroxidase activity and later by the suppression of catalase activity, both involved in maintaining H_2O_2 at low levels. H_2O_2 can then be used as a substrate for the higher activity of Class III peroxidase in the apoplast, which is known to be involved in plant defense against pathogens.

5. SINGLET OXYGEN

Singlet oxygen is the highly reactive, excited state of molecular oxygen that can be formed in a reaction between O_2 and the chlorophyll triplet state [102]. Unlike the formation of H_2O_2 and $O^{\bullet}{}_2{}^-$, the formation of $^1O_2$ is not accompanied by the transfer of an electron to O_2. Instead, one of the unpaired electrons is promoted to a higher energy orbital [103]. Under normal conditions, $^1O_2$ is generated during photosynthesis by the photo activation of photosensitizers, mainly chlorophylls and their precursors [102]. Singlet oxygen is also generated during senescence [25] and under different abiotic stresses [104,105,106]. Similarly to other ROS, $^1O_2$ has a dual effect. As an oxidizing agent it can react with various biological molecules, causing damage and leading to cell death [107]. It can also play a signaling role by activating the expression of different genes [23,108]. As a result of its high reactivity and short lifetime of 3.1 to 3.9 μs in pure water [109], $^1O_2$ is able to interact with molecules mostly in its nearest environment. The diffusion distance of $^1O_2$ has been calculated to be up to 10 nm in a physiologically relevant situation [110]. On the other hand, it was demonstrated that $^1O_2$ is capable of diffusing a distance of over 270 nm in rat nerve cells [111] and that $^1O_2$ produced in the photosynthetic apparatus of *C. reinhardtii* under high light is capable of leaving the thylakoid membrane and reaching the cytoplasm or even the nucleus [112], which makes its role as a signaling molecule feasible.

Significant progress in the investigation of the role of singlet oxygen in signaling was achieved with the use of a conditional fluorescent (*flu*) mutant of *Arabidopsis* that accumulates the photosensitizer protochlorophyllide in the dark and generates singlet oxygen after transfer to light [107,113]. Following illumination with light, a different set of nuclear genes are activated within the *flu* mutant, and they are different from those induced by $O^{\bullet}_2{}^{-}$ and/or H_2O_2, suggesting that singlet oxygen does not act primarily as a toxin but rather as a signal that activates several stress-response pathways [107]. Recently, it was reported that $^1O_2$ could be responsible for increased tolerance to photooxidative stress in *Arabidopsis* plants through the action of β-cyclocitral [114]. β-cyclocitral is a β-carotene derivative produced in high light that is able to induce changes in the expression of a large set of genes, which strongly overlap with the network of genes induced by $^1O_2$ [114]. At the same time it has little effect on the expression of H_2O_2 gene markers. β-cyclocitral-induced reprogramming of gene expression is associated with increased tolerance to photooxidative stress, indicating that β-cyclocitral is a stress signal produced in high light that is able to induce defense mechanisms and represents a likely messenger involved in the $^1O_2$signaling pathway in plants [114]. Further evidence that $^1O_2$ participates in acclimation to photooxidative stress comes from a study with an npq1lut2 double mutant [106]. The npq1lut2 mutant specifically accumulates $^1O_2$ due to its selective loss of lutein and zeaxanthin, which participate in the quenching and scavenging of $^3Chl^*$ and $^1O_2$ [115,116]. Following high light illumination, $^1O_2$ accumulates and modifies the expression of a group of genes encoding chloroplast proteins, leading to a significant change in chloroplast composition and functional modifications. High light induced $^1O_2$ responses were also investigated in an *Arabidopsis* cell suspension culture (ACSC) containing functional chloroplast [117]. An experiment with different fluorescent probes showed that the high light treated cultures emitted fluorescence that corresponded with the production of $^1O_2$. This was accompanied by significant changes in the expression of transcripts specifically upregulated by $^1O_2$, which leads us to conclude that $^1O_2$ plays an important role in the initiation of defense responses to high light.

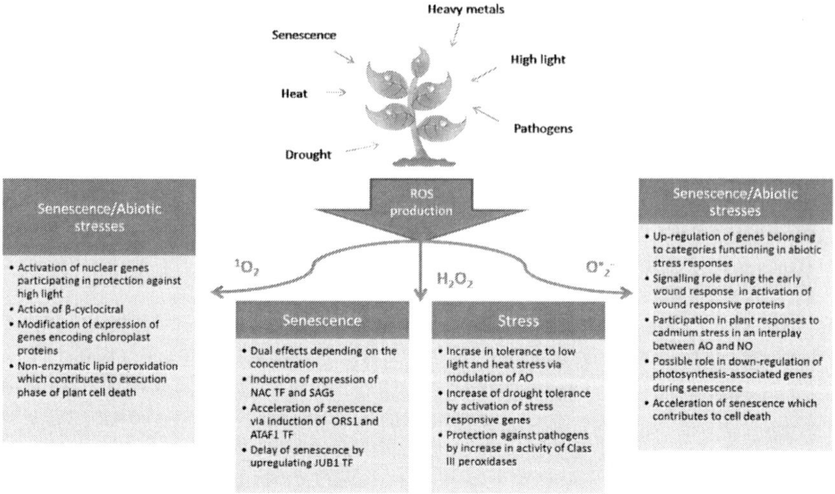

Figure 2. Possible roles of ROS during senescence and abiotic stresses.

When it comes to senescence, our knowledge on the signaling role of the $^1O_2$ is limited by the scarcity of the research relevant to the topic. There is general agreement that ROS production increases during senescence [2,3]. However, the increase in $^1O_2$ is observed simultaneously with that of other ROS, making it difficult to isolate the role of $^1O_2$. It was reported that $^1O_2$ is the main cause of senescence-associated oxidative stress in chloroplasts of sage [118]. However, this was concluded on the basis of the strong degradation of β-carotene and α-tocopherol in drought-stressed plants, which suggests the enhanced formation of singlet oxygen. In addition, in a recent study, a mass generation of singlet oxygen was measured in the early stages of hormone-treated barley but then declined, while in naturally senescing plants there was continuous production of low amounts of singlet oxygen [25]. Together with an increase in $^1O_2$, artificially senescing plants contained oxidative breakdown products of β-carotene such as β-cyclocistral, which is a possible messenger involved in the $^1O_2$ signaling pathway in plants [114], while the same was undetectable in a naturally senescing plant. Similar results were observed in our recent study, where it was shown that during the development and senescence of secondary barley leaves there is a continuous production of low amounts of $^1O_2$ [76]. Another possible role of $^1O_2$ could be its contribution to an increase in lipid peroxidation (LPO), leading to cell death. It is well known that with advancing senescence there is a notable increase in LPO [119,120]. Recently, it has been reported that in optimal growth conditions $^1O_2$ was responsible for more than 80% of the non-enzymatic

LPO in *Arabidopsis* leaf tissues [121]. Lipid peroxidation leads to the generation of free radicals, which can lead to the promotion of senescence [122]. This in turn leads to an increase in lipoxygenase activity, which can further increase LPO and also form $^1O_2$ [123], leading to overproduction of $^1O_2$. Indeed, it was showed that in *Arabidopsis* mutants favoring $^1O_2$ production, photooxidative stress led to a dramatic increase in LPO preceding cell death [121]. Possible roles of ROS during senescence and stress are summarized in Figure 2.

6. CONCLUSIONS

In this review we have given a short overview on the possible role of three classes of ROS during senescence and abiotic stresses. ROS play an important role in different plant processes ranging from plant stress adaptation to defense against pathogen attack. In the ROS family, the signaling roles of H_2O_2 have been most thoroughly studied due to its relative stability and ability to diffuse through membranes. It has been demonstrated that H_2O_2 plays various roles in plant growth, development, and metabolism. It has an important role during the senescence process, where it was shown that it could be used as a signal to promote senescence and during the acquisition of tolerance to different abiotic and biotic stresses. In comparison with H_2O_2, the superoxide anion is less stable and cannot cross the membrane, which makes it less suitable as a signal molecule. Nevertheless, an important signaling role of $O^{\bullet}{}_2^-$ was demonstrated when it was shown that the generation of $O^{\bullet}{}_2^-$ in the absence of H_2O_2 leads to a strong upregulation of the genes that function in abiotic stress responses and during senescence. Finally, in recent years, with the discovery of a *flu* mutant, it was demonstrated that $^1O_2$is capable of activating a set of nuclear genes different from those activated by $O^{\bullet}{}_2^-$ and H_2O_2 and that it plays an important role in plant responses to light.

At present, only the role of H_2O_2 during senescence and abiotic stresses has been extensively studied, while the role of other ROS remains to be further clarified. In recent years there has been significant progress in this area, with development of new techniques and technologies, but still there is no ideal technique that can be applied to a variety of systems and to specific ROS classes. EPR spectrometry can be used to measure specific ROS species but it requires a thorough sample preparation, which prevents the measurement of specific sites of ROS production in tissues. On the other hand, genetically-encoded redox probes can be used to measure ROS production in different cell compartments *in vivo* but they often do not differentiate between different classes of ROS.

Moreover, most of the papers investigate the impact of a group of ROS, while the contribution of individual ROS remains questionable. Significant progress in the future can be made on the signaling roles of $O^{\bullet}{}_2{}^-$ and $^1O_2$ during senescence. Roles of these two classes during this important process have not yet been sufficiently investigated and many questions wait to be answered. Already it has been shown that $O^{\bullet}{}_2{}^-$ upregulates some of the genes that are important during senescence such as WRKY6, but not much more is known. Similarly, the role of $^1O_2$during senescence is limited to its contribution to cell death, while several studies reported that production of this species remains unchanged during natural senescence. Whether this is true or not remains to be seen.

ACKNOWLEDGMENTS

This research was supported by the EU in the frame of Marie Curie Actions (MC-ITN-CropLife; Nr 24394).

The Faculty of Biochemistry, Biophysics, and Biotechnology of Jagiellonian University is a partner of the Leading National Research Center (KNOW), supported by the Ministry of Science and Higher Education.

AUTHOR CONTRIBUTIONS

All authors contributed extensively to the work presented in this paper. Ivan Jajic wrote the manuscript; Tadeusz Sarna and Kazimierz Strzalka supervised the work and gave final corrections and valuable input on how to improve it.

CONFLICTS OF INTEREST

The authors declare no conflict of interest.

REFERENCES

1. Gan, S.; Amasino, R.M. Making sense of senescence (molecular genetic regulation and manipulation of leaf senescence). Plant. Physiol. 1997, 113, 313–319.

2. Lee, S.; Seo, P.J.; Lee, H.J.; Park, C.M. A nac transcription factor ntl4 promotes reactive oxygen species production during drought-induced leaf senescence in arabidopsis. Plant J. Cell Mol. Biol. 2012, 70, 831–844.
3. Prochazkova, D.; Sairam, R.K.; Srivastava, G.C.; Singh, D.V. Oxidative stress and antioxidant activity as the basis of senescence in maize leaves. Plant Sci. 2001, 161, 765–771.
4. Selote, D.S.; Khanna-Chopra, R. Drought acclimation confers oxidative stress tolerance by inducing co-ordinated antioxidant defense at cellular and subcellular level in leaves of wheat seedlings. Physiol. Plant. 2006, 127, 494–506.
5. Silva, E.N.; Ferreira-Silva, S.L.; Fontenele, A.D.V.; Ribeiro, R.V.; Viégas, R.A.; Silveira, J.A.G. Photosynthetic changes and protective mechanisms against oxidative damage subjected to isolated and combined drought and heat stresses in Jatropha curcas plants. J. Plant Phsiol. 2010, 167, 1157–1164.
6. Choudhury, S.; Panda, P.; Sahoo, L.; Panda, S.K. Reactive oxygen species signaling in plants under abiotic stress. Plant Signal. Behav. 2013, 8, e23681.
7. Río, L.A.D.; Pastori, G.M.; Palma, J.M.; Sandalio, L.M.; Sevilla, F.; Corpas, F.J.; Jiménez, A.; López-Huertas, E.; Hernández, J.A. The activated oxygen role of peroxisomes in senescence. Plant Physiol. 1998, 116, 1195–1200.
8. Lushchak, V.I. Adaptive response to oxidative stress: Bacteria, fungi, plants and animals. Comp. Biochem. Physiol. Toxicol. Pharmacol. CBP 2011, 153, 175–190.
9. Ashraf, M. Biotechnological approach of improving plant salt tolerance using antioxidants as markers. Biotechnol. Adv. 2009, 27, 84–93.
10. Mittler, R. Abiotic stress, the field environment and stress combination. Trends Plant Sci. 2006, 11, 15–19.
11. Khanna-Chopra, R.; Selote, D.S. Acclimation to drought stress generates oxidative stress tolerance in drought-resistant than -susceptible wheat cultivar under field conditions. Environ. Exp. Bot. 2007, 60, 276–283.
12. Jubany-Mari, T.; Munne-Bosch, S.; Lopez-Carbonell, M.; Alegre, L. Hydrogen peroxide is involved in the acclimation of the mediterranean shrub, cistus albidus l., to summer drought. J. Exp. Bot. 2009, 60, 107–120.
13. Wang, L.; Su, H.; Han, L.; Wang, C.; Sun, Y.; Liu, F. Differential expression profiles of poplar map kinase kinases in response to abiotic stresses and plant hormones, and overexpression of PtMKK4 improves the drought tolerance of poplar. Gene 2014, 545, 141–148.
14. Suzuki, N.; Koussevitzky, S.; Mittler, R.; Miller, G. Ros and redox signalling in the response of plants to abiotic stress. Plant Cell Environ. 2012, 35, 259–270.
15. Foyer, C.H.; Noctor, G. Oxidant and antioxidant signalling in plants: A re-evaluation of the concept of oxidative stress in a physiological context. Plant Cell Environ. 2005, 28, 1056–1071.
16. Zimmermann, P.; Heinlein, C.; Orendi, G.; Zentgraf, U. Senescence-specific regulation of catalases in Arabidopsis thaliana (L.) heynh. Plant Cell Environ. 2006, 29, 1049–1060.
17. Dat, J.; Vandenabeele, S.; Vranová, E.; van Montagu, M.; Inzé, D.; van Breusegem, F. Dual action of the active oxygen species during plant stress responses. Cell. Mol. Life Sci. 2000, 57, 779–795.
18. Rentel, M.C.; Lecourieux, D.; Ouaked, F.; Usher, S.L.; Petersen, L.; Okamoto, H.; Knight, H.; Peck, S.C.; Grierson, C.S.; Hirt, H.; et al. Oxi1 kinase is necessary for oxidative burst-mediated signalling in arabidopsis. Nature 2004, 427, 858–861.
19. Kumar, N.; Ebel, R.C.; Roberts, P.D. H2O2 degradation is suppressed in kumquat leaves infected with xanthomonas axonopodis pv. Citri. Sci. Hortic. 2011, 130, 241–247.
20. Dias, C.V.; Mendes, J.S.; dos Santos, A.C.; Pirovani, C.P.; da Silva Gesteira, A.; Micheli, F.; Gramacho, K.P.; Hammerstone, J.; Mazzafera, P.; de Mattos Cascardo, J.C. Hydrogen peroxide formation in cacao tissues infected by the hemibiotrophic fungus Moniliophthora perniciosa. Plant Physiol. Biochem. 2011, 49, 917–922.

21. Golldack, D.; Li, C.; Mohan, H.; Probst, N. Tolerance to drought and salt stress in plants: Unraveling the signaling networks. Front. Plant Sci. 2014, 5, 151.
22. Soares, N.C.; Wojtkowska, J.; Jackson, P.A. A proteomic analysis of the wound response in medicago leaves reveals the early activation of a ros-sensitive signal pathway. J. Proteomics 2011, 74, 1411–1420.
23. Laloi, C.; Stachowiak, M.; Pers-Kamczyc, E.; Warzych, E.; Murgia, I.; Apel, K. Cross-talk between singlet oxygen- and hydrogen peroxide-dependent signaling of stress responses in Arabidopsis thaliana. Proc. Natl. Acad. Sci. USA 2007, 104, 672–677.
24. Hideg, E.; Barta, C.; Kalai, T.; Vass, I.; Hideg, K.; Asada, K. Detection of singlet oxygen and superoxide with fluorescent sensors in leaves under stress by photoinhibition or uv radiation. Plant Cell Physiol. 2002, 43, 1154–1164.
25. Springer, A.; Acker, G.; Bartsch, S.; Bauerschmitt, H.; Reinbothe, S.; Reinbothe, C. Differences in gene expression between natural and artificially induced leaf senescence in barley. J. Plant Pysiol. 2015, 176, 180–191.
26. Golemiec, E.; Tokarz, K.; Wielanek, M.; Niewiadomska, E. A dissection of the effects of ethylene, h2o2 and high irradiance on antioxidants and several genes associated with stress and senescence in tobacco leaves. J. Plant Pysiol.2014, 171, 269–275.
27. Rizhsky, L.; Liang, H.; Shuman, J.; Shulaev, V.; Davletova, S.; Mittler, R. When defense pathways collide. The response of arabidopsis to a combination of drought and heat stress. Plant Physiol. 2004, 134, 1683–1696.
28. Rizhsky, L.; Liang, H.; Mittler, R. The combined effect of drought stress and heat shock on gene expression in tobacco. Plant Physiol. 2002, 130, 1143–1151.
29. Prasch, C.M.; Sonnewald, U. Simultaneous application of heat, drought and virus to Arabidopsis thaliana plants reveals significant shifts in signaling networks. Plant Physiol. 2013.
30. Atkinson, N.J.; Lilley, C.J.; Urwin, P.E. Identification of genes involved in the response of arabidopsis to simultaneous biotic and abiotic stresses. Plant Physiol. 2013, 162, 2028–2041.
31. Zhou, M.; Diwu, Z.; Panchuk-Voloshina, N.; Haugland, R.P. A stable nonfluorescent derivative of resorufin for the fluorometric determination of trace hydrogen peroxide: Applications in detecting the activity of phagocyte nadph oxidase and other oxidases. Anal. Biochem. 1997, 253, 162–168.
32. Frahry, G.; Schopfer, P. Nadh-stimulated, cyanide-resistant superoxide production in maize coleoptiles analyzed with a tetrazolium-based assay. Planta 2001, 212, 175–183.
33. Dikalov, S.; Griendling, K.K.; Harrison, D.G. Measurement of reactive oxygen species in cardiovascular studies. Hypertension 2007, 49, 717–727.
34. Gomes, A.; Fernandes, E.; Lima, J.L.F.C. Fluorescence probes used for detection of reactive oxygen species. J. Biochem. Biophys. Methods 2005, 65, 45–80.
35. Benov, L.; Fridovich, I. Is reduction of the sulfonated tetrazolium 2,3-bis (2-methoxy-4-nitro-5-sulfophenyl)-2-tetrazolium 5-carboxanilide a reliable measure of intracellular superoxide production? Anal. Biochem. 2002, 310, 186–190.
36. Swartz, H.M.; Khan, N.; Khramtsov, V.V. Use of electron paramagnetic resonance spectroscopy to evaluate the redox state in vivo. Antioxid. Redox Signal. 2007, 9, 1757–1771.
37. Hideg, E.; Kalai, T.; Hideg, K. Direct detection of free radicals and reactive oxygen species in thylakoids. Methods Mol. Biol. 2011, 684, 187–200.
38. Hideg, É.; Spetea, C.; Vass, I. Singlet oxygen production in thylakoid membranes during photoinhibition as detected by epr spectroscopy. Photosynth. Res. 1994, 39, 191–199.
39. Mubarakshina, M.M.; Ivanov, B.N.; Naydov, I.A.; Hillier, W.; Badger, M.R.; Krieger-Liszkay, A. Production and diffusion of chloroplastic h2o2 and its implication to signalling. J. Exp. Bot. 2010, 61, 3577–3587.
40. Krieger-Liszkay, A.; Kós, P.B.; Hideg, É. Superoxide anion radicals generated by methylviologen in photosystem i damage photosystem ii. Physiol. Plant. 2011, 142, 17–25.

41. Heyno, E.; Mary, V.; Schopfer, P.; Krieger-Liszkay, A. Oxygen activation at the plasma membrane: Relation between superoxide and hydroxyl radical production by isolated membranes. Planta 2011, 234, 35–45.
42. Steffen-Heins, A.; Steffens, B. EPR spectroscopy and its use in planta––A promising technique to disentangle the origin of specific ros. Front. Environ. Sci. 2015, 3.
43. Leister, D. Photosynthesis research protocols. Ann. Bot. 2013, 112, vi–vii.
44. Sandalio, L.M.; Rodríguez-Serrano, M.; Romero-Puertas, M.C.; del Río, L.A. Imaging of reactive oxygen species and nitric oxide in vivo in plant tissues. Methods Enzymol. 2008, 440, 397–409.
45. Fryer, M.J.; Oxborough, K.; Mullineaux, P.M.; Baker, N.R. Imaging of photo-oxidative stress responses in leaves. J. Exp. Bot. 2002, 53, 1249–1254.
46. Flors, C.; Fryer, M.J.; Waring, J.; Reeder, B.; Bechtold, U.; Mullineaux, P.M.; Nonell, S.; Wilson, M.T.; Baker, N.R. Imaging the production of singlet oxygen in vivo using a new fluorescent sensor, singlet oxygen sensor green. J. Exp. Bot. 2006, 57, 1725–1734.
47. Costa, A.; Drago, I.; Behera, S.; Zottini, M.; Pizzo, P.; Schroeder, J.I.; Pozzan, T.; Lo Schiavo, F. H2O2 in plant peroxisomes: An in vivo analysis uncovers a ca(2+)-dependent scavenging system. Plant J. Cell Mol. Biol. 2010, 62, 760–772.
48. Rodriguez-Serrano, M.; Romero-Puertas, M.C.; Zabalza, A.; Corpas, F.J.; Gomez, M.; Del Rio, L.A.; Sandalio, L.M. Cadmium effect on oxidative metabolism of pea (Pisum sativum L.) roots. Imaging of reactive oxygen species and nitric oxide accumulation in vivo. Plant Cell Environ. 2006, 29, 1532–1544.
49. Tarpey, M.M.; Wink, D.A.; Grisham, M.B. Methods for detection of reactive metabolites of oxygen and nitrogen: In vitro and in vivo considerations. Am. J. Physiol. Regul. Integr. Comp. Physiol. 2004, 286, R431–R444.
50. Corpas, F.J.; Fernandez-Ocana, A.; Carreras, A.; Valderrama, R.; Luque, F.; Esteban, F.J.; Rodriguez-Serrano, M.; Chaki, M.; Pedrajas, J.R.; Sandalio, L.M.; et al. The expression of different superoxide dismutase forms is cell-type dependent in olive (Olea europaea L.) leaves. Plant Cell Physiol. 2006, 47, 984–994.
51. Yamamoto, Y.; Kobayashi, Y.; Devi, S.R.; Rikiishi, S.; Matsumoto, H. Aluminum toxicity is associated with mitochondrial dysfunction and the production of reactive oxygen species in plant cells. Plant Physiol. 2002, 128, 63–72.
52. Ragas, X.; Jimenez-Banzo, A.; Sanchez-Garcia, D.; Batllori, X.; Nonell, S. Singlet oxygen photosensitisation by the fluorescent probe singlet oxygen sensor green [registered sign]. Chem. Commun. 2009, 28, 2920–2922.
53. Lukyanov, K.A.; Belousov, V.V. Genetically encoded fluorescent redox sensors. Biochim. Biophys. Acta 2014, 1840, 745–756.
54. Hernandez-Barrera, A.; Quinto, C.; Johnson, E.A.; Wu, H.M.; Cheung, A.Y.; Cardenas, L. Using hyper as a molecular probe to visualize hydrogen peroxide in living plant cells: A method with virtually unlimited potential in plant biology. Methods Enzymol. 2013, 527, 275–290.
55. Rhoads, D.M.; Umbach, A.L.; Subbaiah, C.C.; Siedow, J.N. Mitochondrial reactive oxygen species. Contribution to oxidative stress and interorganellar signaling. Plant Physiol. 2006, 141, 357–366.
56. Sandalio, L.; Rodríguez-Serrano, M.; Romero-Puertas, M.; del Río, L.A. Role of peroxisomes as a source of reactive oxygen species (ROS) signaling molecules. Subcell. Biochem. 2013, 69, 231–255.
57. Rosenwasser, S.; Rot, I.; Sollner, E.; Meyer, A.J.; Smith, Y.; Leviatan, N.; Fluhr, R.; Friedman, H. Organelles contribute differentially to reactive oxygen species-related events during extended darkness. Plant Physiol. 2011, 156, 185–201.
58. Allen, J.F.; Hall, D.O. Superoxide reduction as a mechanism of ascorbate-stimulated oxygen uptake by isolated chloroplasts. Biochem. Biophys. Res. Commun. 1973, 52, 856–862.
59. Asada, K. Production and scavenging of reactive oxygen species in chloroplasts and their functions. Plant Physiol.2006, 141, 391–396.

60. Takahashi, M.A.; Asada, K. Superoxide anion permeability of phospholipid membranes and chloroplast thylakoids. Arch. Biochem. Biophys. 1983, 226, 558–566.
61. Sandalio, L.M.; Fernandez, V.M.; Ruperez, F.L.; Del Rio, L.A. Superoxide free radicals are produced in glyoxysomes. Plant Physiol. 1988, 87, 1–4.
62. Corpas, F.J.; Palma, J.M.; Sandalio, L.M.; Valderrama, R.; Barroso, J.B.; Del Rio, L.A. Peroxisomal xanthine oxidoreductase: Characterization of the enzyme from pea (Pisum sativum L.) leaves. J. Plant Pysiol. 2008, 165, 1319–1330.
63. Del Río, L.A.; Donaldson, R.P. Production of superoxide radicals in glyoxysomal membranes from castor bean endosperm. J. Plant Pysiol. 1995, 146, 283–287.
64. Kaur, G.; Sharma, A.; Guruprasad, K.; Pati, P.K. Versatile roles of plant nadph oxidases and emerging concepts. Biotechnol. Adv. 2014, 32, 551–563.
65. Sagi, M.; Fluhr, R. Production of reactive oxygen species by plant nadph oxidases. Plant Physiol. 2006, 141, 336–340.
66. Torres, M.A.; Dangl, J.L. Functions of the respiratory burst oxidase in biotic interactions, abiotic stress and development. Curr. Opin. Plant Biol. 2005, 8, 397–403.
67. Kwak, J.M.; Mori, I.C.; Pei, Z.M.; Leonhardt, N.; Torres, M.A.; Dangl, J.L.; Bloom, R.E.; Bodde, S.; Jones, J.D.; Schroeder, J.I. Nadph oxidase atrbohd and atrbohf genes function in ROS-dependent aba signaling in arabidopsis. EMBO J. 2003, 22, 2623–2633.
68. Zarepour, M.; Kaspari, K.; Stagge, S.; Rethmeier, R.; Mendel, R.; Bittner, F. Xanthine dehydrogenase atxdh1 from Arabidopsis thaliana is a potent producer of superoxide anions via its nadh oxidase activity. Plant. Mol. Biol. 2010, 72, 301–310.
69. Yesbergenova, Z.; Yang, G.; Oron, E.; Soffer, D.; Fluhr, R.; Sagi, M. The plant mo-hydroxylases aldehyde oxidase and xanthine dehydrogenase have distinct reactive oxygen species signatures and are induced by drought and abscisic acid. Plant. J. 2005, 42, 862–876.
70. McRae, D.G.; Thompson, J.E. Senescence-dependent changes in superoxide anion production by illuminated chloroplasts from bean leaves. Planta 1983, 158, 185–193.
71. Pastori, G.M.; Del Rio, L.A. Natural senescence of pea leaves (an activated oxygen-mediated function for peroxisomes). Plant Physiol. 1997, 113, 411–418.
72. Van Breusegem, F.; Dat, J.F. Reactive oxygen species in plant cell death. Plant Physiol. 2006, 141, 384–390.
73. Zhao, X.; Nishimura, Y.; Fukumoto, Y.; Li, J. Effect of high temperature on active oxygen species, senescence and photosynthetic properties in cucumber leaves. Environ. Exp. Bot. 2011, 70, 212–216.
74. Scarpeci, T.E.; Zanor, M.I.; Carrillo, N.; Mueller-Roeber, B.; Valle, E.M. Generation of superoxide anion in chloroplasts of Arabidopsis thaliana during active photosynthesis: A focus on rapidly induced genes. Plant. Mol. Biol. 2008, 66, 361–378.
75. Robatzek, S.; Somssich, I.E. A new member of the arabidopsis wrky transcription factor family, atwrky6, is associated with both senescence- and defence-related processes. Plant J. Cell Mol. Biol. 2001, 28, 123–133.
76. Jajic, I.; Sarna, T.; Szewczyk, G.; Strzalka, K. Changes in production of reactive oxygen species level in illuminated thylakoids isolated during development and senescence of barley leaf senescence. J. Plant Physiol. 2015. submitted.
77. Jajić, I.; Wiśniewska-Becker, A.; Sarna, T.; Jemioła-Rzemińska, M.; Strzałka, K. EPR spin labeling measurements of thylakoid membrane fluidity during barley leaf senescence. J. Plant Pysiol. 2014, 171, 1046–1053.
78. Niewiadomska, E.; Polzien, L.; Desel, C.; Rozpadek, P.; Miszalski, Z.; Krupinska, K. Spatial patterns of senescence and development-dependent distribution of reactive oxygen species in tobacco (Nicotiana tabacum) leaves. J. Plant Pysiol. 2009, 166, 1057–1068.

79. Kopyra, M.; Gwóźdź, E.A. Nitric oxide stimulates seed germination and counteracts the inhibitory effect of heavy metals and salinity on root growth of lupinus luteus. Plant Physiol. Biochem. 2003, 41, 1011–1017.
80. Romero-Puertas, M.C.; Delledonne, M. Nitric oxide signaling in plant-pathogen interactions. IUBMB Life 2003, 55, 579–583.
81. Quan, L.J.; Zhang, B.; Shi, W.W.; Li, H.Y. Hydrogen peroxide in plants: A versatile molecule of the reactive oxygen species network. J. Integr. Plant Biol. 2008, 50, 2–18.
82. Peng, L.T.; Jiang, Y.M.; Yang, S.Z.; Pan, S.Y. Accelerated senescence of fresh-cut chinese water chestnut tissues in relation to hydrogen peroxide accumulation. J. Plant Phsiol. Mol. Biol. 2005, 31, 527–532. (In Chinese).
83. Kumar, N.; Ebel, R.C.; Roberts, P.D. Antioxidant metabolism of grapefruit infected with xanthomonas axonopodis pv. Citri. Environ. Exp. Bot. 2011, 71, 41–49.
84. Zhang, X.-L.; Jia, X.-F.; Yu, B.; Gao, Y.; Bai, J.-G. Exogenous hydrogen peroxide influences antioxidant enzyme activity and lipid peroxidation in cucumber leaves at low light. Sci. Hortic. 2011, 129, 656–662.
85. Ishibashi, Y.; Yamaguchi, H.; Yuasa, T.; Iwaya-Inoue, M.; Arima, S.; Zheng, S.H. Hydrogen peroxide spraying alleviates drought stress in soybean plants. J. Plant Pysiol. 2011, 168, 1562–1567.
86. Yun, K.-Y.; Park, M.; Mohanty, B.; Herath, V.; Xu, F.; Mauleon, R.; Wijaya, E.; Bajic, V.; Bruskiewich, R.; de los Reyes, B. Transcriptional regulatory network triggered by oxidative signals configures the early response mechanisms of japonica rice to chilling stress. BMC Plant. Biol. 2010, 10, 1–29.
87. Hu, X.; Bidney, D.L.; Yalpani, N.; Duvick, J.P.; Crasta, O.; Folkerts, O.; Lu, G. Overexpression of a gene encoding hydrogen peroxide-generating oxalate oxidase evokes defense responses in sunflower. Plant Physiol. 2003, 133, 170–181.
88. Foyer, C.H.; Noctor, G. Redox homeostasis and antioxidant signaling: A metabolic interface between stress perception and physiological responses. Plant Cell 2005, 17, 1866–1875.
89. Mittler, R.; Vanderauwera, S.; Gollery, M.; van Breusegem, F. Reactive oxygen gene network of plants. Trends Plant Sci. 2004, 9, 490–498.
90. Yang, T.; Poovaiah, B.W. Hydrogen peroxide homeostasis: Activation of plant catalase by calcium/calmodulin. Proc. Natl. Acad. Sci. 2002, 99, 4097–4102.
91. Liao, W.-B.; Zhang, M.-L.; Huang, G.-B.; Yu, J.-H. Hydrogen peroxide in the vase solution increases vase life and keeping quality of cut Oriental × Trumpet hybrid lily "manissa". Sci. Hortic. 2012, 139, 32–38.
92. Khandaker, M.M.; Boyce, A.N.; Osman, N. The influence of hydrogen peroxide on the growth, development and quality of wax apple (Syzygium samarangense, [blume] merrill & LM. Perry var. Jambu madu) fruits. Plant Physiol. Biochem. 2012, 53, 101–110.
93. Bieker, S.; Riester, L.; Stahl, M.; Franzaring, J.; Zentgraf, U. Senescence-specific alteration of hydrogen peroxide levels in Arabidopsis thaliana and oilseed rape spring variety Brassica napus L. Cv. Mozart. J. Integr. Plant Biol. 2012, 54, 540–554.
94. Balazadeh, S.; Wu, A.; Mueller-Roeber, B. Salt-triggered expression of the ANAC092-dependent senescence regulon in Arabidopsis thaliana. Plant Signal. Behav. 2010, 5, 733–735.
95. Balazadeh, S.; Kwasniewski, M.; Caldana, C.; Mehrnia, M.; Zanor, M.I.; Xue, G.-P.; Mueller-Roeber, B. ORS1, an H2O2-responsive nac transcription factor, controls senescence in Arabidopsis thaliana. Mol. Plant. 2011, 4, 346–360.
96. Wu, A.; Allu, A.D.; Garapati, P.; Siddiqui, H.; Dortay, H.; Zanor, M.I.; Asensi-Fabado, M.A.; Munne-Bosch, S.; Antonio, C.; Tohge, T.; et al. JUNGBRUNNEN1, a reactive oxygen species-responsive NAC transcription factor, regulates longevity in Arabidopsis. Plant Cell 2012, 24, 482–506.
97. Garapati, P.; Xue, G.P.; Munne-Bosch, S.; Balazadeh, S. Transcription factor ATAF1 in arabidopsis promotes senescence by direct regulation of key chloroplast maintenance and senescence transcriptional cascades. Plant Physiol. 2015.

98. Chen, H.J.; Wu, S.D.; Huang, G.J.; Shen, C.Y.; Afiyanti, M.; Li, W.J.; Lin, Y.H. Expression of a cloned sweet potato catalase SPCAT1 alleviates ethephon-mediated leaf senescence and H2O2 elevation. J. Plant Pysiol. 2012, 169, 86–97.
99. Prasad, T.K.; Anderson, M.D.; Martin, B.A.; Stewart, C.R. Evidence for chilling-induced oxidative stress in maize seedlings and a regulatory role for hydrogen peroxide. Plant Cell 1994, 6, 65–74.
100. Gao, Y.; Guo, Y.-K.; Lin, S.-H.; Fang, Y.-Y.; Bai, J.-G. Hydrogen peroxide pretreatment alters the activity of antioxidant enzymes and protects chloroplast ultrastructure in heat-stressed cucumber leaves. Sci. Hortic. 2010, 126, 20–26.
101. Sielewiesiuk, J. Why there are photodamages to photosystem II at low light intensities. Acta Physiol. Plant. 2002, 24, 399–406.
102. Krieger-Liszkay, A. Singlet oxygen production in photosynthesis. J. Exp. Bot. 2005, 56, 337–346.
103. Gill, S.S.; Tuteja, N. Reactive oxygen species and antioxidant machinery in abiotic stress tolerance in crop plants. Plant Physiol. Biochem. 2010, 48, 909–930.
104. Pospíšil, P.; Šnyrychová, I.; Nauš, J. Dark production of reactive oxygen species in photosystem II membrane particles at elevated temperature: EPR spin-trapping study. Biochim. Biophys. Acta 2007, 1767, 854–859.
105. Morker, K.H.; Roberts, M.R. Light as both an input and an output of wound-induced reactive oxygen formation in Arabidopsis leaves. Plant Signal. Behav. 2011, 6, 1087–1089.
106. Alboresi, A.; Dall'osto, L.; Aprile, A.; Carillo, P.; Roncaglia, E.; Cattivelli, L.; Bassi, R. Reactive oxygen species and transcript analysis upon excess light treatment in wild-type Arabidopsis thaliana vs. a photosensitive mutant lacking zeaxanthin and lutein. BMC Plant. Biol. 2011, 11, 62.
107. Op den Camp, R.G.L.; Przybyla, D.; Ochsenbein, C.; Laloi, C.; Kim, C.; Danon, A.; Wagner, D.; Hideg, É.; Göbel, C.; Feussner, I.; et al. Rapid induction of distinct stress responses after the release of singlet oxygen in arabidopsis. Plant Cell 2003, 15, 2320–2332.
108. Kim, C.; Meskauskiene, R.; Apel, K.; Laloi, C. No single way to understand singlet oxygen signalling in plants. EMBO Rep. 2008, 9, 435–439.
109. Krasnovsky, A.A., Jr. Singlet molecular oxygen in photobiochemical systems: IR phosphorescence studies. Membr. Cell Biol. 1998, 12, 665–690.
110. Sies, H.; Menck, C.F.M. Singlet oxygen induced DNA damage. Mutat. Res./DNAging 1992, 275, 367–375.
111. Skovsen, E.; Snyder, J.W.; Lambert, J.D.C.; Ogilby, P.R. Lifetime and diffusion of singlet oxygen in a cell. J. Phys. Chem. B 2005, 109, 8570–8573.
112. Fischer, B.B.; Krieger-Liszkay, A.; Hideg, E.; Snyrychova, I.; Wiesendanger, M.; Eggen, R.I. Role of singlet oxygen in chloroplast to nucleus retrograde signaling in chlamydomonas reinhardtii. FEBS Lett. 2007, 581, 5555–5560.
113. Wagner, D.; Przybyla, D.; Op den Camp, R.; Kim, C.; Landgraf, F.; Lee, K.P.; Wursch, M.; Laloi, C.; Nater, M.; Hideg, E.; et al. The genetic basis of singlet oxygen-induced stress responses of Arabidopsis thaliana. Science 2004, 306, 1183–1185.
114. Ramel, F.; Birtic, S.; Ginies, C.; Soubigou-Taconnat, L.; Triantaphylides, C.; Havaux, M. Carotenoid oxidation products are stress signals that mediate gene responses to singlet oxygen in plants. Proc. Natl. Acad. Sci. USA 2012, 109, 5535–5540.
115. Dall'Osto, L.; Lico, C.; Alric, J.; Giuliano, G.; Havaux, M.; Bassi, R. Lutein is needed for efficient chlorophyll triplet quenching in the major LHCII antenna complex of higher plants and effective photoprotection in vivo under strong light. BMC Plant. Biol. 2006, 6, 32.
116. Dall'Osto, L.; Cazzaniga, S.; Havaux, M.; Bassi, R. Enhanced photoprotection by protein-bound vs. free xanthophyll pools: A comparative analysis of chlorophyll b and xanthophyll biosynthesis mutants. Mol. Plant. 2010, 3, 576–593.
117. González-Pérez, S.; Gutiérrez, J.; García-García, F.; Osuna, D.; Dopazo, J.; Lorenzo, Ó.; Revuelta, J.L.; Arellano, J.B. Early transcriptional defense responses in arabidopsis

cell suspension culture under high-light conditions. Plant Physiol. 2011, 156, 1439–1456.
118. Munné-Bosch, S.; Jubany-Marí, T.; Alegre, L. Drought-induced senescence is characterized by a loss of antioxidant defences in chloroplasts. Plant Cell Environ. 2001, 24, 1319–1327.
119. Berger, S.; Weichert, H.; Porzel, A.; Wasternack, C.; Kühn, H.; Feussner, I. Enzymatic and non-enzymatic lipid peroxidation in leaf development. Biochim. Biophys. Acta 2001, 1533, 266–276.
120. Zhang, M.P.; Zhang, C.J.; Yu, G.H.; Jiang, Y.Z.; Strasser, R.J.; Yuan, Z.Y.; Yang, X.S.; Chen, G.X. Changes in chloroplast ultrastructure, fatty acid components of thylakoid membrane and chlorophyll a fluorescence transient in flag leaves of a super-high-yield hybrid rice and its parents during the reproductive stage. J. Plant Pysiol. 2010, 167, 277–285.
121. Triantaphylides, C.; Krischke, M.; Hoeberichts, F.A.; Ksas, B.; Gresser, G.; Havaux, M.; van Breusegem, F.; Mueller, M.J. Singlet oxygen is the major reactive oxygen species involved in photooxidative damage to plants. Plant Physiol.2008, 148, 960–968.
122. Arora, A.; Sairam, R.K.; Srivastava, G.C. Oxidative stress and antioxidative system in plants. Curr. Sci. 2002, 82, 1227–1238.
123. Thompson, J.E.; Legge, R.L.; Barber, R.F. The role of free radicals in senescence and wounding. New Phytol. 1987, 105, 317–344.

CHAPTER 9

Characterization of Grain Nutritional Quality in Wheat

S. A. Mallick • Kousar Azaz • Moni Gupta • Vikas Sharma • B. K. Sinha

Division of Biochemistry and Plant Physiology, Faculty of AgricultureSher-e-Kashmir University of Agricultural Sciences and TechnologyJammuIndia

ABSTRACT

Ten Indian wheat varieties viz. RSP-566, RSP-561, PBW-396, HD-2687, C-306, PBW-175, RSP-81, PBW-550, DBW-17 and WH-542 were characterized for grain nutritional quality parameters viz., macronutrients (viz. starch, protein, protein fractions, sugars, fat), essential elements (calcium, phosphorus, iron and zinc), carotenoids, antioxidant and antinutritional parameters (phytic acid, total phenol, polyphenol oxidase (PPO) and trypsin inhibitor). RSP-561 possessed highest starch content, total protein, albumin, globulin, microelements (iron and zinc) and lowest antinutritional phytic acid and its grain contained second highest values of gluten (gliadin + glutenin), calcium, carotenoids and antioxidant contents and second lowest in antinutritional total phenol, PPO, trypsin inhibitor compositions. HD-2687 showed highest content of albumin, gliadin and total phenol constituents besides highest starch and total protein content. PBW-175 had highest sugar, calcium and carotenoids. However, antinutritional trypsin inhibitor, total phenol and PPO were found lowest in RSP-566, PBW-550 and RSP-81 respectively. The finding of this study concludes that on the basis of overall nutritional status, RSP-561 genotypes can be selected as one of best genotypes.

Keywords: Grain quality Starch Protein fractions Carotene Antinutritional factors

Among cereals, wheat is the most important crop in terms of production and consumption. World nutrition mostly depends on wheat and wheat products viz. chapati, bread, biscuit, pasta and fermented products, as the people all over the world consume wheat product(s) in one of these forms (Agrawal and Gupta 2006). A balanced food containing enough calories, balanced proteins and micronutrients with low antinutritional components is needed for the proper growth and development. Around three billion people throughout the world face problem of malnutrition due to micronutrient deficiency (Anonymous 2000). So, emphasis is being given to improve wheat grain nutrition quality to meet the challenge of world malnutrition. Therefore, along with optimal level of starch and protein, adequate level of essential elements like calcium, phosphorus, iron, zinc and carotenoids and antioxidant are needed to fight against world malnutrition. High protein and starch are important for growth and energy whereas gluten, a complex protein made of glutenin (Gln) and gliadin (Gld), is essential for water and gas retention ability for making loaf and chapati. Water soluble albumin and globulin improve biological value of protein and are considered as factors for nutritional superiority (Stehno et al. 2008). The divalent iron, calcium and zinc, which are directly involved in haemoglobin biosynthesis, ossification, and brain development respectively, are feared to be deficient with the increase of phytic acid level as they are made unavailable due to chelating action by phytic acid (Ekhlom et al. 2003), whereas trypsin inhibitor inhibits protein digestion and polyphenol produces brown coloured compound on oxidation by polyphenol oxidase (PPO) which makes flour and its product brown. Knowledge of nutritional status of wheat genotypes is being exploited for breeding programme for enhancing nutritional quality. Therefore, a detailed study of nutritional status of diverse wheat genotypes was undertaken to characterize these genotypes in basis of quality traits.

This experiment was conducted during *rabi* (2010–2011) at Chatha farm, Sher-e-Kashmir University of Agricultural Sciences and Technology-Jammu, with ten leading wheat cultivars namely RSP-566, RSP-561, PBW-396, HD-2687, C-306, PBW-175, RSP-81, PBW-550, DBW-17 and WH-542. Sugars were extracted from harvested powdered grain sample by washing with hot 80 % ethanol. Starch was then extracted from sugar free residue by treating with 52 % cold (4 °C) perchloric acid. Both total sugar and starch were estimated colorimetrically by phenol sulphuric acid method (Dubois et al. 1956). Starch content was obtained by multiplying the glucose value by factor of 0.9. Reducing sugar was estimated from ethanol extract by the method of Miller (1972). Total crude protein was determined by micro-kjeldahl method (Sadasivam and Manickam 1992). Different protein fractions were extracted following the method of Dvoracek et al. (2001) and then estimated by the method of Lowry et al. (1951). Albumin (A) and globulin (G) were extracted from grain samples by dissolving with distilled water and 0.5 N NaCl solution respectively. Gln and Gld were then extracted from residue step wise by homogenizing with 5 ml of Na-Borate buffer pH 10.0, 70 % ethanol and 5 mM β-mercapto-ethanol (β-ME) gradually. Fat content was estimated following petroleum ether extraction (AOAC 1990).

Micro-elements were extracted by digesting wheat grain powder with di-acid mixture (nitric acid:perchloric acid 4:1 v/v). Zinc and iron were determined by atomic absorption spectrophotometer (HITACHI, Z-2300, Japan). Calcium and phosphorus content were estimated titrimetrically (Cheng and Bray 1951) and colorimetrically (Koenig and Johnson 1942). Carotenoids were estimated colorimetrically from ethanol extracted samples following Jensen (1978). For antioxidant assay, methanol (80 %) extracted flour sample was assayed by auto-oxidation of carotene and linoleic acid coupled reaction followed by the method of Emmons and Peterson (1999).

Antinutritional phytic acid was extracted from powdered grain sample with 3 % trichloroacetic acid and then precipitated as ferric salt and the iron content was estimated colorimetrically after developing colour with potassium thiocyanate. Phytate phosphorus content was calculated from this value assuming a constant 4Fe:6P molecular ratio in the precipitate (Wheeler and Ferrel 1971). Total polyphenol content was estimated colorimetrically from methanol extracted sample (Schanderi 1970). PPO was extracted from sample with ice cold 0.1 M Tris–HCl buffer (pH 7.5) containing 5×10^{-3} m mercapto-ethanol and PPO activity was assayed by measuring the rate of increased absorbance at 410 nm caused due to oxidation of substrate catechol (Augustin et al. 1985). Trypsin inhibitor was extracted in distilled water and estimated spectrophotometrically using benzyl-DL-arginine para-nitroanilide Manjunath et al. (1983).

All measurements were made on samples drawn in triplicates. The data were analyzed by analysis of variance (ANOVA) and CD was calculated at $P \leq 0.05$.

Grain quality parameters for all the genotypes are shown in Table 1. HD-2687 showed highest values in starch, total protein, gluten and albumin but lowest in sugar, whereas RSP-561 possessed starch and total protein contents at par with HD-2687, had maximum water soluble protein (albumin + globulin = 3.13 %) and high gluten content (9.55 %) on the other hand PBW-175 revealed highest levels of total and reducing sugar and fat compositions, but had lowest starch content. RSP-566 recorded maximum values in globulin and starch (at par with HD-2687 and RSP-561) with moderate levels in sugars and proteins. Starch, besides energy and palatability provider also maintains the viscosity of flour to increase extensibility of the dough, an important factor for bakery products. In our study HD-2687, RSP-561 and RSP-566 with maximum starch content are potential candidates for bakery industry. Protein is an essential component for bread/or chapati making quality of wheat. High protein was reported to improve nutritional as well as bread/chapati making quality (Balla et al. 2011) and proteins assist to improve quality of food (Stehno et al. 2008). RSP-561 which possessed highest in albumin + globulin value was considered best for nutritionally rich protein. On the other hand, glutenin and Gld, the gluten constituents, responsible for gas and water retention capacity for better loaf/chapati making property were maximum in HD-2687 and RSP-561.

Table 1. Grain nutritional quality parameters in different wheat varieties

Variety	Starch (%)	Total protein (%)	Total sugar (%)	Reducing sugar (%)	Albumin (%)	Globulin (%)	Glutenin (%)	Gliadin (%)	Fat (%)
RSP-566	72.33	11.58 (3.55)	4.01 (2.25)	3.08 (2.02)	1.27 (1.51)	1.82 (1.68)	4.37 (1.68)	4.14 (2.27)	1.29
RSP-561	74.15	12.60 (3.68)	4.03 (2.26)	3.32 (2.08)	1.36 (1.54)	1.77 (1.67)	5.30 (1.89)	4.25 (2.29)	1.32
PBW-396	70.95	10.71 (3.42)	3.34 (2.08)	2.25 (1.80)	1.21 (1.49)	1.68 (1.63)	3.66 (1.63)	4.18 (2.28)	1.58
HD-2687	74.30	12.83 (3.72)	3.01 (2.03)	2.32 (1.88)	1.49 (1.58)	1.56 (1.59)	5.01 (1.85)	4.78 (2.40)	1.45
C-306	69.50	11.18 (3.49)	4.73 (2.39)	3.55 (2.18)	1.43 (1.56)	1.56 (1.59)	5.02 (1.85)	3.20 (2.05)	1.54
PBW-175	66.15	12.06 (3.61)	5.80 (2.47)	3.83 (2.20)	1.46 (1.57)	1.44 (1.54)	5.08 (1.87)	4.07 (2.25)	1.71
RSP-81	66.47	11.85 (3.58)	5.10 (2.45)	4.32 (2.31)	1.24 (1.50)	1.50 (1.57)	5.07 (1.87)	4.02 (2.24)	1.55
PBW-550	68.90	12.33 (3.65)	4.86 (2.43)	3.05 (2.01)	1.46 (1.57)	1.42 (1.53)	5.08 (1.87)	4.25 (2.29)	1.59
DBW-17	67.80	11.00 (3.46)	3.85 (2.21)	2.91 (1.97)	1.45 (1.57)	1.56 (1.59)	4.29 (1.67)	3.72 (2.17)	1.27
WH-542	66.17	11.72 (3.57)	4.01 (2.25)	3.05 (2.14)	1.41 (1.55)	1.47 (1.56)	4.52 (1.70)	3.67 (2.16)	1.39
CD(0.05)	**2.340**	**0.240**	**0.420**	**0.200**	**0.025**	**0.016**	**0.170**	**0.140**	**0.11**

Within parentheses are transformed values

Besides enough calories and quantity/quality of protein, a balanced level of micronutrients is essential for proper growth and development. Table 2 depicts that grain calcium varied from 25.42 to 76.67, phosphorus 41.57–63.94, iron 3.82–4.45 and zinc 2.50–3.95 mg g^{-1}respectively. Carotenoids and antioxidant were highest in PBW-396 and PBW-175 respectively. RSP-561 had maximum iron and zinc content and was second highest in calcium and carotenoids. It is therefore considered most potential genotype that can overcome micronutrient deficiency in consumers.

Table 2. Essential elements, carotenoids and antioxidant composition of the wheat varieties

Variety	Essential elements (mg 100 g^{-1} dry wt)				Carotenoids (μg 100 g^{-1})	Antioxidant (%)
	Calcium	Phosphorus	Iron	Zinc		
RSP-566	68.00	45.70	3.95	2.75	66.00	15.71
RSP-561	71.75	58.96	4.45	3.95	105.67	16.68
PBW-396	76.67	43.01	3.82	2.69	118.00	17.06
HD-2687	68.50	55.51	4.03	3.42	87.00	11.78
C-306	57.33	57.73	3.83	3.01	89.33	14.98
PBW-175	40.94	55.26	4.07	2.62	93.00	17.75
RSP-81	25.42	48.89	4.12	2.57	68.00	10.45
PBW-550	48.50	63.94	4.36	3.52	96.00	12.64
DBW-17	27.42	54.16	4.39	2.50	72.00	9.88
WH-542	28.00	41.57	4.18	2.53	65.33	12.56
CD (0.05)	0.52	2.82	0.08	0.13	7.35	0.49

Table 3. Antinutritional parameters in grains of the wheat varieties

Variety	Total phenol (mg g^{-1})	PPO (unit min^{-1} 100 g^{-1} grain)	Phytic acid (mg 100 g^{-1} grain)	Trypsin inhibitor (TIU)
RSP-566	3.93	3.63	0.53	66.33
RSP-561	3.88	3.59	0.38	98.33
PBW-396	4.94	6.75	0.42	135.67
HD-2687	5.14	6.73	0.35	114.00
C-306	4.68	8.63	0.46	318.83
PBW-175	4.9	6.74	1.19	368.33
RSP-81	3.89	6.93	0.66	394.09
PBW-550	5.07	3.36	0.65	150.05
DBW-17	4.24	4.78	1.20	151.02
WH-542	4.77	4.52	1.60	114.02
CD(0.05)	0.32	0.50	0.03	16.99

Data for antinutritional parameters revealed that phytic acid varied from 0.35 (HD-2687) to 1.60 mg 100 g^{-1} (WH-542), total phenol 38.80 (RSP-81) to 51.44 mg 100 g^{-1} (HD-2687), PPO 3.36 (PBW-550) to 10.63 unit min^{-1} 100 g^{-1} (PBW-396) and trypsin inhibitor 66.33 (RSP-566) to 394 TIU (RSP-81) respectively (Table 3). Antinutritional factors cause negative impact directly or indirectly to consumer's health and hence should be lowered down or removed from wheat grains to improve quality. Reducing antinutritional parameters has become an important criteria of breeding programme to improve wheat grain quality (Guttieri et al. 2006). In the present study, RSP-561 and RSP-566 had low total phenol and PPO and relatively low phytic acid and trypsin inhibitor values. So, both RSP-561 and RSP-566 are considered to have less antinutritional effect with better nutritional quality.

Out of the ten genotypes evaluated for nutritional RSP-561 possessed maximum levels of starch protein, iron, zinc, soluble protein fraction, gluten, calcium, carotenoids and antioxidant compositions and low phytic acid and total phenol, PPO and trypsin inhibitor. Hence, RSP-561 is considered the most nutritionally efficient genotype that can be used in the breeding programmes and processing industry for consumer.

REFERENCES

1. Agrawal, P. K., & Gupta, H. S. (2006). Enhancement of nutritional quality of cereals using biotechnological options. In P. S. Kendurkar, G. P. Srivastava, M. Mohan & Vajpeyi (Eds.), *Proceeding of ICPHT* (pp. 48–58).
2. Anonymous. (2000). Fourth Report on the World Nutrition Situation, ACC/SCN (Administrative committee on coordination, subcommittee on Nutrition) and International Food Policy Research Institute. United Nations, Geneva.
3. AOAC. (1990). *Official Methods of Analysis, Association of Official Analytical Chemists* (15th edn.). AOAC International, USA.
4. Augustin, M. A., Ghazil, H. M., & Hussain, H. (1985). Polyphenol oxidase from guava (*Psidium guaava* L.). *Journal of the Science of Food and Agriculture, 36*, 1259.
5. Balla, K., Rakszegi, M., Li, Z., Bekes, F., Bencze, S., & Veisz, O. (2011). Quality of winter wheat in relation to heat and drought shock after anthesis. *Czech Journal of Food Science, 29*, 117–128.
6. Cheng, K. L., & Bray, R. H. (1951). Determination of calcium and magnesium in soil and plant material. *Journal of Soil Science, 72*, 449–458.
7. Dubois, M., Gilles, K. A., Hamilton, J. K., Rober, P. A., & Smith, F. (1956). Colorimetric estimation of carbohydrates by phenol sulphuric acid method. *Analytical Chemistry, 28*, 350–356.
8. Dvoracek, V., Moudary, I., & Curn, V. (2001). Studies of protein fractions in grain of split wheat *(Triticum split* L.) and common wheat (*Triticum aestivum* L.). *Scientific Agricultural Biochemica, 32*, 287–305
9. Ekhlom, P., Liisav, M., Maija, Y., & Lisa, J. (2003). The effect of phytic acid and some natural chelating agents on the solubility of mineral elements in oat bran. *Food Chemistry, 80*, 165.
10. Emmons, C. L., & Peterson, D. M. (1999). Antioxidant activity and phenolics of oat groats and hulls. *Cereal Chemistry, 76*, 902–906.
11. Guttieri, M., Peterson, K., & Souza, E. J. (2006). Milling and baking quality of low phytic acid wheat. *Crop Science, 46*, 2403–2408.
12. Jensen, A. (1978). Chlorophyll and carotenoids. In A. Hellebust & J. S. Crargie (Eds.), *Hand book of phytological methods* (pp. 59–70). London: Cambridge University Press.
13. Koenig, R., & Johnson, C. (1942). Colorimetric determination of phosphorus in biological material. *Industrial and Engineering Chemistry, Analytical Edition, 14*, 155–156.
14. Lowry, O. H., Rosebrough, N. J., Farr, A. L., & Randall, R. J. (1951). Protein measurement with the Folin Phenol reagent. *Journal of Biological Chemistry, 193*, 265–275.

15. Manjunath, N. H., Veerabhadrappa, P. S., & Virupaksha, T. K. (1983). Estimation of trypsin inhibitors. *Phytochemistry, 22*, 2349–2357.
16. Miller, G. L. (1972). Use of dinitrosalicylic acid reagent for determination of reducing sugar. *Analytical Chemistry, 31*, 426–428.
17. Sadasivam, S., & Manickam, A., (1992). Estimation of nitrogen by Micro-Kjeldahl. In*Biochemical methods for agricultural sciences* (pp. 34–37). New Delhi: Wiley Eastern Limited.
18. Schanderi, S. H. (1970). *Methods in food analysis* (p. 709). New York: Academic Press.
19. Stehno, Z., Dvoracek, V., & Dotlacil, L. (2008). Wheat protein fractions in relation to grain quality characters of the cultivars. In *Proceedings of 11th International Wheat Genetics Symposium* (Vol. 2, pp. 556).
20. Wheeler, E. L., & Ferrel, R. E. (1971). A method for phytic acid determination in wheat and wheat fractions. *Cereal Chemistry, 48*, 312–320.

CHAPTER 10

Evaluating Postharvest Characteristics of Apantu (Local False Horn) Plantain for Harvest Indices Determination

Dzomeku B. M.[1]* , Sarkordie-Addo J.[2] , Darkey S. K[1] Bam R.K and Wuensche J[3]

[1]*CSIR-Crops Research institute, P. O. Box 3785, Kumasi, Ghana.*

[2]*Crop and soil Science Department, College of Agriculture and Natural Resources, Kwame Nkrumah University of Science and Technology, Kumasi, Ghana.*

[3] *University of Hohenheim, Stutgart, Germany.*

ABSTRACT

Plantain (Musa *sp*., genome AAB) is a major locally grown starchy staple food highly consumed by Ghanaians with a per capita consumption of 84.4 kg. However, the maturity index of the fruit has been dependent on various characteristics including angularity, fruit tip drying, etc. A physiological study was conducted on the fruit characteristics of Apantu at harvest to determine the appropriate maturity index to be used by smallholders. The results were subjected to correlative analysis and revealed a strong positive correlation between age at harvest and pulp to peel ratio. The number of leaves declined with age while pulp to peel ratio, bunch weight, peel weight and fruit weight showed positive correlation with age. The low correlations observed between the age at harvest and peel thickness, peel weight, density of fruit and pulp thickness makes them unreliable indices, possibly due to the difference in environmental conditions. Generally this is true since smallholders rely upon rainfed irrigation and physiological characteristics are greatly influenced by the watering component of the environmental. It is therefore recommended that farmers use bunch ages to determine harvest maturity. This could be done through tagging of plants at flower emergence due to the strong correlation it had with pulp:peel ratio. Further testing on each variety is needed to make tangible conclusions on some of the maturity indices to be adopted or must be backed with other variables.

Keywords: Musa, Apantu, physiology, postharvest, maturity indices.

INTRODUCTION

Plantain (AAB subgroup) is one of the earliest crops domesticated by man and grown in organized agriculture in the humid forest zones of West and Central Africa. Plantain makes up the fourth most important world food item after cereals in terms of the gross value of production. They have been identified to be of great socio-economic and nutritional significance in growing regions (Dzomeku et al., 2011). In the developing world plantain is identified as the fourth most important food commodity after rice, wheat and maize (FAO, 2010, Wattanachaiyingcharoen and Boonyanuphap, 2003).

Despite the high value of plantain, poor soils,growing pest threats and disease pressures have affected production, the most notable being the fungal disease - Black Sigatoka (*Mycosphaerella fijiensis*) (IITA, 1992;Stover and Simmonds, 1987; Swennen, 1990). Yield losses due to the disease are highly significant ranging from 20 to 50%. Under very severe conditions yield losses may be as high as 80%. All the plantain landraces in Ghana are susceptible to black Sigatoka disease and other pests except cooking bananas (Dzomeku et al., 2006). The fear that the disease could wipe out the susceptible cultivars, research efforts have been to introduce developed hybrids in readiness for any eventuality. New hybrids, resistant or tolerant to the Black Sigatoka disease were introduced into the country as a long term solution. A large deployment project was executed with four hybrids.

Apantu, a False Horn plantain is the most popular plantain cultivar in Ghana. Apantu cultivation has become a feature of great socioeconomic importance in Ghana from the point of view of food security and job creation. It belongs to the non-traditional sector of the rural economy, where it is used mainly to shade cocoa and is also an essential component of the diet. More than 90% of the cultivated area in Ghana belongs to smallholder farming system. In the agricultural sector, plantain is ranked fourth in Ghana (FAO, 2010) and contributes about 13.1% to the Agricultural Gross Domestic Product (AGDP). Its per capita consumption of 84 kg (SRID-MOFA, 2010) is higher than all other starchy staples. A total of 359,865 ha of land area in Ghana is used to cultivate plantain producing an annual average of 3.7 million tonnes of fruits, of which more than 95% is sold on the domestic market (SRID-MOFA, 2010). Apantu production is concentrated in the three agro-ecological zones namely rain forest, moist semi-deciduous forest and forest-savanna transition. The rainfall pattern is bi-modal from March to July as the major rainy season and August to November as the minor season. Although Apantu plays a significant role in the farming systems in Ghana, there has not been any conscious effort to develop any maturity indices for the crop. Maturity indices of plantain under rainfed production could be unreliable as they are often influenced by the environment. Physical, biochemical and physiological parameters are used to define the maturity stage for harvesting of fruits (Jha et al., 2006). Huda (2003) and Robinson and Saúco (2010) described that the banana fruits having the stage of three quarters round was considered as commercial maturity. The common maturity indices used by most plantain farmers for assessing for harvest in view of the absence of universal criteria, are any combination of the following: (i) By experience and judged largely by the visual appearance of the hanging bunch and particularly by the angularity of individual fingers (Palmer, 1971). (ii) Fruits harvested when the fingers of the first hand on the bunch show signs of ripening or yellowing or when the finger tips turn black (Dadzie, 1994b, c). (iii) In commercial plantations, fruits destined for distant markets are harvested at a stage known as 'three quarters full', when the fingers are still clearly angular. For local markets fruits are often harvested when fingers are full or rounded (iv) Usually coloured ribbons are used in commercial plantations to provide information regarding bunch age. (v) Fruit diameter (or caliper grade of fruit) and fruit length may be used as criteria to determine when to harvest (Dadzie and Orchard, 1997). Dadzie and Orchard (1997) stated again that while it is advantageous if the maturity index is non-destructive so that every fruit can be evaluated, it is important that the indices can be measured in a rapid, simple and inexpensive way (Reid, 2002). The method of evaluation should be simple and easily replicable, not requiring any complex scientific methodology so that farmers with basic literacy in mathematics can evaluate their plantain production and maturity time. In other jurisdictions, various instrument based techniques are used to measure maturity and ripeness. It is important to note that if the fruits are very mature at harvest, particularly following a heavy rain, peel splitting can occur.

It is evident that under rainfed conditions, the maturity indices are influenced by the environment. In the dry season angularity may be deceptive as an index of maturity. Also the soil nutrient and soil moisture could also influence the maturity indices. The objective of this study therefore was to critically evaluate the physiological characteristics of Apantu with maturity indices.

MATERIALS AND METHODS

Field samples of Apantu (False-horn) plantain were harvested at 5- day intervals of 70 to 105 days after flower emergence. The fruits were harvested from the Crops Research Institute plantain orchard at Fumesua near Kumasi in the rainy and dry seasons. Data was taken on 20 plants and replicated three times. In all, a total of 480 plants were harvested from the orchard. The field is located at 1° 31'W and 6° 43'N. The soil type is the sandy-loam (Aeronosols). The annual rainfall ranges between 1400 and 1700 mm. Annual temperature range of 22 to 33°C can be recorded on this field all year round.

In this study, 60 plants per maturity were tagged at flower emergence to determine their age after flowering. At sampling ages fresh samples were harvested in the cool of the morning to conform to postharvest standards of keeping the fruit at a minimum achievable temperature to preserve harvest quality before reaching the pack house, market or the consumer. This study was carried out on the field right after harvesting and continued in the laboratory before any deterioration. Studies were carried out within 48 h of harvesting. The methods of Dadzie and Orchard (1997) and Gowen (1995) were used. All measurements were limited to the fingers of the second hand on each bunch sampled. However, when there were not enough samples, the third hand was included. Some agronomic data were recorded prior to harvesting to assess any correlation between them and observations at harvest. Some of these were; (a) plant height; (b) pseudostem girth; (c) number of functional leaves; (d) bunch weight; (e). total number of fingers, and (f) number of hands.

At harvest, bunch weight was determined by weighing individual bunches with a balance. Number of hands was obtained by counting the number of hands on each bunch. Number of fingers was obtained by simply counting the number of fingers (per hand) on each bunch. Fruit weight was determined by weighing individual fruit on a weighing scale. Fruit length was determined by measuring the outer curve of individual fruit with a tape measure from the distal end to the point at the proximal end where the pulp is judged to terminate. Fruit girth or circumference was determined by measuring individual fruit with a tape measure at the widest midpoint of each fruit. Volume of fruit was obtained by direct volume displacement or by weighing fruit under water (Kushman and Pope, 1968; Kushman et al., 1966). Pulp and peel weight were determined after fingers have been hand-peeled and peel and pulp weighed separately with Mettler electronic balance usually to 2 decimal places). Pulp and peel were separated, weighed individually and expressed as pulp to peel ratio. Hand peeling of each fruit after cutting transversely at the midpoint was done, and the peel and pulp separately measured with a pair of calipers.

The data was subjected to ANOVA and the standard deviation was emphasized to reveal or reduce significant errors between duplication.

RESULTS AND DISCUSSION

Maturity indices must meet two requirements for producers, locations and seasons (years): (i) Minimum acceptable eating quality, and (ii) Long storage life (Dadzie and Orchard, 1997).

The False Horn plantain fruit diameter recorded throughout the year ranged between 33.4 and 44.2 mm.

They all fall within the minimum acceptable limits indicating that all were mature for consumption and compliant to African standards. With visual observation of fruit angularity, the ages at which fruits were harvested displayed significant change from angular, at 70 days after flower emergence (AFE) to rounded, at 95 days AFE, a visual indication of maturity.

There was a positive correlation between pulp-to peel ratio and bunch age (Figure 1). This corroborated well with the results of Dadzie and Orchard (1997) on hybrid cooking bananas as the ratio of pulp to peel increases with age at 1.27 (70 days) to 1.63 (95 days); a strong indicator of increasing maturity and readiness for harvest. The only challenge with the use of this index is the destructive approach. A finger must be removed from the bunch to determine if the requirement is met. This approach could serve as an entry point for pathogens as a wound would be created in the process of removal of the finger. However, smearing of petroleum jelly to prevent entry of pathogens through the wounded portion is recommended.

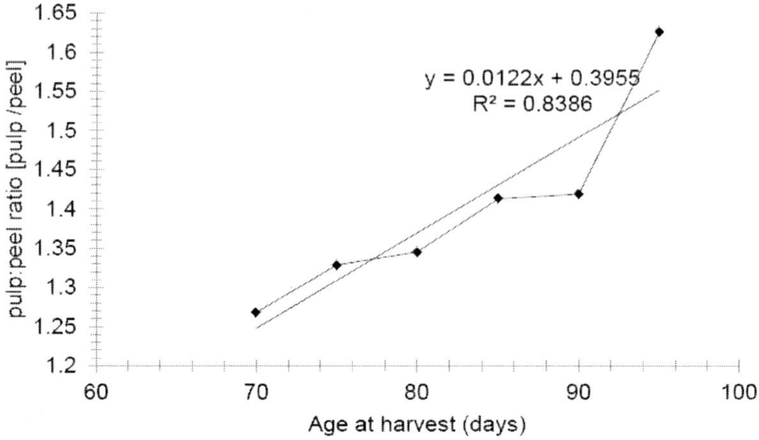

Figure 1. Relationship of pulp: peel ratio of Apantu with Bunch age

There was a decline in the number of functional leaves as the fruit approached maturity (Figure 2). This was not surprising as this is directly linked to translocation of nutrients as into fruits during fruit -filling as well as the high susceptibility of the landrace to diseases. The most destructive of these diseases being Black Sigatoka caused by *M. fijiiensis* as indicated by Dzomeku et al. (2006), IITA (1992).

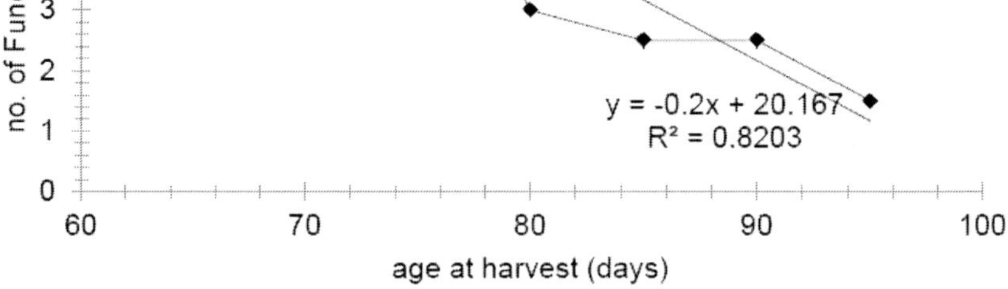

Figure 2. Relationship of functional leaves of Apantu with maturity.

The relationship between age and the weight of the fruit (Figure 3), showed a clear increase possibly due to the fruit filling. Plantain is a determinate plant and after flowering, the physiological process of nutrient translocation is for fruit filling. There is also translocation of photosynthase into the fruit without any new leaf production. Evidence shows that fruit filling is a critical stage of the plantain crop when abiotic factors can influence the yield. Turner and Thomas (1998) reported that, plantains are sensitive to soil water deficits; expanding tissues such as growing fruit are among the first to be affected.

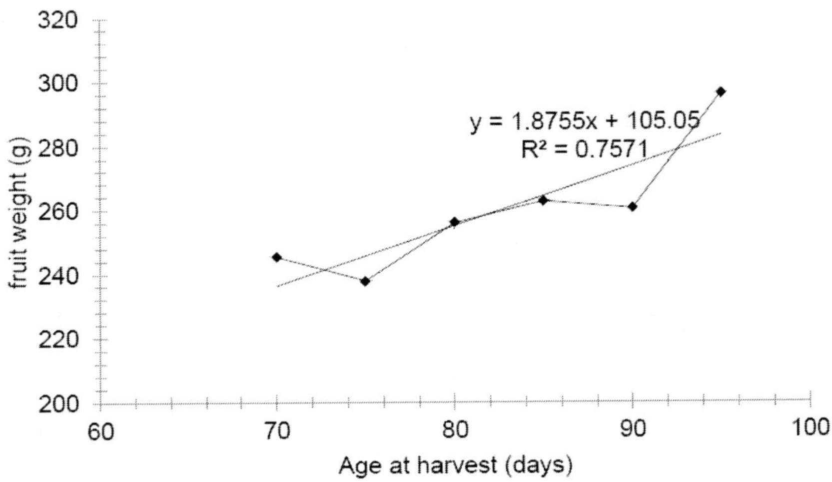

Figure 3. Relationship of fruit weight to bunch age of Apantu.

Bunch weight weakly correlated to number of fingers (Figure 4). The varying growth and developmental differences of the various fingers translated into physiological difference manifested in the weights. The environmental conditions under rainfed systems often show variations. It has been observed that the number of hands on a bunch as well as the number of fingers could be influenced by soil nutrient and rainfall.

There was a low correlation observed between the ages at harvest and peel thickness (Figure 5). This correlation does not make the parameter a reliable maturity index for Apantu. There is normally a decline in fruit density due to the reduction of its displacement power (weight) and this was reflected in the test. The low correlations observed between the age at harvest and peel thickness, peel weight, density of fruit and pulp thickness makes them unreliable indices (Figure 6).

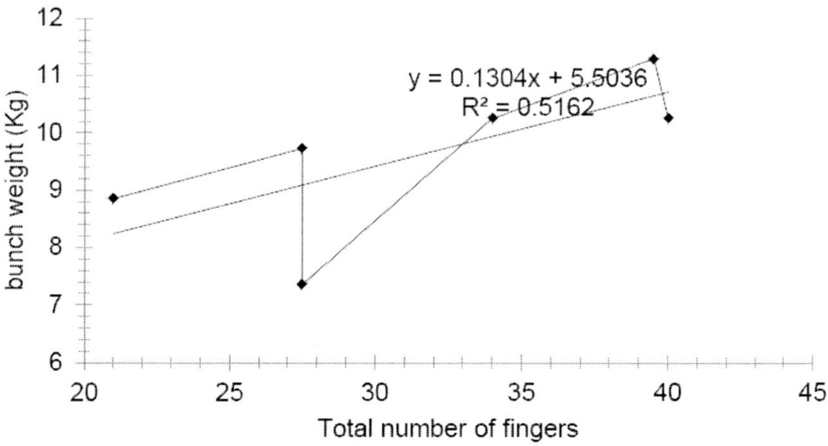

Figure 4. Relationship between number of fingers and bunch weight of Apantu.

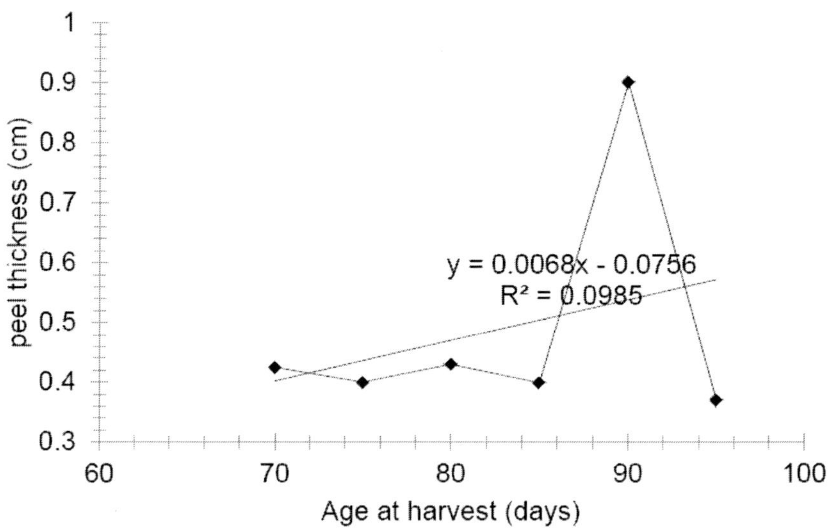

Figure 5. Correlation between bunch maturity and peel thickness of Apantu.

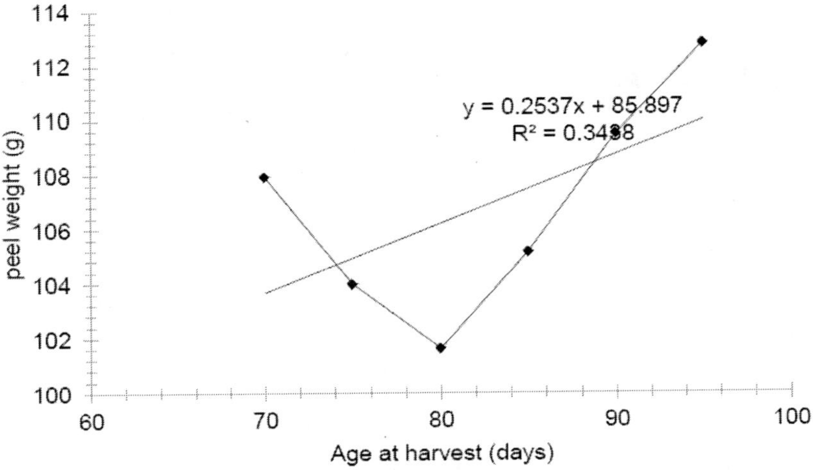

Figure 6. The relationship between bunch maturity and peel weight of Apantu bunches.

They can, however; be indicators of other environmental conditions within which the fruit grew. Most of these parameters are cultivar dependent and as such cannot be blindly compared to other cultivars in literature. The study has shown that rainfed plantain production cannot in anyway bring about the full potential of the crop. Productivity is affected because of the erratic nature of the rain coupled with the soil nutrient. The critical stage of growth of the crop is at flowering. Water deficit at flowering severely affects fruit filling. Plantains (*Musa* spp.) rarely attain their full genetic potential for yield due to limitations imposed by water limiting the plants photosynthesis. Plantains are known to be particularly sensitive to changes in the environment. Plantain growth and yield is known to decrease drastically when the interval between watering was increased with the soil moisture falling below 66% of total available soil moisture (Robinson and Bower, 1988). Under severe soil moisture stress, physiological and biochemical processes occur. Physiologically stomata close and this affects photosynthesis. It is worth noting that if the fruit is too matured at harvest, particularly following a heavy rain, peel splitting can occur. In the commercial banana industry, maturity is estimated through fruit diameter which differs with the standards used the world over; a difference that may be expressed in the following ways: (a) Total thirty-seconds of an inch (e.g. grade 42 which is 110/32 inch); (b) The number of thirty-seconds of an inch above thirty-two (e.g. grade 10 which is 110/32 inch); (c) Millimeters (e.g. grade 42 is 33 mm using 0.794 for each thirty-second of an inch). In Central and South America, grade is expressed as the number of thirty-seconds of an inch above 1, whereas millimeters are used in the Caribbean and Africa (Stover and Simmonds, 1987). It is evident that under rainfed conditions these parameters may vary according to the seasons. During the major harvesting seasons (June to February) angularity would be a very pronounced index unlike in the harvesting months of March to May.

CONCLUSIONS

The peel to pulp is the singular most important indicator of fruit maturity that can be used on the field. However due to its destructive nature, it cannot be extensively used. In combination with the other parameters measured, the age of bunch is the most reliable stand-alone parameter and is recommended for extensive promotion and use by smallholder farmers.

CONFLICT OF INTERESTS

The authors have not declared any conflict of interests.

REFERENCES

1. Dadzie BK, Orchard JE (1997). Routine Post-Harvest Screening of Banana/Plantain Hybrids: Criteria and Methods, an INIBAP Technical Guidelines 2.
2. Dadzie BK (1998) Post-harvest characteristics of black Sigatoka resistant banana, cooking banana and plantain hybrids. An INIBAP Technical guideline. International Plant Genetic Resources Institute Via delle Sette Chiese, 142 00145 Rome, Italy.
3. Dadzie BK (1994b). Quarterly report for the INIBAP/FHIA/NRI (ODA Holdback) project on post-harvest cooking banana and plantain characterisation (October - December, 1994).
4. Dadzie BK (1994c). Six monthly report for the INIBAP/FHIA/NRI (ODA Holdback) project on post-harvest cooking banana and plantain characterisation (April - September, 1994).
5. Dzomeku, BM, Dankyi AA, Darkey SK (2011). Socioeconomic Importance of Plantain Cultivation in Ghana. J. Anim. Plant Sci. 21(2):269-273.
6. Dzomeku BM, Osei-Owusu, M, Ankomah AA, Akyeampong E, Darkey SK (2006). Sensory Evaluation of Some Hybrid Cooking Bananas in Ghana. J. Appl. Sci. 6(4):835-837.
7. FAO-Food and Agriculture Organization (2010). FAOSTAT Rome, Italy: http://faostat. external.fao.org/. Accessed on 17th April, 2010.
8. Gowen SR (1995). Bananas and plantains. Chapman & hall, London pp. 383-402.

9. Huda ARF (2003). Effect of various treatment on banana ripening. Ahfad J. 20(1):17-19.
10. IITA (International Institute of Tropical Agriculture) (1992). Sustainable food production in Sub-Saharan Africa, 1. IITA, Ibadan, Nigeria, 208 p.
11. Jha, S, Kingsly A, Chopra S (2006). Physical and mechanical Properties of mango during growth and storage for determination of maturity. J. Food Eng. 72(1):73-76.
12. Kushman LJ, Pope DT (1968). Procedure for determining intercellular space of roots and specific gravity of sweetpotato root tissue. HortScience 3(1):44-45.
13. Kushman LJ, Pope, DT Monroe RJ (1966). Estimation of intercellular space and specific gravity of five varieties of sweetpotato root.
14. Ortiz, R, Vuylsteke D (1996). Improving plantain and banana-based system. In: Ortiz, R. and Akoroda, M.O., (Ed.) Plantain and Banana Production and Research in West and Central Africa. Proceedings of a Regional Workshop 23-27 September, 1995.
15. Palmer JK (1971). The banana. Pp. 65-105 in The Biochemistry of fruits and their products. Vol. 2. (A. C. Hulme, ed.). Academic Press, London.
16. Reid MS (2002). Maturation and Maturity Indices. In: Kader AA (ed) Postharvest Technology of Horticultural Crops, 3rd. Ed., Univ. California, Oakland, CA, USA: ANR Publication 3311, Pp. 55-62.
17. Robinson JC, Bower JP (1988). Transpiration from banana leaves in the subtropics. Sci. Hortic. 37(1-2):129-143.
18. Stover RH, Simmonds NW (1987). Banana (3rd edition). John Wiley & Sons, Inc. New York 468 p.
19. Robinson JC, Saúco VG (2010). Bananas and plantains, In Morphological characteristics and plant development. (ed. J. C. Robinson) CABI Oxford.
20. SRID-MOFA, (Statistical, Research and Information Directorate of Ministry of Food and Agriculture (2010). Agriculture in Ghana. Facts and Figures 58 p.
21. Turner, D W, Thomas D S (1998). Leaf gas exchange of droughted and irrigated banana cv. Williams (Musa spp.) growing in hot, arid conditions. J. Hortic. Sci. Biotechnol. 73(3):419-429
22. Wattanachaiyingcharoen D, Boonyanuphap J (2003). Production Management of Musa (AAB group) "Kluai Nam Wa" for Export and Agro-industry Using GIS (Geographic Information System). Research report (In Thai, with English abstract). The National Science and Technology Development Agency (NSTDA). Bangkok. 114 p.

CHAPTER 11

Physiological Responses to Humic Substances as Plant Growth Promoter

Luciano P Canellas* and Fábio L Olivares

Núcleo de Desenvolvimento de Insumos Biológicos para Agricultura, Universidade Estadual do Norte Fluminense Darcy Ribeiro, Av Alberto Lamego 2000, Campos dos Goytacazes 28013-602, Brazil

ABSTRACT

Humic substances (HS) have been widely recognized as a plant growth promoter mainly by changes on root architecture and growth dynamics, which result in increased root size, branching and/or greater density of root hair with larger surface area. Stimulation of the H^+-ATPase activity in cell membrane suggests that modifications brought about by HS are not only restricted to root structure, but are also extended to the major biochemical pathways since the driving force for most nutrient uptake is the electrochemical gradient across the plasma membrane. Changes on root exudation profile, as well as primary and secondary metabolism were also observed, though strongly dependent on environment conditions, type of plant and its ontogeny. Proteomics and genomic approaches with diverse plant species subjected to HS treatment had often shown controversial patterns of protein and gene expression. This is a clear indication that HS effects of plants are complex and involve non-linear, cross-interrelated and dynamic processes that need be treated with an interdisciplinary view. Being the humic associations recalcitrant to microbiological attack, their use as vehicle to introduce beneficial selected microorganisms to crops has been proposed. This represents a perspective for a sort of new biofertilizer designed for a sustainable agriculture, whereby plants treated with HS become more susceptible to interact with bioinoculants, while HS may concomitantly modify the structure/activity of the microbial community in the rhizosphere compartment. An enhanced knowledge of the effects on plants physiology and biochemistry and interaction with rhizosphere and endophytic microbes should lead to achieve increased crop productivity through a better use of HS inputs in Agriculture.

Keywords: Biochemical effects Biofertilizers Inoculation technology Plant growth-promoting bacteria Sustainable agriculture technology

INTRODUCTION

Climate warming and changes in global precipitation patterns, particularly drought, are already affecting crop production [1]. Moreover, edaphic constraints from inherently low soil fertility and/or soil degradation from unsustainable farming practices also contribute to the limited crop yield [2]. Based on these challenges, there has been a call for a second green revolution with a goal of enhancing crop yields, improving soil fertility through better management practices [3, 4], breeding crops with greater tolerance to edaphic stresses [5] and by development of new inputs based on optimization of biological/microbiological process [6]. A new generation of crops adapted to low/reduced input systems will not only enable people in some of the poorest parts of the world to provide themselves with adequate food, but it will also play a vital role in the high-input systems of the developed world, by reducing agrochemical inputs and their associated economic and environmental costs [7]. The main adaptation of plants to natural low fertility soils includes anatomical changes at the root system, such as production of lateral roots and root hair with consequent increase of either root length or surface area. Geometrical changes in roots can be linked to increased rhizodeposition by exudation of organic anions and enzymes, rhizosphere acidification and association with specific microbes [7]. All of these processes could be mediated by soluble humic substances (Figure 1).

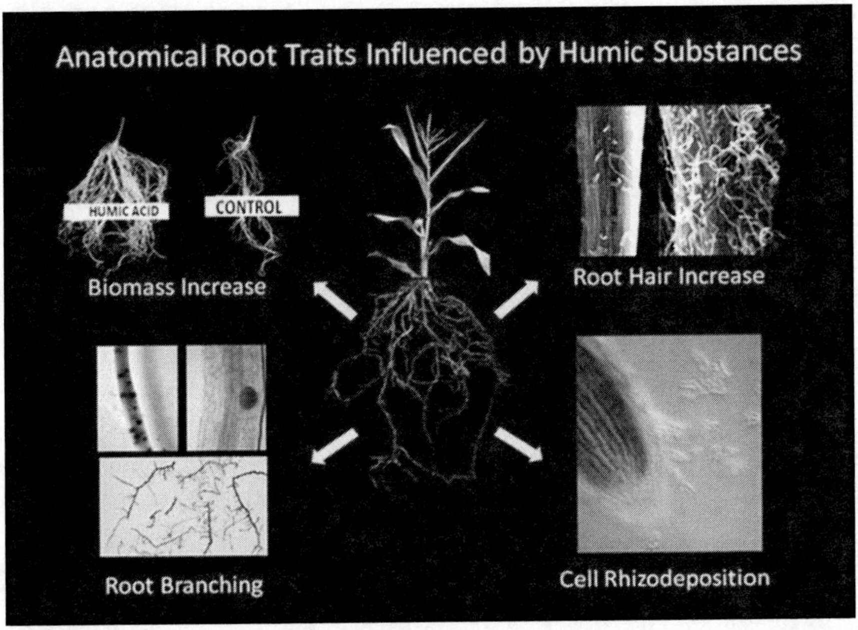

Figure 1. ROOT TRAITS OF IMPORTANCE IN ADAPTATION TO LOW FERTILITY SOILS THAT COULD BE INFLUENCED BY HS. Root architecture, root hair, root exudate, enzyme production.

Natural organic matter (NOM) is a product of plant and animal tissue decay, and, together with the biota, is of pivotal importance in the global carbon cycle. The relationship between NOM and crop production is consistent with the results of over a hundred years of modern agricultural research and with thousands of years of on-farm observation and experimentation [8]. This relationship is based on the critical influence of NOM on soil properties, processes and functions. Humus, otherwise referred to as HS, is the NOM comprising up to 80% of soil organic matter. Because of the beneficial effects that HS have on the physical, chemical, and biological properties of soil, their role in sustaining plant growth is recognized.

The direct use of soluble HS as a plant growth promoter is not a novelty, but the increase of HS market as plant stimulators has attracted the attention of new producers, businesses and farmers interested in knowing how such dilute concentrations can bring many benefits to sustainable production. The most described effect of HS is the promotion of plant root system. However, primary plant metabolism, diverse and complex enzymatic machinery related with a plethora of cell processes and more recent changes on secondary metabolism by HS are being increasingly documented.

The root system plays a central role in the acquisition of water and nutrients in a natural heterogeneous soil environment and how plants promote changes in its rhizosphere for defence, improve nutrient mineralization and select microbial community. These processes could be modulated by HS. It is not a surprise that HS can modify plant root growth and architecture since the genesis and soil use history is profoundly marked by HS formation. During the conquest of land, the salt stress ceased to exist, and plants switched to H^+ as the master ion [9]. Brundrett [10]) suggests that, as plants colonized the land, they would have faced powerful selection pressure to increase the surface area of their absorptive surfaces in soil to parallel that occurring in their photosynthetic organs. Interception of light and CO_2 fixation would have thereby become in balance with nutrients and water uptake from soils. How HS pressed this selection is not very clear but its influence is presumable.

Recent advances of knowledge on humus chemistry have emphasize that small and heterogeneous molecules are randomly associated in hydrophobic-hydrophilic phases which are either contiguous to or contained in each other [11]. Their environmental reactivity is dictated by their hydophobic/hydrophilic ratio [3] and a detailed molecular characterization of HS would eventually allow a structure-activity relationship between humic molecules and their activity inside plant cells [12, 13, 14]. Manufactured technologies for controlling HS activity can now take a boost. Powerful methods on molecular biology along with new innovative methods for conservation of agricultural diversity demand a new generation of agrochemicals acting as plant growth promoters based on the physiological effects of natural humic matter. The objective to be achieved is the improvement of crop productivity by relating structure and conformation of HS to plant physiological and biochemical activity.

Here we consider some of the main mechanisms involved in root growth promotion, absorption of water and nutrients, and interaction with beneficial soil microbes mediated by soluble HS applied directly to plants at low concentration. The implications on primary and secondary plant metabolism will be also briefly discussed.

REVIEW

Changes in root architecture and morphology by HS

The main evidence of biological activity of HS is lateral root induction (Figure 2A).

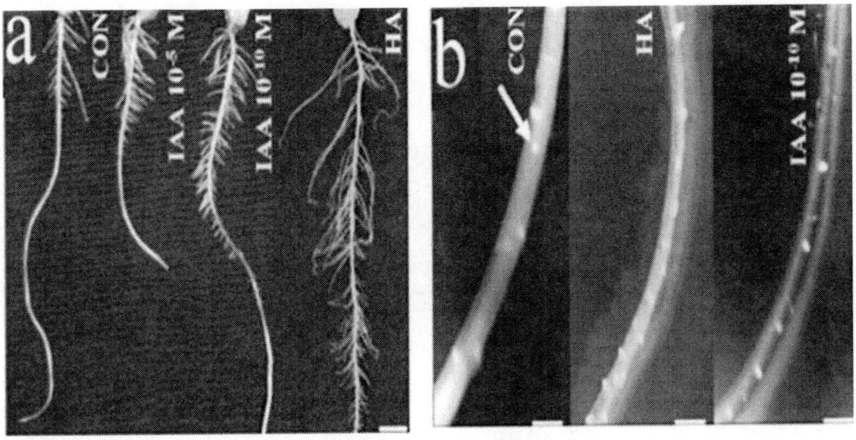

Figure 2. INDUCTION OF LATERAL ROOT EMERGENCE (A) AND MITOTIC SITES (B) BY HUMIC ACIDS (HA) AND AUXINS (IAA). See altered patter in elongation zone of maize seedlings treated w ith HA **(A)**. Densely meristematic cells coloured by hematoxylin (**B**, adapted f rom Zandonadi et al. [15]).

The elongation and differentiation zone of roots include small, dense meristematic cells that are in continuous metabolic activity and are susceptible to lateral root formation. HS were found to have marked effect on the emergence of lateral roots and the hyperinduction of sites of lateral root emergence upon HS treatments have been observed (Figure 2B; [15, 16]).

Auxin, the first hormone discovered in plants, is one of the most important morphogenic compounds that shape the whole plant body. A well-known natural auxin is indole-3-acetic acid (IAA), which is endogenously produced in the apical meristem of the shoot and can be rapidly transported to the roots via the phloem. Lateral roots originate from mature, non-dividing pericycle cells within the parent root. Auxin signals trigger groups of pericycle cells to re-enter the cell cycle and establish lateral root mitotic sites [17]. Auxin perception and signalling pathways are also essential to complete auxin's biological function [18]. This does not necessarily imply that root morphology is only under the control of the endogenous auxin. Indeed, the experimentally observed lateral root induction by exogenous IAA could be seen as a normal phenomenon since auxin is commonly present in natural soil environment. Hager [19] showed massive evidence that the final target of auxin action is the plasma membrane H^+-ATPase, that excretes H^+ ions into the cell wall compartment. The auxin-enhanced H^+ pumping lowers the pH of cell walls, activates pH-sensitive enzymes and proteins within the wall, and initiates cell-wall loosening and extension growth. This is the basis of the acid growth theory (Figure 3).

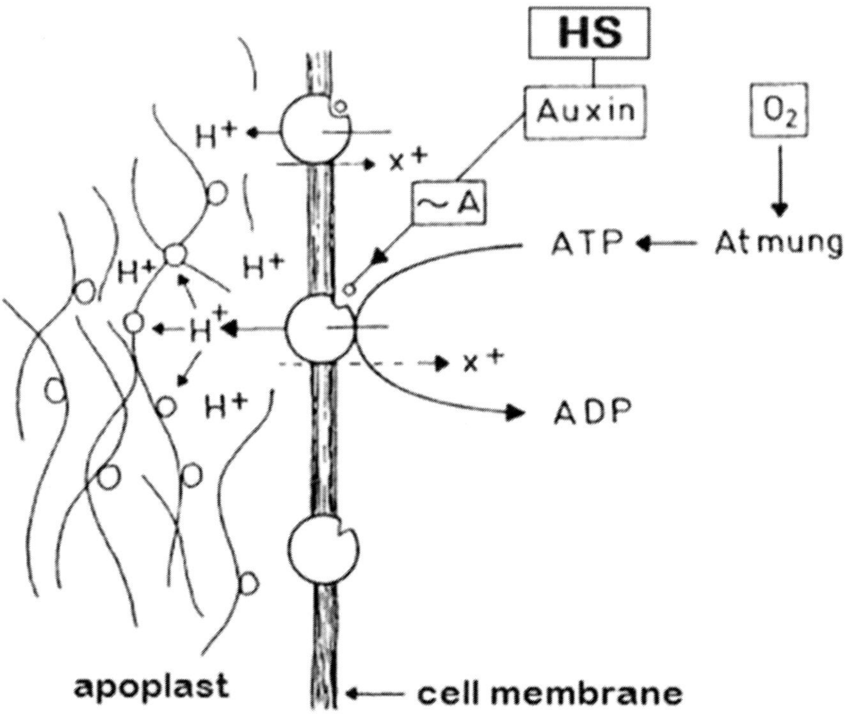

Figure 3. AUXIN OR AUXIN-LIKE SUBSTANCES PRESENT IN HS IN ACTIVE FORM (~A) ACTIVATES H^+-PUMPING ATPASE AT PLASMA MEMBRANE. This event leads to an increase in the activity of enzymes that soften the cell wall and thus trigger cell elongation according acid growth theory (Adapted from [19]).

Plasma membrane vesicles isolated from maize roots treated with HS derived from vermicompost exhibited a clear stimulation of the vanadate-sensitive ATPase activity [16]. The presence of intrinsic small bioactive molecules such as auxin entrapped into the HS supramolecular arrangement might be related to both induction of lateral root emergence and H^+-ATPase activation. Auxin-like compounds were also detected by immunoassay in HS extracted from other sources [20]. *DR5* auxin synthetic promoter fused to the β-glucuronidase (*GUS*) reporter gene (DR5::GUS) has been used as a tool to visualize auxin responses in tissues and mark auxin signalling in lateral root primordia at all developmental stages. The first evidence using DR5::GUS tools that HS caused activation of expression of this gene reporter in lateral roots, in a way comparable to exogenously applied auxins, was provided by Trevisan et al. [21]. These results were further corroborated by employing tomato (cv. Micro-Tom) seedlings expressing *DR5:GUS* (Figure 4) and low auxin-sensitive *diageotropica* (*dgt*) mutant [22].

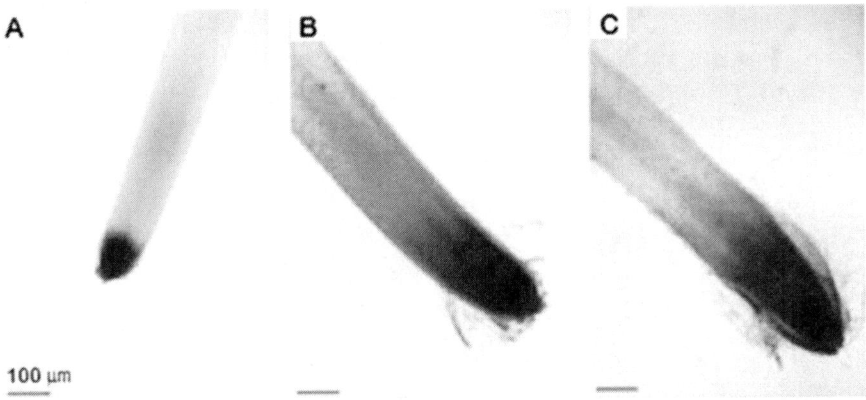

Figure 4. VISUALIZATION OF B-GLUCORONIDASE (GUS) ACTIVITY IN THE ROOT OF DR5:GUS TRANSGENIC TOMATO. Seedlings were grown for 4 days in a 2-mM $CaCl_2$ medium and then incubated for 4 days in water **(A)**, 10^{-6} M indole acetic acid **(B)** and 2.5 mM C of humic substances isolated from cattle manure vermicompost after free lipid removal **(C)** (Adapted from Canellas et al. [22]).

Quaggiotti et al. [23] provided unequivocal evidence of the expression induced by HS from the major isoform of the maize plasma membrane H^+-ATPase (MHA2), which was characterized by Frias et al. [24]. An about two-fold increase of mRNA levels of this isoform was observed in seedlings treated with HS. The HS-induced synthesis of new H^+-ATPases enhanced not only H^+-extrusion, but also the number of H^+-ATPase in a given membrane area. Despite the extension of plasma membrane during growth, these changes resulted in the maintenance of the membrane potential, the energy-dependent transport of solutes to and from the cell and the turgor [19]. Zandonadi et al. [15] showed that different HS isolated from different sources as well as low auxin concentrations (10^{-10} and 10^{-15} M) promoted a differential activation not only of the plasma membrane but also of vacuolar H^+-ATPases and H^+-pyrophosphatase. This represents a complementary view of the acid growth mechanism in which a concerted activation of the plasmalemma and tonoplast H^+ pumps plays a key role in the root cell expansion process, which is driven by environment-derived molecules endowed with auxin activity, such as HS. On the other hand, activation of PM H^+-ATPase also improves plant nutrition by enhancing the electrochemical proton gradient that drives ion transport across cell membranes via secondary transport systems [25]. It is not surprising to find nutrient accumulation in plants treated with HS as observed by Baldotto et al. [26] in pineapples, whereby the increase of N, P, K, Ca and Mg in leaves was 52%, 71%, 50%, 58% and 59% larger than control, respectively.

Root hairs are specialized root epidermal cells of higher plants whose functions are water absorption and anchorage. Application of HS caused increase in root hair length and density and an increase in cell proliferation in the root ground tissue (Figure 5).

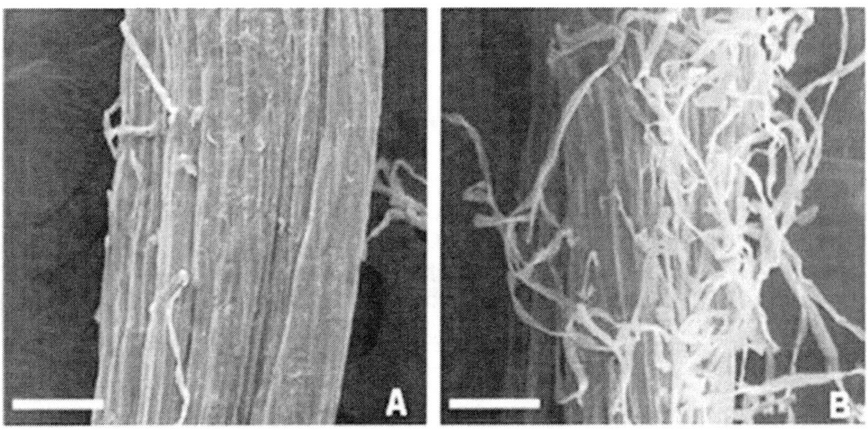

Figure 5. SCANNING ELECTRON MICROSCOPY (SEM) OF MAIZE ROOT SEGMENTS TREATED WITH $CACL_2$ OR $CACL_2$ WITH HUMIC-LIKE ACIDS FROM VERMICOMPOST. (A) 2-mM $CaCl_2$. **(B)** 2-mM $CaCl_2$ plus 50 mg L^{-1} humic-like acids. Bars = 100 μm (Adapted from Canellas et al. [27]). Other evidence of root hair induction by HS can be observed in SEM of wheat [28].

Zandonadi et al. [29] showed that HS can induce nitric oxide (NO) in sites of lateral root emergence. NO is a bioactive molecule that is involved in numerous plant physiological processes including root development among others [30]. NO was also shown to be involved in *Arabidopsis* root hair formation in both initiation and elongation processes [31, 32]. The application of HS on roots of cucumber plants caused a primary increase in NO accumulation and it was associated by Mora et al. [33] with the expression of the following morphological root changes: (1) increase in the number of secondary roots that were measured in the medium region of root system, (2) increase in root thickness, and (3) increase in root fresh weight.

Roots exude an enormous range of compounds into the surrounding soil. The quantity and quality of root exudates are a function of plant species, age of an individual plant, and external factors, such as biotic and abiotic stressors [34]. Root exudates play a decisive role in solubilizing and mobilizing nutrients in soil, thereby enhancing nutrient uptake by plants, toxic element immobilization and protection against toxicity, and substrate availability for microbial activity in the rhizosphere [35]. Canellas et al. [36] observed changes in profile exudation of organic acids by maize seedlings treated with HS (Figure 6). Puglisi et al. [37] also reported an enhancement of organic acid exudation in maize seedlings following HS treatment. Several factors can be involved in the increase of anion extrusion by seedlings treated with HS, including response to apoplast acidification following an enhanced activity of plasma membrane H^+-ATPase.

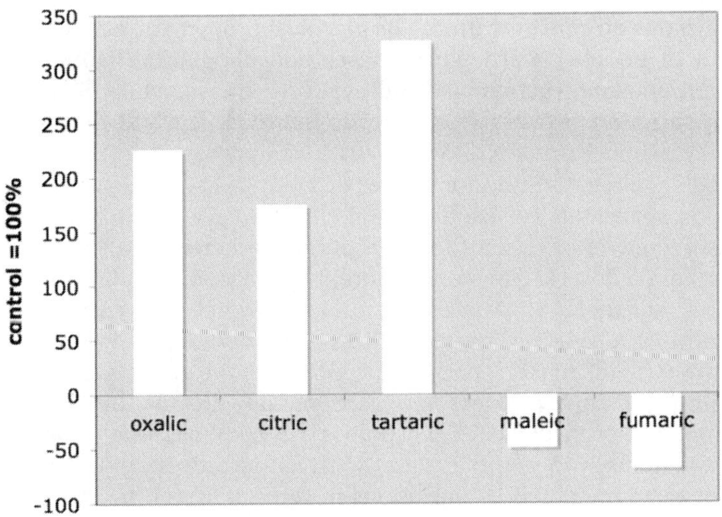

Figure 6. DIFFERENTIAL PROFILE OF ORGANIC ACIDS EXUDED FROM ROOTS OF MAIZE SEEDLINGS CONSIDERING CONTROL PLANTS (100%).

The disruption of HS associations in solution by the action of organic acids such as those exuded by plants and soil biomass has been repeatedly indicated [11]. It is thus conceivable that molecules released from humic superstructures may then access cell membranes and induce different physiological responses. The assembling and disassembling behavior of HA molecular domains and their dynamic release of different molecular constituents [38] may then explain why humic compounds had been previously found to possess hormone-like activities similar to those of auxins, gibberellins, cytokinins and polyamines [39]. The ultimate physiological response would involve a large level of cell activity and tissue differentiation, which result in root growth.

Changes on primary and secondary metabolism

Primary metabolism refers to a biochemical basic process essential for plant survival, growth and reproduction. The involved molecules are called primary metabolites and are generally constituted by proteins, carbohydrates and lipids. Glycolysis is a basic metabolic process with crucial importance in plants because it is the predominant pathway that 'fuels' plant respiration. Nardi et al. [40] showed that HS affected the enzyme activities related to glycolysis and the tricarboxylic acid cycle (TCA) in different ways, depending on humic molecular size, molecular characteristic and concentration. The glycolysis enzymatic activities studied by Nardi et al. [40] were glucokinase, phosphoglucose isomerase, PPi-dependent phosphofructokinase and pyruvate kinase, while those involved in the respiration process were cytrate synthase, malate dehydrogenase and the cytosolic form of $NADP^+$-isocitrate dehydrogenase. The stimulation increased significantly for all enzymes at 1 mg $C \cdot L^{-1}$ for either a soil HS or its fraction separated by size exclusion chromatography. It was also found as a positive effect of HS on the main photosynthetic metabolism in maize leaves, where a decrease in starch content was accompanied by an increase in soluble sugars [41]. Moreover, sucrose, which is synthesized from carbohydrates produced by photosynthesis via the reductive pentose phosphate (Calvin-Benson) pathway, represents the starting compound for the respiratory pathway, and its increase may justify the 40% activity improvement of rubisco (D-ribulose-1,5-bisphosphate carboxylase/oxygenase activity) in plants treated with leonardite HS [41]. This change appeared to be mediated by variations in the activity of the main enzymes involved in carbohydrate metabolism. According to Merlo et al. [42], in most plants, starch and sucrose are the principal end products of photosynthesis. The cytoplasmatic fructose-1,6-bisphosphate and sucrose phosphate synthase appear to be important enzymes in the control of sucrose formation, whereas ADP-glucose pyrophosphorylase has a key role in the starch biosynthesis pathway. These authors found that leaf starch content decreased in plants treated with HS, whereas the level of soluble sugars concomitantly increased. The decrease of starch was accompanied by an enhanced activity of amylase, whereas the activity of ADP glucose pyrophosphorylase was not affected. Activities of invertases and sucrose synthase were stimulated by HS treatment in apical tissues and inhibited when HS were applied to basal tissues. HS may change both the level and percent distribution of sugars of maize leaves, by affecting enzyme activities involved in carbohydrate metabolism. More recently, Canellas et al. [6] verified that free carbohydrate content in leaf extracts was 60% lower than control in maize plants treated with HS, showing reduction of fructose, glucose and starch. Finally, the induction of leaf invertase activity by HS was also verified by Concheri et al. [28] in wheat root seedlings.

The N metabolism is the basis of amino acids, proteins, enzymes and nucleotide formation. The enhancement of N uptake/assimilation and N metabolism in plants treated with HS had been well documented. Piccolo et al. [43] showed that HS increased NO−3 uptake by barley seedlings to the same extent previously observed by Albuzio et al. [44]. Quaggiotti et al. [23] in a seminal experiment did not find a relationship between uptake and high/or low affinity of NO−3 cell transporter to HS stimuli. This appears to confirm the hypothesis proposed by Nardi et al. [45] that stimulation of nitrate uptake by HS cannot be explained by an effect on the primary transport of solutes, but it should instead decrease pH at the root surface, thus facilitating the H^+/NO−3 symport. As cited earlier, Quaggiotti et al. [23] elegantly demonstrated the induction of MHA2 plasma membrane H^+-ATPase isoform by HS. Besides NO−3uptake, HS also induce changes on N cell metabolism. For example, Ertani et al. [41] evaluated, in maize treated by HS, the activities of glutamine synthetase (GS EC 6.3.1.2) and glutamate synthase (GOGAT EC 1.4.7.1), which are key enzymes involved in N assimilation. They found a 65% increase in GS root activities and GOGAT enzyme activity increased by 176% in the roots and 204% in leaves, in comparison to control. The influence of soil HS on NH+4 uptake was studied by Panuccio et al. [46] in two different coniferous species. They measured the activity of key enzymes also involved in N metabolism, such as malate dehydrogenase (MDH; EC 1.1.1.37), glutamate dehydrogenase (GDH; EC 1.4.1.3) and phosphoenolpyruvate carboxylase (PEPC; EC 4.1.1.31), and observed enzyme activities increased concomitantly with ammonium uptake but not in nitrate uptake.

These results revealed the role of HS in primary plant metabolism and in enzyme activity modulation. Carletti et al. [47] reported for the first time a proteomic approach in a study on maize seedlings treated with HS. These authors isolated plasma membrane proteins, including triosophosphate isomerase (TIM; EC 5.3.1.1), glyceraldehyde-3-P dehydrogenase (GAPD; EC 1.2.1.12), phosphoglycerate mutase (PGAM; EC 5.4.2.1), fructose biphosphate aldolase (FBA; EC 4.1.2.13), 2-phosphoglycerate dehydratase (enolase, ENO; EC 4.2.1.11), phosphoglucomutase (PGM; EC 5.4.2.2), and 3-phosphoglycerate kinase (PGK; EC 2.7.2.3), and verified whether these were down-regulated by the HS treatment. They found that the isolated proteins were all down-regulated by the HS treatment, except for TIM, which showed a 69% increase, and GAPD, which maintained the same level of expression in treated and untreated samples. Moreover, they identified two adjacent protein spots as cytosolic aconitate hydratase (ACO), one showing down regulation (spot 230, −77%) and the other up regulation (spot 224, +79%). This result is suggestive of a post-translational modification of the enzyme. The expression of ACO may induce changes in root excretion and participate in the rhizosphere cross-talk between plant and soil. This was observed by Canellas et al. [36] and Puglisi et al. [37], who showed an increase in acid citric exudation by maize seedlings when treated with different HS. Citric acid is a component of root exudates and plays an important role in disaggregating humic supramolecular structures [45].

The proteomic approach used to assess the effect of HS on plant metabolism was followed by a genomic assay in *Arabidopsis thaliana* treated with HS [48]. Down and up regulation behaviours were also observed, thereby demonstrating that HS affect plant physiology by means of complex transcriptional networks. From the overall transcriptomic results, we may hypothesize that HS exert their function through a multifaceted mechanism of action, partially connected to their well-demonstrated auxin activity, but involving also IAA-independent signalling pathways.

The Pandora box was open and elucidation of complex interactions between plant and HS should be done step by step and should grant more reliability on the recommendation of doses and times of applications to obtain plant growth stimulation based on HS physiological effects. For instance, Jannin et al. [49] used microarray analysis to evaluate genes involved in physiological response of *Brassica napus* to HS and indentified over expression in nine clusters, which covered the major metabolic plant functions: carbon and photosynthesis, general cell metabolism, fatty acids, nitrogen/sulphur, phytohormones, plant development, senescence, responses to stress and transport of ions and water. HS can thus induce carbon and nitrogen metabolism and a plethora of cell process can be down or up regulated [48].

Schiavon et al. [50] showed for the first time a relevant effect of HS on secondary plant metabolism. Plants produce a remarkably diverse array of over 100,000 low-molecular-mass natural products, also known as secondary metabolites that are distinct from the components of primary metabolism since they are generally non-essential for the basic metabolic processes. According to Dixon [51], most of the secondary metabolites are derived from the isoprenoid, phenylpropanoid, alkaloid or fatty/polyketide pathways and are produced to improve plant defence against biotic and abiotic stress. Schiavon et al. [50] showed that HS treatment enhanced the expression of the phenylalanine (tyrosine) ammonia-lyase (PAL/TAL; EC

4.3.1.5) that catalyzes the first committed step in the biosynthesis of phenolics, by converting phenylalanine to trans-cinnamic acid and tyrosine to p-coumaric acid. The expression of PAL/TAL was accompanied by leaf phenol accumulation. The authors concluded that the stimulatory effects of HS on plant secondary metabolism provide an innovative approach to explore plant responses to stress. The influence of HS on enzyme activities linked to cell protection has been observed and explained as induction of esterase activity [52] and protection against oxidative stress [53]. The door was open to drive the use of HS as stimulator of secondary metabolites in improving medicinal, aromatics, spices plants and reducing the use of synthetic agrochemical as plant protector.

HS and plant growth-promoting bacteria

By definition, HS are assemblies of heterogeneous compounds that are insoluble in water and recalcitrant to microbial activity and represent the stable part of earth C cycle [3]. However, since HS affect both plant primary and secondary metabolism including changes on exudation profile, it is pertinent to consider that HS may interfere with microorganism community in the rhizosphere. Puglisi et al. [37]; [54]) showed that the addition of HS had a significant effect on the amount of bioavailable C deposited by maize plant roots, thus resulting in a significant change in the structure of soil microbial communities. More recently, Puglisi et al. [55] used DGGE analysis and showed that the influence of HS on microorganism diversity reaches the bulk soil beyond the rhizosphere zone. Plants thus select their microbial community in order to improve their physiological processes involved in defence against pathogens, and mineralization and solubilization of nutrients.

Plant growth-promoting bacteria (PGPB) are a wide range of microorganisms that induce plant growth by several processes including biological N_2 fixation, increase of nutrient availability in the rhizosphere, enlargement of root surface area, and enhancement of beneficial symbioses for the host [56]. Relationships between PGPB and their hosts can be either rhizospheric or endophytic. We proposed a new biofertilizer concept based on the combination of HS and endophytic diazotrophic bacteria. Such biofertilizer implies an increase of endophytic interactions associated with plant host and protection of the bioinoculant in the HS hydrophobic domains which act as carriers. Different from rhizobia-legume symbioses, endophytic PGPB interaction do not have specialized root hair infection mechanisms, which are induced by chemotaxis responses to flavonoid compounds present in legume exudates [57]. The infection of endophytic PGPB is opportunistic since it exploits the natural openings occurring during the lateral root emergence that is significantly promoted by HS. The result is an increased population of endophytic bacteria associated to plants. Furthermore, highly efficient PGPB are selected in the laboratory by allowing maximum expression of their beneficial traits such as auxin (and other hormones) production, phosphate solubilization, antibiosis, etc. However, the introduction of selected PGPB to crop field represents a surviving challenge for incoming microorganisms which have to face the competition with natural microbiota adapted to infection niches in the host rhizosphere.

Piccolo [11] postulated that hydrophobic humic components deriving from plant degradation and microbial activity are able to randomly incorporate more polar molecules and hence protect them against degradation. Spaccini et al. [58] showed that the organic compounds released in soils during mineralization of fresh maize residues were stabilized against microbial degradation by surrounding hydrophobic components. Hydrophobic protections by HS against degradation of labile organic compound were demonstrated by Spaccini et al. [59] and in a long-term experiment by Piccolo et al. [60]. Therefore, we assume that selected PGPB may be protected from immediate degradation when packed in the hydrophobic HS aggregates. However, this encapsulation is dynamic and the disaggregating activity of organic acids exuded by plants upon metastable hydrophobic and hydrophilic humic associations should increase the chances of PGPB to colonize plants [11]. Moreover, the heterogeneous molecular composition of humic matter allows adsorption on root surfaces that, in turn, enables a greater interaction with root surfaces of bacteria associated to HS thereby increasing the microbial population intimately associated to plants (Figure 7).

The anatomical changes deriving from an increase of both area and length of root hair favour bacteria association with root surface. The mechanism of endophytic penetration of bacteria is crucial for their activity in plants [61] and represents the selective advantage for bacteria competing for colonization sites in plants. The bacterial population with largest attachment to roots surfaces is more likely to develop biofilms, thereby increasing their activity, persistency and overall probability to colonize the inner tissue of the inoculated plant (Figure 7). The observations of such structural interactions between bacteria and plants in the presence of HS represent the explanation for the positive and consistent effects of the combined use of humic acids and PGPB under field conditions.

In fact, it was found that maize colonization by *Herbaspirillum seropedicae* was increased by the presence of 20 $mg \cdot L^{-1}$ of different HS (Figure 8). In addition, HS with large hydrophobic/hydrophilic ratio, as measured by ^{13}C-CPMAS-NMR, induced slightly greater plant colonization by bacteria, confirming the importance of the humic chemical composition in carrying selected microbes to roots and contribute to their colonization capacity [62].

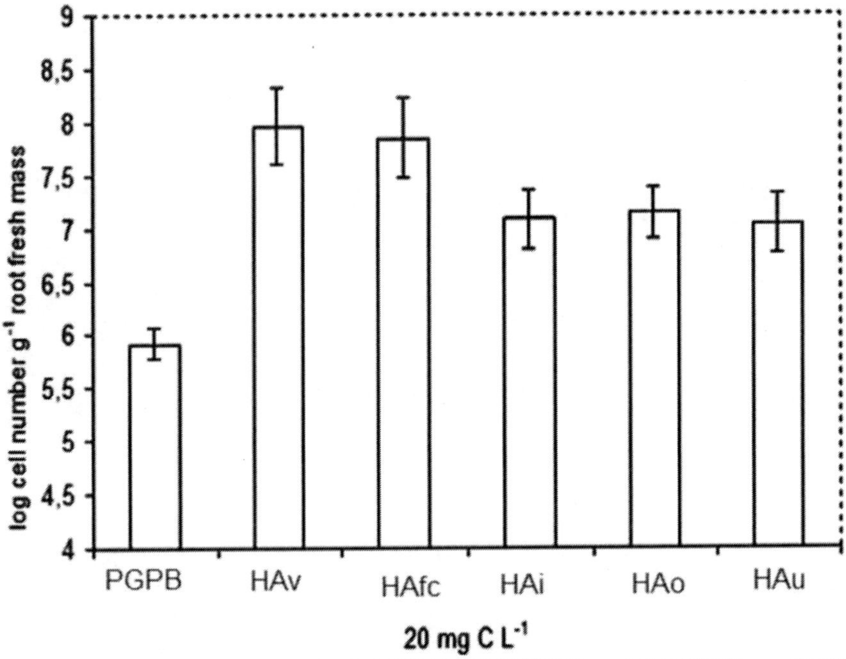

Figure 7. PLANT GROWTH-PROMOTING BACTERIA (PGPB) COLONIZATION BY MOST PROBABLE NUMBER METHOD OF FRESH ROOT TISSUES OF MAIZE SEEDLINGS. After 7 days of inoculation with a suspension of 10^9 cell · mL^{-1} of *Herbaspirillum seropedicae* and in the presence of 20 mg C L^{-1} of different humic acids (HA): HAv, vermicompost; HAfc, from filter cake a residue of sugarcane factory, HAi, inceptisol; HAo, oxisol; and HAu, ultisol.

A combined application of HS and bacteria was performed for the first time with sugarcane plantlets [63]. It is a common practice in sugarcane micropropagated biofactories to perform a disinfection procedure to seeds that decreases the natural population of microorganisms and may be an opportunity to introduce selected microbiota. In this line, Marques Júnior et al. [63] showed an increased stimulation of root growth following infection of sugarcane seeds treated with humic matter isolated from vermicompost (Figure 9). Another technological opportunity is seed coating. Increase of maize growth after seed coating and infection improved by HS was experimentally reported by Conceição et al. [64].

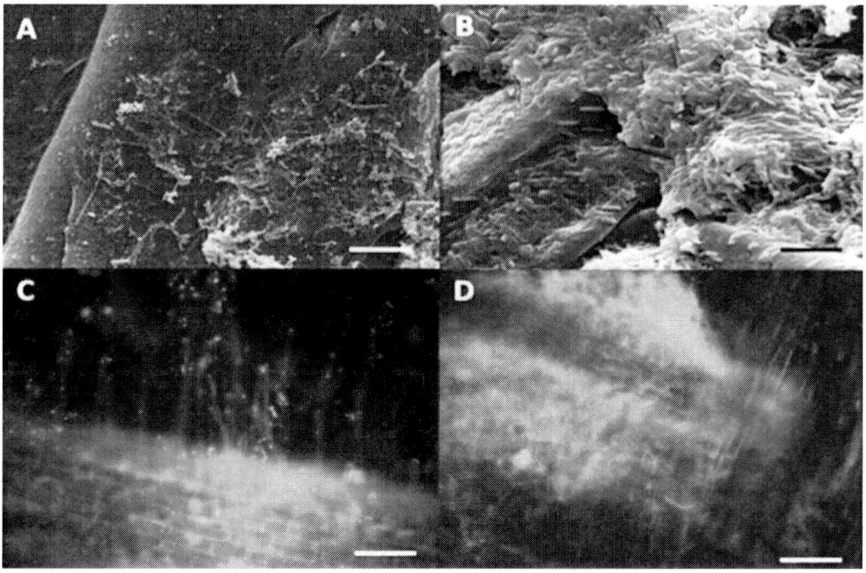

Figure 8. SEM OF ***Herbaspirillum seropedicae*** HRC54 CELL ATTACHMENT AND EPIFLUORESCENCE MICROSCOPY OF ***gfp*** -LINKED ***H. seropedicae*** RAM 10 OVER ROOT SURFACE. SEM comparing *H. seropedicae* HRC54 cell attachment on the root surface of sugarcane **(A)** without and **(B)** with 20 mg C L^{-1} of humic acids (HA). **(C)** Epifluorescence microscopy of *gfp*-linked *H. seropedicae* RAM 10 colonizing the root hair zone of maize roots as a single cells and **(D)** differentiation zone of maize root as matured biofilm, both with 20 mg C L^{-1} of HA.

Field experiments using low fertilizer input (20 kg N-urea) and foliar application of both HS from vermicompost and a plant growth-promoting bacteria strain of *H. seropedicae* (Z67) were carried out [6]. It was found, at early stages (7 and 45 days old) in greenhouse, that *H. seropedicae* in mixture with HS from vermicompost activated plant metabolic processes such as enhancement of plasma membrane H^+-ATPase activity, alteration of both sugar and N metabolism and increase of net photosynthesis. The number of viable bacterial cells was larger in root tissues when inoculation was performed in the presence of soluble HS. Foliar application of endophytic diazotrophic bacteria in mixture with HS increased maize grain production by 65% under field conditions. Again in field experiments, a 30% yield increase of sugarcane stems was obtained by foliar application of *H. seropedicae* in mixture with humic acid (Figure 9), without any negative effect for the agroindustrial sugar content (data not shown).

Figure 9. TECHNOLOGICAL APPLICATION FOR NON-LEGUMINOUS PLANTLETS. These are under greenhouse, seed pieces, sugarcane crop under field condition and maize seeds covered by combined use of humic acid and plant grow th promoting bacteria.

Zaler [65] proposed the use of vermicompost as suitable plant growth media in horticulture instead of peat, whose indiscriminate use menaces endangered bog ecosystems worldwide. The benefits of vermicompost to field crops are well documented [66] and are often attributed to its large nutrient potential, content of humified molecules possessing hormone-like activity [67], and presence of bioactive microorganisms. The introduction of a mixed inoculum of PGPB strains of *H. seropedicae* (HRC 54) and *Burkholderia silvatlantica* (UENF 101) at the initial stage of the vermicomposting process revealed the increase of total N and water-soluble phosphorus in the mature vermicompost [68], thus showing that the addition of beneficial microorganisms enables the strengthening of the growth substrate. Moreover, in a study on the organic matter molecular composition of different vermicomposts, a significant relationship was found between the presence of long-chain hydrophobic compounds in compost and the preservation of PGPB during 1 year after inoculation [62].

Concluding remarks

Humic substances (HS) are the major fraction of the soil organic matter which represent the final stage of a complex interaction between non-living organic matter and microbial communities. The critical influence of HS on the physical, chemical, and biological properties of soil has been extensively demonstrated and their role in sustaining plant growth recognized. More recently, a number of experimental data has shown that HS and its different fractions may affect plant growth and development, involving specific structural and physiological responses to HS applications. Remarkable effects on root architecture, such as induction of lateral root and root hair may be accompanied by changes in the biochemistry of energy generation and transport system across plasma membranes. Primary and secondary plant metabolisms of above- and below-ground tissues are recognized to be affected by HS. Thus, a systematic and coordinated research combining different approaches in functional genomics (e.g. transcriptomic and metabolomic) may help to overcome the complexity if HS influence on plant biology, and allow developing new technologies to increase plant growth based on humic matter. Technological research have already shown the benefits of combining the application of plant growth-promoting microbes with humic substances, the latter acting as a transporter for microbial delivery under field conditions. This synergy of effects favours the increase in population and activity of selected microbes and consequent response to plant growth promotion, thus opening opportunities to develop a new generation of biofertilizers for sustainable agricultural systems.

ACKNOWLEDGEMENTS

The work in NUDIBA laboratory was supported by Conselho Nacional de Desenvolvimento Científico e Tecnológico (CNPq), Fundação de Amparo à Pesquisa do Estado do Rio de Janeiro (FAPERJ), Instituto Nacional de Ciência e Tecnologia (INCT) para a Fixação Biológica de Nitrogênio, Internacional Foundation of Science (IFS) and OCWP. We are also grateful to editor and three anonymous reviewers for their constructive comments on the original version of the manuscript, which helped to improve it greatly. Lastly, we apologize to those authors whose work could not be discussed because of the space limitation.

COMPETING INTERESTS

The authors declare that they have no competing interests.

AUTHORS' CONTRIBUTIONS

LPC and FLO contributed equally to this review and both authors read and approved the final manuscript.

REFERENCES

1. Clair SB, Lynch JP: The opening of Pandora's Box: climate change impacts on soil fertility and crop nutrition in developing countries. Plant Soil. 2010, 335: 101-115.
2. Lal R: Anthropogenic influences on world soils and implications to global food security. Adv Agron. 2007, 93: 69-93.
3. Piccolo A: The nature of sol organic matter and innovative soil management to fight global changes and maintain agricultural productivity. Carbon sequestration in Agricultural soils: a multidisciplinary approach to innovative methods. Edited by: Piccolo A. 2012, Heidelberg: Springer, 1-20.
4. Sanchez PA: Ecology—soil fertility and hunger in Africa. Science. 2002, 295: 2019-2020.
5. Lynch JP: Roots of the second green revolution. Aust J Bot. 2007, 55: 493-512. 10.1071/BT06118.
6. Canellas LP, Martínez Balmori D, Médici LO, Aguiar NO, Campostrini E, Rosa RC, Façanha A, Olivares FL: A combination of humic substances and *Herbaspirillum seropedicae* inoculation enhances the growth of maize (*Zea mays* L.). Plant Soil. 2013, 366: 119-132.
7. Brown LK, George TS, Dupuy LX, White PJ: A conceptual model of root hair ideotypes for future agriculture environments: what combination of traits should be targeted to cope with limited P availability?. Ann Bot. 2013, 112: 317-330.
8. Herrick JE, Wander MM: Relationships between soil organic carbon and soil quality in cropped and rangeland soils: the importance of distribution, composition and soil biological activity. Soil process and the carbon cycle. Edited by: Lal R, Kimble JM, Follett RF, Stewart BA. 1997, Boca Raton: CRC, 405-425.
9. Felle H: pH as a signal and regulator of membrane transport. Handbook of Plant Growth pH as the Master Variable. Edited by: Rengel Z. 2002, New York: Marcel Dekker, 118-141.
10. Brundrett MC: Coevolution of roots and mycorrhizas of land plants. New Phytol. 2002, 154: 275-304.
11. Piccolo A: The supramolecular structure of humic substances: a novel understanding of humus chemistry and implications in soil science. Adv Agron. 2002, 75: 57-134.
12. Canellas LP, Spaccini R, Piccolo A, Dobbss LB, Okorokova-Facanha AL, Santos GA, Olivares FL, Facanha AR: Relationships between chemical characteristics and root growth promotion of humic acids isolated from Brazilian oxisols. Soil Sci. 2009, 174: 611-624.

13. Nebbioso A, Piccolo A: Basis of a humeomics science: chemical fractionation and molecular characterization of humic biosuprastructures. Biomacromolecules. 2011, 12: 1187-1199.
14. Nebbioso A, Piccolo A: Advances in Humeomic: enhanced structural identification of humic molecules after size fractionation of a soil humic acid. Ana Chim Acta. 2012, 720: 77-90.
15. Zandonadi DB, Canellas LP, Façanha AR: Indolacetic and humic acids induce lateral root development through a concerted plasmalemma and tonoplast H^+ pumps activation. Planta. 2007, 225: 1583-1595.
16. Canellas LP, Olivares FL, Okorokova-Façanha AL, Façanha AR: Humic acids isolated from earthworm compost enhance root elongation, lateral root emergence, and plasma membrane H^+-ATPase activity in maize roots. Plant Physiol. 2002, 130: 1951-1957.
17. Casimiro I, Marchant A, Bhalerao RP, Beeckman T, Dhooge S, Swarup R, Graham N, Inzé D, Sandber G, Casero PJ, Benett M: Auxin transport promotes Arabidopsis lateral root initiation. Plant Cell. 2001, 13: 843-852.
18. Correa-Aragunde N, Lanteri ML, García-Mata C, Have A, Laxalt AM, Graziano M, Lamattina L: Nitric Oxide Functions as Intermediate in Auxin, Abscisic Acid, and Lipid Signaling Pathways. Nitric Oxide in Plant Growth Plant Cell Monogr. Edited by: Lamattina L, Polacco JC. 2006, Berlin Heidelberg: Springer-Verlag, 21-6
19. Hager A: Role of the plasma membrane H^+-ATPase in auxin-induced elongation growth: historical and new aspects. J Plant Res. 2003, 116: 483-505.
20. Muscolo A, Cutrupi S, Nardi S: IAA detection in humic substances. Soil Biol Biochem. 1998, 30: 1199-1201.
21. Trevisan S, Pizzeghello D, Ruperti B, Francioso O, Sassi A, Palme K, Quaggiotti S, Nardi S: Humic substances induce lateral root formation and the expression of the early auxin-responsive IAA19 gene and DR5 synthetic element in Arabidopsis. Plant Biol. 2010, 12: 604-614.
22. Canellas LP, Dantas DJ, Aguiar NO, Peres LEP, Zsögön A, Olivares FL, Dobbss LB, Façanha AR, Nebbioso A, Piccolo A: Probing the hormonal activity of fractionated molecular humic components in tomato auxin mutants. Ann Appl Biol. 2011, 159: 202-211.
23. Quaggiotti S, Ruperti B, Pizzeghello D, Francioso O, Tugnoli V, Nardi S: Effect of low molecular size humic substances on nitrate uptake and expression of genes involved in nitrate transport in maize (*Zea mays* L.). J Exp Bot. 2004, 55: 803-813.
24. Frias I, Caldeira MT, Perez CJR, Navarro AJP, Culianez MFA, Kuppinger O, Stransky H, Pages M, Hager A, Serrano R: A major isoform of the maize plasma membrane H_-ATPase: characterization and induction by auxin in coleoptiles. Plant Cell. 1996, 8: 1533-1544.
25. Morsomme P, Boutry M: The plant plasma membrane H^+-ATPase: structure, function and regulation. Biochim Biophys Acta. 2000, 1465: 1-16.
26. Baldotto LEB, Baldotto MA, Giro VB, Canellas LP, Olivares FL, Bressan-Smith B: Performance of 'vitória' pineapple in response to humic acid application during acclimatization. R Bras Ci Solo. 2009, 33: 979-990.
27. Canellas LP, Piccolo A, Dobbss LB, Spaccini R, Olivares FL, Zandonadi DB, Facanha AR: Chemical composition and bioactivity properties of size-fractions separated from a vermicompost humic acid. Chemosphere. 2010, 78: 457-466.
28. Concheri G, Nardi S, Reniero F, Dell'Agnola G: The effects of humic substances within the Ah horizon of a Calcic Luvisol on morphological changes related to invertase and peroxidase activities in wheat roots. Plant Soil. 1996, 179: 65-72.
29. Zandonadi DB, Santos MP, Dobbss LB, Olivares FL, Canellas LP, Binzel ML, Okorokova-Facanha AL, Facanha AR: Nitric oxide mediates humic acids-induced root development and plasma membrane H^+-ATPase activation. Planta. 2010, 231: 1025-1036.
30. Lamattina L, García-Mata C, Graziano M, Pagnussat GC: Nitric oxide: the versatility of an extensive signal molecule. Annu Rev Plant Biol. 2003, 54: 109-136.

31. Lombardo MC, Graziano M, Polacco JC, Lamattina L: Nitric oxide functions as a positive regulator of root hair development. Plant Signaling Behav. 2006, 1: 28-33.
32. Lombardo MC, Lamattina L: Nitric oxide is essential for vesicle formation and trafficking in Arabidopsis root hair growth. J Exp Bot. 2012, 63: 4875-4885.
33. Mora V, Baigorri R, Bacaicoa V, Zamarreño A, García-Mina JM: The humic acid-induced changes in the root concentration of nitric oxide, IAA and ethylene do not explain the changes in root architecture caused by humic acid in cucumber. Environ Exp Bot. 2012, 76: 24-32.
34. Baldri DV, Vivanco JM: Regulation and function of root exudates. Plant Cell Environ. 2009, 32: 666-681.
35. Aullakh MS, Wassmann R, Bueno C, Kreuzweisser J, Rennenberg H: Characterization of root exudates at different growth stages of ten rice (*Oryza sativa* L.) cultivars. Plant Biol. 2001, 3: 139-148.
36. Canellas LP, Teixeira Junior LRL, Dobbss LB, Silva CA, Medici LO, Zandonadi DB, Façanha AR: humic acids crossinteractions with root and organic acids. Ann Appl Biol. 2008, 153: 157-166.
37. Puglisi E, Fragoulis G, Del Re AA, Spaccini R, Piccolo A, Gigliotti G, Said-Pullicino D, Trevisan M: Carbon deposition in soil rhizosphere following amendments with compost and its soluble fractions, as evaluated by combined soil–plant rhizobox and reporter gene systems. Chemosphere. 2008, 73: 1292-1299.
38. Šmejkalová D, Piccolo A: Aggregation and disaggregation of humic supramolecular assemblies by NMR diffusion ordered spectroscopy (DOSY-NMR). Environ Sci Technol. 2008, 42: 699-706.
39. Nardi S, Carletti P, Pizzeghello D, Muscolo A: Biological activities of humic substances. Biophysico-chemical processes involving natural nonliving organic matter in environmental systems. Vol 2, part 1: fundamentals and impact of mineral-organic biota interactions on the formation, transformation, turnover, and storage of natural nonliving organic matter (NOM). Edited by: Senesi N, Xing B, Huang PM. 2009, Hoboken: Wiley, 305-339.
40. Nardi S, Muscolo A, Vaccaro S, Baiano S, Spaccini R, Piccolo A: Relationship between molecular characteristics of soil humic fractions and glycolytic pathway and krebs cycle in maize seedlings. Soil Biol Biochem. 2007, 39: 3138-3146.
41. Ertani A, Francioso O, Tugnoli V, Righi V, Nardi S: Effect of commercial lignosulfonate-humate on *Zea mays* L. Metabolism J Agri Food Chem. 2011, 59: 11940-11948.
42. Merlo L, Ghisi R, Rascio N, Passera C: Effects of humic substances on carbohydrate metabolism of maize leaves. Can J Plant Sci. 1991, 71: 419-425.
43. Piccolo A, Nardi S, Concheri G: Structural characteristics of humic substances as related to nitrate uptake and growth regulation in plant systems. Soil Biol Biochem. 1992, 24: 373-380.
44. Albuzio A, Ferrari G, Nardi S: Effects of humic substances on nitrate uptake and assimilation in barley seedlings. Can J Soil Sci. 1986, 66: 73l-736l.
45. Nardi S, Pizzeghello D, Gessa C, Ferrarese L, Trainottic L, Casadoro G: A low molecular weight humic fraction on nitrate uptake and protein synthesis in maize seedlings. Soil Biol Biochem. 2000, 32: 415-419.
46. Panuccio MR, Muscolo A, Nardi S: Effect of humic substances on nitrogen uptake and assimilation in two species of pinus. J Plant Nutr. 2001, 24: 693-704.
47. Carletti P, Masi A, Spolaore B, Polverino De Laureto P, De Zorzi M, Turetta L, Ferretti M, Nardi S: Protein expression changes in maize roots in response to humic substances. J Chem Ecol. 2008, 34: 804-818.
48. Trevisan S, Botton A, Vaccaro S, Vezzaroa A, Quaggiotti S, Nardia S: Humic substances affect Arabidopsis physiology by altering the expression of genes involved in primary metabolism, growth and development. Environ Exp Bot. 2011, 74: 45-55.
49. Jannin L, Arkoun M, Ourry A, Laîné P, Goux D, Garnica M, Fuentes M, Francisco SS, Baigorri R, Cruz F, Houdusse F, Garcia-Mina JM, Yvin JC, Etienne P: Microarray

analysis of humic acid effects on *Brassica napus* growth: involvement of N, C and S metabolisms. Plant Soil. 2012, 359: 297-319.

50. Schiavon M, Pizzeghello D, Muscolo A, Vaccaro S, Francioso O, Nardi S: High molecular size humic substances enhance phenylpropanoid metabolism in maize (*Zea mays* L.). J Chem Ecol. 2010, 36: 662-669.
51. Dixon RA: Natural products and plant disease resistance. Nature. 2001, 411: 843-847.
52. Muscolo A, Felici M, Concheri G, Nardi S: Effect of earthworm humic substances on esterase and peroxidase activity during growth of leaf explants of *Nicotiana plumbaginifolia*. Biol Fertil Soils. 1993, 15: 127-131.
53. Garcia AC, Santos LA, Guridi F, Sperandi MVL, Castro RN, Berbara RLL: Vermicompost humic acids as an ecological pathway to protect plant agaisnt oxidative stress. Ecol Eng. 2012, 47: 203-208.
54. Puglisi E, Fragoulis G, Ricciuti P, Cappa F, Spaccini R, Piccolo A, Trevisan M, Crecchio C: Effects of a humic acid and its size-fractions on the bacterial community of soil rhizosphere under maize (Zea mays L.). Chemosphere. 2009, 77: 829-837.
55. Puglisi E, Pascazio S, Suciu N, Cattani I, Fait G, Spaccini R, Crecchio C, Piccolo A, Trevisan M: Rhizosphere microbial diversity as influenced by humic substance amendments and chemical composition of rhizodeposits. J Geochem Expl. 2013, 129: 82-94.
56. Vessey JK: Plant growth promoting rhizobacteria as biofertilizers. Plant Soil. 2003, 255: 571-586.
57. Sugiyama A, Yazaki K: Root exudates of legume plants and their involvement in interactions with soil microbes. Secretions and exudates in biological systems. Edited by: Vivanco JM, Baluška F. 2012, Heidelberg: Springer, 27-48.
58. Spaccini R, Piccolo A, Haberhauer G, Gerzabek MH: Transformation of organic matter from maize residues into labile and humic fractions of three European soils as revealed by ^{13}C distribution and CPMAS-NMR spectra. Eur J Soil Sci. 2000, 51: 583-594.
59. Spaccini R, Piccolo A, Conte P, Haberhauer G, Gerzabek MH: Increased soil organic carbon sequestration through hydrophobic protection by humic substances. Soil Biol Biochem. 2002, 34: 1839-1851.
60. Piccolo A, Spaccini R, Nieder R, Richter J: Sequestration of a biologically labile organic carbon in soils by humified organic matter. Clim Change. 2004, 67: 329-343.
61. James EK, Olivares F: Infection and colonization of sugarcane and other graminaceous plants by endophytic bacteria. Crit Rev Plant Sci. 1998, 17: 77-119.
62. Martinez-Balmori D, Olivares FL, Spaccinic R, Aguiar KP, Araújo MF, Aguiar NO, Guridi F, Canellas LP: Molecular characteristics of vermicompost and their relationship to preservation of inoculated nitrogen-fixing bacteria. J Anal Appl Pyrol. 2013,
63. Marques Júnior RB, Canellas LP, Silva LG, Olivares FL: Promoção de enraizamento de microtoletes de cana-de-açúcar pelo uso conjunto de substâncias húmicas e bactérias diazotróficas endofíticas. Rev Bras Ci Solo. 2008, 32: 1121-1128.
64. Conceição PM, Vieira HD, Canellas LP, Marques Júnior RB, Olivares FL: Corn seed coating with humic acids and endophytic diazotrophic bacteria. Pesq Agropec Bras. 2008, 43: 545-548.
65. Zaler JG: Vermicompost in seedling potting media can affect germination, biomass allocation, yields and fruit quality of three tomato varieties. Eur J Soil Biol. 2007, 43: S332-S336.
66. Arancon NQ, Edwards CA, Bierman P, Metzger PJ, Lee S, Welch C: Effects of vermicomposts on growth and marketable fruits of field-grown tomatoes, peppers and strawberries. Pedobiologia. 2003, 47: 731-735.
67. Aguiar NO, Olivares FL, Novotny EH, Dobbss LB, Martizez-Balmori D, Santos-Júnior LG, Chagas JG, Façanha AR, Canellas LP: Bioactivity of humic acids isolated from vermicomposts at different maturation stages. Plant Soil. 2013, 362: 161-174.

68. Busato JG, Silva LL, Aguiar NO, Canellas LP, Olivares FL: Changes in labile phosphorus forms during maturation of vermicompost enriched with phosphorus-solubilizing and diazotrophic bacteria. Bioresour Technol. 2012, 110: 390-395.

CHAPTER 12

Formation of (2E)-4-Hydroxy-2-nonenal and (2E)-4-Hydroxy-2-hexenal by Plant Enzymes: A Review Suggests a Role in the Physiology of Plants

Harold W. Gardner*

Wild Edges Nature Preserve, Carlisle, PA, USA (formerly ARS, USDA)

ABSTRACT

It is demonstrated that (3Z)-nonenal (NON) and (3Z)-hexenal (HEX) are oxidized in a cascade by lipoxygenase (LOX) and hydroperoxide peroxygenase (HP peroxygenase) into (2E)-4-hydroxy-2- nonenal (HNE) and (2E)-4-hydroxy-2-hexenal (HHE), respectively. In turn, HNE inactivates LOX terminating the cascade. The hydroxy-alkenals produced serve to inhibit plant pathogens, which initiated the cascade. In addition to LOX, other unknown oxygenases may be involved in the cascade.

Keywords: (3Z)-Nonenal, (3Z)-Hexenal, (2E)-4-Hydroxy-2-nonenal, (2E)-4-Hydroxy-2-hexenal, (2E)-4-Hydroperoxy-2-hexenal, Lipoxygenase, Hydroperoxide Peroxygenase, Hydroperoxide Lyase, (2E)-4-Hydroperoxy-

2-Nonenal, 3, 4-Epoxynonanal, (2E)-4-Oxo-2-nonenal, Glycine max, Vica faba, Plant Pathogen

1. INTRODUCTION

Using a cloned hydroperoxide lyase (HP lyase) preparation from alfalfa Nordermeer et al. [1] reported that (3Z)- nonenal (NON) autoxidized to (2E)-4-hydroperoxy-2-nonenal (HPNE), and (3Z)-hexenal autoxidized to (2E)-4- hydroperoxy-2-hexenal (HPHE) in the absence of lipoxygenase (LOX) and hydroperoxide peroxygenase (HP peroxygenase). They reduced HPNE and HPHE by $NaBH_4$ to (2E)-4-hydroxy-2-nonenal (HNE) and (2E)-4-hydroxy-2-hexenal (HHE), respectively, for analysis. As first reported by us, there certainly exists autoxidation of NON and HEX, but very feebly, as carefully determined and discussed in the publication using Vica faba preparations [2] [1], as well as, Glycine max (soybean) preparations [3] . On the basis of the work of Noordermeer et al. [1] the so-called "autoxidation-only theory" was adopted by Schneider et al. [4] [5] evidently without examining the proof of enzyme action involved. Unfortunately, the "autoxidation-only theory" may have prejudiced all future research.

2. DISCUSSION

In the work of Noordermeer et al. [1] the concentration of NON and (3Z)-hexenal (HEX) was 0.025 mM, which was incubated for 1 to 1.5 hr. In some experiments this resulted in HNE as $NaBH_4$-reduced HPNE at a maximum of about 50% conversion from NON. In research with Vica faba membrane preparations [2] the concentration of NON was 40-fold higher at 1 mM, and the conversion to HNE was almost complete after only 5 min. at 23°C giving about 45% yield of HNE. As shown by Gardner and Hamberg [2] , the membrane preparation inactivated by boiling water for 5 min. resulted in nearly undetectable NON conversion to HNE. If significant HPNE was formed in the heat-inactivated preparation, it would be partially detected by our method [3] [6] . Other experiments used either triphenylphosphite or triphenylphosphine reduction to detect HPNE [2] [3] . The disappearance of NON in heat-inactivated preparation was attributed to reaction as Schiff base with inactivated protein. With soybean Glycine max preparations [3] the concentration used was 40- to 80-fold higher (1 to 2 mM) than used by Noordermeer et al. [1] . With a 9,300 g supernatant of soybean and 2 mM NON conversion was nearly complete at 25°C within 5 min. at 15% yield of

HNE, whereas the preparation inactivated at 80°C for 15 min. was inactive. That the soybean activity mainly resided in the 200,000 g membrane fraction of soybean was shown by a 19% conversion of 1 mM NON within 15 min; whereas, the soluble fraction gave only a 3.3% conversion. Actually, the soluble fraction converted NON and HEX into mainly (2E)-nonenal and (2E)-hexenal, respectively. Heat-inactivated membrane fraction gave only 1.2% yield of HNE after 15 min., and buffer afforded only 0.7% HNE. If one considers these latter two values to be mainly due to autoxidation, extrapolation to one hr. gives 0.048 mM and 0.028 mM HNE, respectively. Of course, the conditions are different than used by Noordermeer et al. [1] . In our experience it is completely logical that Noordermeer et al. [1] detected autoxidation of NON under their conditions of low NON concentration and long incubation times. Water or buffer alone should cause autoxidation of 0.025 mM NON in 1 to 1.5 hr. In the discussion to follow there is a strong indication of an enzymatic component to the oxidation of NON and HEX.

Early experiments with soybean, Glycine max, preparations (Triton-100R, pH 7.5 10,000 Xg) (9S, 10E, 12Z)- 9-hydroperoxy-10, 12-octadecadienoic acid (9S-HPOD) was converted into the expected HP lyase product, 9-oxononanooic acid, but only traces of nonenals [7] . The absence of nonenals was accounted for as the appearance of HNE. (13S, 9Z, 11E, 15Z)-13-Hydroperoxy-9, 11, 15-octadectrienoic acid (13S-HPOT) as a substrate produced HEX, which peaked after only 2 min. reaction and then rapidly decreased to undetectable. In later reports HEX was shown to be converted into (2E)-4-hydroperoxy-2-hexenal (HPHE) and (2E)-4-hydroxy-2-hex-enal (HHE) by soybean preparations [3] , as well as, Vica faba membrane preparation [2] . Several reports were published concerning HP peroxygenase [8] - [11] . Since NON could be both a LOX and a HP peroxygenase mimic substrate, it seemed plausible that a dual process was at work. Chemically 3, 4-epoxy-nonanal was rearranged to form HNE [12] . Thus, a dual pathway involving LOX and HP peroxygenase was postulated [13] . In Hamberg's lab initial experiments with NON and a soybean LOX-1 (Sigma Type V) preparation were frustrating indicating a slow autoxidation was occurring (this result was clarified in later research discussed below). Hamberg suggested a membrane preparation of Vicia faba known to contain HP peroxygenase and LOX. This membrane preparation proved that NON was converted to HNE enzymically with heat-inactivated membrane producing much less conversion (very little autoxidation) [2] . The supernatant fraction was active, but much less so, and buffer only was ineffective [2] .

In this work we isolated both HPNE and trapped 3, 4-epoxynonanal with $NaBH_4$ reduction furnishing stable 3, 4-epoxynonanol [2] . It was also shown that HEX was converted to HHE. In addition to $NaBH_4$ trapping of 3,

4-epoxy-nonanal, it was demonsrated that HP peroxygenase was involved by use of several experiments. Preincubation with H_2O_2 is known to inactivate HP peroxygenase, and such a H_2O_2 preincubation experiment produced about 50% less HNE as triphenylphosphite reduced HPNE. And, the HP peroxygenase co-substrate, oleic acid, reduced the conversion of NON to HNE with the simultaneous appearance of 9, 10-epoxyoctadecanoic acid. Also, the addition of [13-$^{18}O_2$]-(13S, 9Z, 12E)-hydroperoxy-9, 12-octadecadienoic acid ([13-$^{18}O_2$]-13S- HPOD) both inhibited HNE formation and labeled HNE with ^{18}O. Takamura and Gardner [3] preincubated a 9,300 g supernatant from soybean with H_2O_2 and demonstrated that NON was converted into HPNE by TLC of the product, which was isolated from TLC, reduced by triphenylphosphine, and rechromatographed to correspond to a HNE standard. Thus, it was clearly demonstrated that the Vicia faba and Glycine max membrane preparations produced HNE from HPNE by HP peroxygenase. Table 1 summarizes the conversions of substrates to products by various preparations.

As discussed briefly above, it was demonstrated that soybean Glycine max preparations were giving similar results [3] using both HEX and NON. The pH optimum of HNE and HHE formation at 9.5 was suspiciously similar to soybean LOX-1 pH optimum. HEX was much less active, which one would expect with LOXs given the ⍰3 double bond of HEX. In further work [14] it was demonstrated that soybean LOX-1 activity and NON oxidation activity co-eluted by DEAE Sephadex column chromatography. It required a relatively large amount of LOX-1 to produce mainly (2E)-4-oxo-2-nonenal and less HPNE. This was reminiscent of the formation of 13-oxo-octadecadienoic acid and (13S, 9Z, 12E)-hydroperoxy-9,12-octadecadienoic acid (13S-HPOD) with LOX under partial anaerobic conditions [15] [16] . The latter report implied that HPNE was inefficient at forming the Fe^{3+} form of LOX. It seemed that the reaction needed a lipid hydroperoxide to "prime-the-pump" of native Fe^{2+} [17] . By adding a small amount of 13S-HPOD, NON was efficiently oxidized to HPNE. That the reaction was caused by LOX-1 was demonstrated by showing that the recovered HNE was 83% S-configuration as expected. NON was also oxidized by other LOX isozymes isolated by column chromatography. LOX-3 gave HPNE at 55.5% and 71.4% 4S-stereoconfiguration at pH 8.6 and 6.5, respectively. LOX-2 resulted in HPNE with 60.8% and 86.5% 4S-stereoconfiguration at pH 8.6 and 6.5, respectively. This result with LOX-2 was close to the 75.1% 13S-HPOD obtained at pH 6.6 with linoleic acid oxidation [18] . Soybean LOX-3 afforded nearly racemic R, S-9-and 13-HPOD products from linoleic acid, which did not thoroughly coincide with the HPNE result.

Yet other studies showed that HP lyase product of 13S-HPOD, 12-oxo-9Z-dodecenoic acid, was converted into 12-oxo-9-hydroxy-10E-dodecenoic acid by both soybean and alfalfa seedling preparations [19] . Nakashima et al. [20] identified 12-oxo-9-hydroxy-10E-dodecenoic acid as a glyceride component of galactolipid. Noordermeer et al. [1] reported that 12-oxo-9Z-dodecenoic acid was autoxidized to both 12-oxo-9-hydroxy-10E-dodecenoic acid and 12-oxo-11-hydroxy-9Z-dodecenoic acid.

Counter-intuitively, HNE inhibits soybean LOX-1 [21] probably by reacting with essential amino acid (s). LOX has three histidines as ligands to the iron active site [22] [23] It is known that histidine is particularly susceptible to reaction with HNE [24] . It seems certain this inhibition is a feed-back mechanism. That is, after wounding by a plant pathogen LOX produces HPOD/HPOTs, which are both substrates and promoters of hydroxy-alkenal formation. In turn, the build-up of hydroxy-alkenals then turns off LOX oxidation of linoleic acid, linolenic acid, and alkenals preserving the plant but inhibiting invaders. Deighton and coworkers [25] [26] showed high concentrations of HNE (4 to 20 nmol · g^{-1}) in fruit of sweet pepper infected by Botrytis cinerea. In certain experiments HNE was as high as 30 nmol · g^{-1} (Deighton, N., personal communication, unpublished). Interestingly, the senescence hormone, methyl jasmonate, up-regulated 13-LOXs and HP lyase in barley leaves leading to 15 nmol · g^{-1} of endogenous HHE after 48 hrs [27] .

This leads to the issue of S, R-configuration of the hydroxyl group of HNE. In the Gardner and Hamberg paper [2] with Vicia faba at pH 6.5 the configuration of HPNE was racemic. HNE was slightly more R, and$NaBH_4$-trapped 3, 4-epoxynonanol was 61% R. With crude 10,000 g soybean preparations the result was about 65% 4S-HNE (Gardner, H.W., unpublished). Future research should be pursued to better understand the role of LOX isozymes and HP peroxygenase in stereo-selectivity. Since pH 6.7 buffer was used with the V. faba preparation, it is plausible that more than one LOX isozyme was present. LOXs with optimal activity at neutral pHs are known as type-2 LOXs, and these enzymes often result in HPOD with a mix of regio-and stereo-isomers. At pH 6.5, even soybean LOX-1 gave a mix of regio-and stereo-isomers [28] . In membrane preparations of Vicafaba and Glycine max it is likely that endogenous lipid provides the hydroperoxide to “prime the LOX pump”, thus starting the oxidative cascade. Scheme 1 summarizes the hypothesis that the oxidative cascade to 4-HNE and 4-HHE plays a role in the physiology of plants.

Table 1. Summary of NON and HEX oxidations.

Substrate	Product	References	Enzymes
NON	HPNE	2, 3, 14	LOX + H_2O_2-HPP, LOX-1
NON	HNE	2, 3	LOX + HPP
HEX	HHE	2, 3	LOX + HPP

LOX: Lipoxygenase/Oxygenase; HPP: Hydroperoxide peroxygenase; H_2O_2-HPP: H_2O_2 inhibited Hydroperoxide Peroxygenase, LOX-1: Soybean Lipoxygenase-1 isolate.

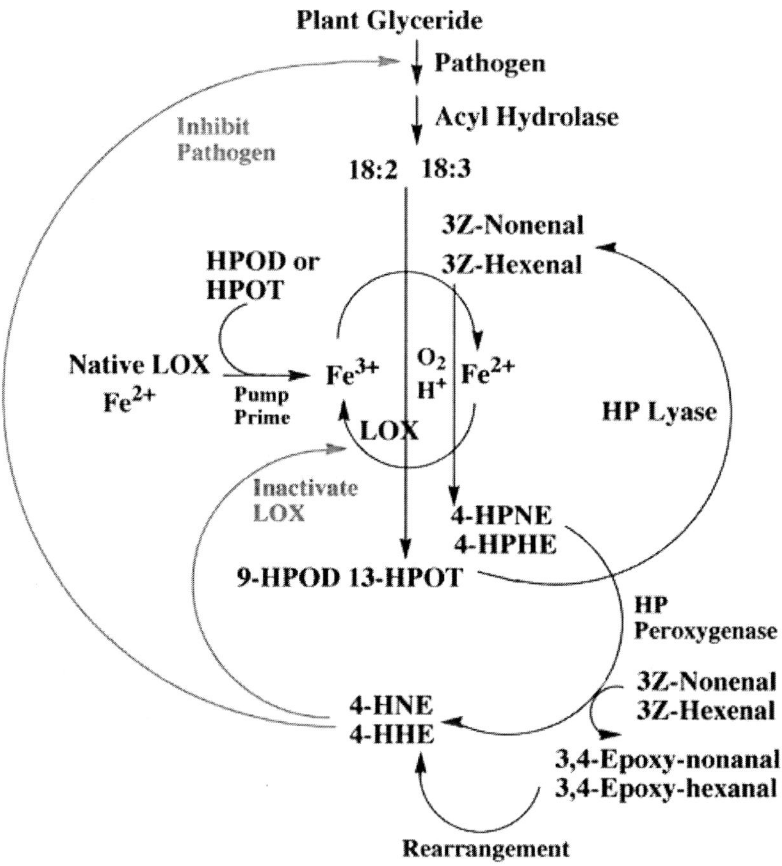

Scheme 1. Oxidative cascade of plants.

It was shown adequately that HP peroxygenase plays a role in the cascade. Although soybean LOX-1 is capable of forming HPNE with the expected S stereo-configuration, the oxygenase portion of the cascade is not totally clear. With Vica faba membrane preparations [2] the apparent inaction of LOX was shown by the inaction of LOX inhibitors, 5, 8, 11, 14-

ecosatetraynoic acid and nordihydroguaiaretic acid. Also, in the soybean membrane preparations Takamura and Gardner [3] showed that there was about 50% inhibition by LOX inhibitors, 5, 8, 11, 14-eicosatetraynoic acid and salicylhydroxamate at pH 6.5, but not at pH 9.5 where LOX-1 has optimum activity. The lack of effect by LOX inhibitors was a confounding result that suggests additional research is needed. It is plausible that an aldehyde oxygenase also may be involved.

CITE THIS PAPER

Harold W. Gardner (2016) Formation of (2*E*)-4-Hydroxy-2-nonenal and (2*E*)-4-Hydroxy-2-hexenal by Plant Enzymes: A Review Suggests a Role in the Physiology of Plants. *Advances in Enzyme Research*,**04**,56-61.

REFERENCES

1. Noordermeer, M.A., Fuessner, I., Kolbe, A., Veldink, G.A. and Vliegenthart, J.F.G. (2000) Oxygenation of (3Z)-Alkenals to 4-Hydroxy-(2E)-Alkenals in Plant Extracts: A Nonenzymatic Process. Biochemical and Biophysical Research Communications, 277, 112-116.
2. Gardner, H.W. and Hamberg, M. (1993) Oxygenation of (3Z)-Nonenal to (2E)-4-Hydroxy-2-nonenal in the Broad Bean (Vica faba L.). Journal of Biological Chemistry, 268, 6971-6977. [Citation Time(s):11]
3. Takamura, H. and Gardner, H.W. (1996) Oxygenation of (3Z)-Alkenal to (2E)-4-Hydroxy-2-alkenal in Soybean Seed (Glycine max L.). Biochimica et Biophysica Acta, 1303, 83-91.
4. Schneider, C., Tallman, K.A,, Porter, N.A. and Brash, A.R. (2001) Two Distinct Pathways of Formation of 4-Hydroxy- nonenal Mechanisms of Nonenzymatic Transformation of the 9- and 13-Hydroperoxides of Linoleic Acid to 4-Hydroxyalkenals. The Journal of Biological Chemistry, 276, 20831-20838.
5. Schneider, C., Porter, N.A. and Brash, A.R. (2008) Routes to 4-Hydroxynonenal: Fundamental Issues in the Mechanisms of Lipid Peroxidation. The Journal of Biological Chemistry, 283, 15539-15543.
6. Grechkin, A.N., Mukhtarova, L.S. and Hamberg, M. (2005) Thermal Conversions of Trimethylsilyl Peroxides of Linoleic and Linolenic Acid. Chemistry and Physics of Lipids, 138, 93-101.
7. Gardner, H.W., Weisleder, D. and Plattner, R.D. (1991) Hydroperoxide Lyase and Other Hydroperoxide-Metabolizing Activity in Tissues of Soybean, Glycine max. Plant Physiology, 97, 1059-1072.
8. Hamberg, M., and Fahlstadius, P. (1992) On the Specificity of a Fatty Acid Epoxygenase in Broad Bean (Vicia faba L.). Plant Physiology, 99, 987-995.
9. Hamberg, M. and Hamberg, G. (1990) Hydroperoxide-Dependent Epoxidation of Unsaturated Fatty Acids in the Broad Bean (Vica faba L.). Archives of Biochemistry and Biophysics, 283, 409-416.
10. Blee, E. and Schuber, F. (1990) Efficient Epoxidation of Unsaturated Fatty Acids by a Hydroperoxide-Dependent Oxygenase. The Journal of Biological Chemistry, 265, 12887-12894.

11. Blee, E. and Schuber, F. (1990) Stereochemistry of the Epoxidation of Fatty Acids Catalyzed by Soybean Peroxygenase. Biochemical and Biophysical Research Communications, 173, 1354-1360.
12. Gardner H.W., Bartelt, R.J. and Weisleder, D. (1992) A Facile Synthesis of 4-Hydroxy-2(E)-nonenal. Lipids, 27, 686-689.
13. Gardner, H.W., Weisleder, D. and Plattner, R.D. (1991) An Enzyme Preparation from Seeds of Soybean Glycine max Converts 9S-Hydroperoxide of Linoleic Acid into 4-Hydroxy-trans-2-nonenal. Plant Physiology Meeting, Albuquerque, NM, 28 July-1 August 1991. [Citation Time(s):1]
14. Gardner, H.W. and Grove, M.J. (1998) Soybean Lipoxygenase-1 Oxidizes 3Z-Nonenal. Plant Physiology, 116, 1359- 1366.
15. Garssen, G.J., Vliegenthart, J.F.G. and Boldingh, J. (1971) An Anaerobic Reaction between Lipoxygenase, Linoleic Acid and Its Hydroperoxide. Biochemical Journal, 122, 327-332.
16. Verhagen, J., Veldink, G.A., Egmond, M.R., Vliegenthart, J.F.G., Boldingh, J. and van der Star, J. (1978) Steady-State Kinetics of the Anaerobic Reaction of Soybean Lipoxygenase-1 with Linoleic Acid and 13-L-hydroperoxylinoleic Acid. Biochimica et Biophysica Acta (BBA)—Lipids and Lipid Metabolism, 529, 369-379.
17. Smith, W.L. and Lands, W.E.M. (1972) Oxygenation of Unsaturated Fatty Acids by Soybean Lipoxygenase. The Journal of Biological Chemistry, 247, 1038-1047. [Citation Time(s):1]
18. Fukushige, H., Wang, C., Simpson, T.D., Gardner, H.W. and Hildebrand, D.F. (2005) Purification of Linoleic Acid Hydroperoxides Generated by Soybean Seed Lipoxygenases 2 and 3. Journal of Agricultural and Food Chemistry, 53, 5691-5694.
19. Gardner, H.W. (1998) 9-Hydroxy-traumatin, a New Metabolite of the Lipoxygenase Pathway. Lipids, 33, 745-749.
20. Nakashima, A., von Reuss, S.H., Tasaka, H., Nomura, M., Mochizuki, S., Iijima, Y., Aoki, K., Shibata, D., Boland, W., Takabayashi, J. and Matsui, K. (2013) Traumatin- and Dinortraumatin-Containing Galactolipids in Arabidopsis: Their Formation in Tissue-Disrupted Leaves as Counterparts of Green Leaf Volatiles. The Journal of Biological Chemistry, 288, 26078-26088.
21. Gardner, H.W. and Deighton, N. (2001) Effect of 4-Hydroxy-2(E)-nonenal on Soybean Lipoxygenase-1. Lipids, 36, 623-628.
22. Boyington, J.C., Gtaffney, B.J. and Amzel, L.M. (1993) The Three-Dimensional Structure of an Arachidonic Acid 15-Lipoxygenase. Science, 260, 1482-1486.
23. Minor, W., Steczko, J., Stec, B., Otwinowski, Z., Bolin, J.T., Walter, R. and Axelrod, B. (1996) Crystal Structure of Soybean Lipoxygenase L-1 at 1.4 Angstrom Resolution. Biochemistry, 35, 10687-10701.
24. Uchida, K. and Stadtman, E.R. (1992) Modification of Histidine Residues in Proteins by Reaction with 4-Hydroxyno- nenal. Proceedings of the National Academy of Sciences of the United States of America, 89, 4544-4548.
25. Deighton, N., Muckenschnabel, I., Goodman, B.A. and Williamson, B. (1999) Lipid Peroxidation and the Oxidative Burst Associated with Infection of Capsicum annuum by Botrytis cinerea. The Plant Journal, 20, 485-492.
26. Adams, L.K., Benson, E.E., Staines, H.J., Bremmer, D.H., Millam, S. and Deighton, N. (1999) Effects of the Lipid Peroxidation Products 4-Hydroxy-2-nonenal and Malondialdehyde on the Proliferation and Morphogenetic Development of in Vitro Plant Cells. Journal of Plant Physiology, 155, 376-386.
27. Kohlmann, M., Bachmann, A., Weichert, H., Kohbe, A., Balkenhohl, T., Wasternack, C. and Feussner, I. (1999) Formation of Lipoxygenase-Pathway-Derived Aldehydes in Barley Leaves upon Methyl Jasmonate Treatment. European Journal of Biochemistry, 260, 885-895.
28. Gardner, H.W. (1989) Soybean Lipoxygenase-1 Enzymically Forms Both (9S)- and (13S)-Hydroperoxides from Linoleic Acid by a pH-Dependent Mechanism. Biochimica et Biophysica Acta (BBA)—Lipids and Lipid Metabolism, 1001, 274-281.

NOTES

[1]In the publication of Gardner, H.W. Hamberg, M. [2] we erroneously named "3, 4-epoxynonenal" in the Abstract, which should be corrected to read 3, 4-epoxynonanal.

CHAPTER 13

Rescue of Heavy Metal Effects on Cell Physiology of The Algal Model System *Micrasterias* by Divalent Ions

Stefanie Volland[a], Elisabeth Bayer [a], Verena Baumgartner [a], Ancuela Andosch[a], Cornelius Lütz [b], Evelyn Sima[a], Ursula Lütz-Meindl [a,*]

[a] *Plant Physiology Division, Cell Biology Department, University of Salzburg, Hellbrunnerstraße 34, 5020 Salzburg, Austria*

[b] *Institute of Botany, Faculty of Biology, University of Innsbruck, Sternwartestraße 15, 6020 Innsbruck, Austria*

ABSTRACT

Recent studies have shown that metals such as copper, zinc, aluminum, cadmium, chromium, iron and lead cause severe dose-dependent disturbances in growth, morphogenesis, photosynthetic and respiratory activity as well as on ultrastructure and function of organelles in the algal model system *Micrasterias denticulata* (Volland et al., 2011, 2012; Andosch et al., 2012). In the present investigation we focus on amelioration of these adverse effects of cadmium, chromium and lead by supplying the cells with different antioxidants and essential micronutrients to obtain insight into metal uptake mechanisms and subcellular metal targets. This seems particularly interesting as *Micrasterias* is adapted to extremely low-concentrated, oligotrophic conditions in its natural bog environment.

The divalent ions of iron, zinc and calcium were able to diminish the effects of the metals cadmium, chromium and lead on *Micrasterias*. Iron showed most ameliorating effects on cadmium and chromium in short- and long-term treatments and improved cell morphogenesis, ultrastructure, cell division rates and photosynthesis. Analytical transmission electron microscopic (TEM) methods (electron energy loss spectroscopy (EELS) and electron spectroscopic imaging (ESI)) revealed that chromium uptake was decreased when *Micrasterias* cells were pre-treated with iron, which resulted in no longer detectable intracellular chromium accumulations. Zinc rescued the detrimental effects of chromium on net-photosynthesis, respiration rates and electron transport in PS II. Calcium and gadolinium were able to almost completely compensate the inhibiting effects of lead and cadmium on cell morphogenesis after mitosis, respectively. These results indicate that cadmium is taken up by calcium and iron transporters, whereas chromium appears to enter the algae cells via iron and zinc carriers. It was shown that lead is not taken up into *Micrasterias* at all but exerts its adverse effects on cell growth by substituting cell wall bound calcium. The antioxidants salicylic acid, ascorbic acid and glutathione were not able to ameliorate any of the investigated metal effects on the green alga *Micrasterias* when added to the culture medium.

Keywords: Amelioration of metal effects, Antioxidants, Green algae, Heavy metals, Ions

INTRODUCTION

Metals are necessary components of all ecosystems and occur naturally in the earth's crust (Pinto et al., 2003). They appear in a wide range of oxidative states and coordination numbers, influencing their chemical characteristics and thus their bioavailability and toxicity (Pinto et al., 2003; Verbruggen et al., 2009). Certain metals such as iron (Fe), copper (Cu) and zinc (Zn) are considered essential nutrients to plants and are needed for photosynthesis and as cofactors for many enzymes (e.g. Kovacik et al., 2010; Shanmugam et al., 2011). Plants take up essential elements from their surroundings, but they are also able to accumulate elements, which have no known biological function, such as heavy metals like cadmium (Cd), chromium (Cr) or lead (Pb) (Mendoza-Cozatl and Moreno-Sanchez, 2005; Peralta-Videa et al., 2009). These nonessential metals are able to enter plant cells via metal transporters and carriers for the uptake of essential metals (Clemens, 2001; Shanker et al., 2005).

Aquatic environments are particularly exposed to increasing amounts of industrial and agricultural wastes (Kovacik et al., 2010). They may contain Cd, Cr and Pb which are toxic to most organisms at low concentration and have serious negative effects on plant growth, development and photosynthesis (di Toppi and Gabbrielli, 1999; Panda and Choudhury, 2005; Sacan et al., 2007; Peralta-Videa et al., 2009). Experimental amelioration of heavy metal effects by addition of antioxidants or essential ions provides insight into uptake and distribution mechanisms as well as on physiological and sub-structural targets of metals and increases our understanding on possibilities to limit damage to an aquatic ecosystem.

Antioxidants and certain essential micronutrients have the ability to inhibit heavy metal uptake, to contribute to detoxification or to decrease damage to plant cells. Salicylic acid (SA) is commonly known throughout the plant kingdom as regulator for physiological processes and as stress hormone during biotic and abiotic stress (Metraux, 2002; Belkhadi et al., 2010). Several studies demonstrate successful amelioration of heavy metal damage by SA (Guo et al., 2009; Belkhadi et al., 2010). The protective function of SA during heavy metal stress is not fully understood, but seems to mainly derive from its activity as antioxidant, scavenging reactive oxygen species (ROS) (Shah, 2003; Ahmad et al., 2011). The generation of ROS, either directly through Haber–Weiss reactions, or as a consequence of the metals toxicity, is the primary response of plants to heavy metal stress (Yadav, 2010). Other antioxidants such as glutathione (GSH) and ascorbic acid (AA) are also involved in the quenching of ROS, generally via the glutathione-ascorbate cycle (El-Naggar and El-Sheekh, 1998; Noctor et al., 2012; Bielen et al., 2013). GSH is a tripeptide and an indispensable small molecule to higher plants, with multiple functions in biosynthetic pathways, metal detoxification, antioxidant biochemistry and redox homeostasis (Zechmann et al., 2008; Noctor et al., 2012). Both AA and GSH were also found to be capable of diminishing heavy metal effects in the alga Chlorella (El-Naggar and El-Sheekh, 1998).

Among essential ions particularly Fe, Zn and calcium (Ca) may affect uptake and toxicity of heavy metals in plants and algae. Ca and Fe are able to reduce the uptake of Cd (Peralta-Videa et al., 2009). Ca has been shown only recently, to rescue Cd damage on photosynthesis and ultrastructure in the alga *Micrasterias* (Andosch et al., 2012) but was also found to have ameliorating functions on Cd effects in higher plants (Choi et al., 2001; Suzuki, 2005; Wan et al., 2011). Ca also has a positive effect on cell number and size of Pb treated protonema cells and reduced typical cell malformations found under Pb influence (Krzeslowska et al., 2004). Fe and Zn have been reported to ameliorate toxic effects and uptake of Cr in plant cells (Mallick et al., 2010; Branzini et al., 2012). All these ameliorating effects seem to arise mainly from chemical similarities of essential and toxic ions and their competition for carrier uptake into the plant cell (di Toppi and Gabbrielli, 1999; Shanker et al., 2005). This is also supported by experiments with gadolinium (Gd), a well-known Ca-channel blocker which diminishes Cd uptake, suggesting a rescue mechanism via Cd–Ca exchange (Hinkle et al., 1987).

In the present study we investigate ameliorating effects of signaling molecules, antioxidants and essential ions (AS, GSH, AA, Fe, Ca, Zn and Gd) on impact of the heavy metals Cd, Cr and Pb on the alga *Micrasterias denticulata* in order to obtain insight into heavy metal uptake mechanisms and intracellular targets. Previous publications have shown severe dose-dependent effects of different metals on growth, morphogenesis, photosynthetic and respiratory activity as well as on ultrastructure and function of organelles in the unicellular fresh-water alga *Micrasterias* (Volland et al., 2011, 2012; Andosch et al., 2012) which has been employed as a cell biological model since many years (e.g. Meindl, 1993; Oertel et al., 2004; Eder and Lütz-Meindl, 2008; Affenzeller et al., 2009; etc.).

By considering the fact that *Micrasterias* inhabits oligotrophic peat bog ponds and is adapted to extremely nutrient-depleted, low concentrated aquatic environments the hypotheses to be tested in the present study were the following: (1). Do metals such as Cd, Cr and Pb enter *Micrasterias* cells via natural transport systems, such as Ca, Fe or Zn channels like in higher plants? (2). Does addition of antioxidants or micronutrients, though unusually elevating the concentration of the algal environment, prevent damage by the metals? As *Micrasterias* belongs to a group of algae (Streptophyta) which are closest relatives of higher plants (Wodniok et al., 2011) the results of this study are not only relevant for our cell physiological understanding of heavy metal uptake but also in respect to an evolutionary point of view.

MATERIAL AND METHODS

Chemicals

All chemicals were purchased from Sigma–Aldrich (Vienna, Austria), Alfa Aesar (Karlsruhe, Germany) or Carl Roth (Karlsruhe, Germany) unless stated differently.

Cell cultures

Micrasterias denticulata cells were grown in liquid Desmidiacean culture medium (Schlösser, 1982) in Erlenmeyer flasks under semi-sterile conditions. The medium contained a substantial amount of soil extract providing good pH buffering properties. Cells were kept at 20 ± 1 °C at a photoperiod of 14 h light:10 h dark. Every 4–6 weeks the cells were sub-cultured. 3–4 week old cultures during exponential growth were used for experiments (for detailed culture conditions see Meindl and Lütz, 1996; Affenzeller et al., 2009).

Light microscopy

To capture both, short-term metal effects on cell growth and morphogenesis and long-term impact on cell division rates, viability, ultrastructure and metabolic functions *Micrasterias* cells were exposed to short- and long-term incubations. For short-term treatments dividing cells (15–75 min after mitosis) were selected from cultures and were exposed to nutrient solutions containing different metals. For each metal the highest concentration was chosen which the cells were able to survive during a test series (see also our previous publications (Volland et al., 2011, 2012; Andosch et al., 2012). The following concentrations were used: 15 µM $CdSO_4$ (Cd), 1 mM $K_2Cr_2O_7$(Cr) or 40 µM $Pb(NO_3)_2$ (Pb) respectively. For long-term treatments cell cultures were treated with 600 nM Cd,10 µM Cr and 5 µM Pb. After 4 h of incubation (short-term) or 21 days (long-term) of treatment, the effects on the cells were examined with a Univar light microscope (Reichert, Vienna, Austria) and documented with a Canon Powershot A620 camera (Tokyo, Japan).

The following agents were used for rescue experiments of heavy metals impact on *Micrasterias* cells: cultures were pre-treated with 20 µM salicylic acid $C_7H_6O_3$ (SA), 20 µM L-ascorbic acid $C_6H_8O_6$ (AA), 300 nM reduced L-glutathione $C_{10}H_{17}N_3O_6S$ (GSH), 100 µM Fe-EDTA (Fe), 300 nM $ZnSO_4$ (Zn) or 1 mM $CaSO_4$ (Ca) for 1 week to allow acclimatization of the cells to the substances. Dividing cells were then collected and incubated for 4 h, or cultures were treated for 21 days with heavy metals and ameliorating substances simultaneously, before the effects were examined. For rescue treatments with gadolinium (Gd), cells from untreated cultures were exposed to 15 µM Cd and 40 µM Pb either together with 3 µM Gd or after 30 min pre-treatment with Gd. All used concentrations of rescue substances were previously determined to have no negative effect on cell development, morphology and photosynthesis when applied to *Micrasterias* cells alone. Heavy metals and rescue substances were added to the cell cultures only once at the given time points.

Preparation methods for transmission electron microscopy and analytical TEM

Micrasterias cells after selected short- and long-term treatments (see above) were investigated in the TEM.

High pressure freeze fixation of treated cells and controls were done in a Leica EMPACT high-pressure freezer. Cryo-substitution was performed in a Leica EM AFS (Leica Mikrosysteme GmbH, Vienna, Austria) as described by Meindl et al. (1992) and Aichinger and Lütz-Meindl (2004). After freeze substitution cells were infiltrated and embedded in Agar low viscosity resin (Agar Scientific, Essex, U.K.) and polymerized for 16–24 h at 60 °C.

For structural analysis, ultrathin sections of 40–60 nm were cut on a Leica UC7 ultramicrotome (Leica Microsystems GmbH, Vienna, Austria) and were placed on formvar-coated copper grids for conventional imaging. For EELS measurements hexagonal narrow mesh copper grids were used. Sections were examined in a LEO 912 AB transmission electron microscope (Zeiss, Oberkochen, Germany) with in-column energy filter, operated with a LaB_6 cathode and an acceleration voltage of 80 kV for conventional imaging and 120 kV for EELS. Micrographs and EELS were recorded with a slow scan dual speed CCD camera Sharpeye (Tröndle, Moorenweis, Germany), operated by iTEM software (Soft Image System, Münster, Germany).

Intracellular metal localization via EELS and ESI

For the present study *Micrasterias* cells treated with 10 µM Cr plus 100 µM Fe, with 5 µM Pb for 21 days and with 40 µM Pb for 4 h respectively, were investigated.

For electron energy-loss spectroscopy (EELS) magnifications between ×25,000 and ×40,000 were chosen. The measurement area for EELS was defined by a 100 µm spectrometer entrance aperture and 5–7 integration cycles were taken per measurement. Cr and Fe were detected at the $L_{2,3}$ edge at an electron loss of 570 eV and 708 eV respectively. Illumination angles between 1 and 1.6 mrad, exposure times between 2 and 5 s and a spectrum magnification of 200× was used. For the detection of Pb via the $M_{4,5}$ edge at a high energy loss of 2484 eV, a micrograph of the spectrum was captured as previously described by Zheng et al. (2012). The Pb $M_{4,5}$ edge was then identified by superimposing an intensity profile over an image of the spectrum. Element maps (ESI) were taken with the three-window power-law method at a lower magnification of ×8,000.

Measurement of photosynthetic activity

In order to determine the physiological status of treated *Micrasterias* cells, photosynthetic activity was measured by oxygen turnover (production/consumption) and by fast chlorophyll fluorescence (for method see Affenzeller et al., 2009).

Photosynthetic oxygen measurements

For rescue experiments cells were pre-treated for 1 week with 20 µM AA, 20 µM SA, 300 nM GSH, 100 µM Fe and 300 nM Zn followed by a 21 days incubation in 600 nM Cd and 10 µM Cr. Additionally cell cultures treated only with 5 µM Pb for 21 days were measured. Roughly 2000 cells were used for each run with 3–4 light/dark cycles, which were repeated 3 times each. Oxygen turnover as indicator for photosynthetic activity was measured by a Hansatech (King' Lynn, UK) polarographic oxygen electrode and µM oxygen/h/mg chlorophyll were determined. Illumination was set to 200 µM photons m^2 s^{-1} to enable comparison with in earlier experiments (Andosch et al., 2012) and the temperature was kept constant at 20 °C. After each measurement an aliquot of the suspension was removed for a later determination of the total chlorophyll content by pigment extraction (Porra et al., 1989).

Fast chlorophyll fluorescence

To assay the efficiency of photosystem II (PS II) activity a Handy PEA (Hansatech, King' Lynn, UK) was used as described by Affenzeller et al. (2009). Five drops of the cell suspensions treated as described above were pipetted on pieces of filter paper and incubated in darkness for 20 min. A minimum of 7 parallel measurements was performed. During this incubation the cells were kept moist in the sample holders. PS II activity was expressed as F_v/F_m (variable over maximum fluorescence as an introduced parameter) and the shape of the fluorescence induction curves was used to demonstrate the differences in fast energy conversion after the treatments in comparison to the controls, as has been described in detail by Strasser et al. (1995).

Cell vitality

The percentage of living cells was determined by analyzing cell plasmolysis. Per treatment 50 cells were collected and the nutrient solution was substituted by 500 mM sorbitol (for method see Andosch et al., 2012). Cells not undergoing plasmolysis within 20 min sorbitol exposure were assumed dead and counted in a binocular (Nikon, Chiyoda-ku, Japan). Cell vitality-assays were carried out with 600 nM Cd, 600 nM Cd + 100 µM Fe, 600 nM Cd + 300 nM Zn and 10 µM Cr, 10 µM Cr + 100 µM Fe, 10 µM Cr + 300 nM Zn, 5 µM Pb after 21 days treatment. For rescue experiments cells were pre-treated for 1 week with the respective rescue agent prior to the experiment. All experiments were done in triplicate. 150 cells were analyzed in total. A Student' *t*-test was done to determine statistical significance.

Cell division rates

Dividing *Micrasterias* cells were selected and grown at culture conditions for 2 days in order to obtain interphase cells of the same age. Cell division rates of alga cells treated with 600 nM Cd + 100 µM Fe, 600 nM Cd + 300 nM Zn and 10 µM Cr + 100 µM Fe, 10 µM Cr + 300 nM Zn plus controls were examined over the course of 21 days. Division rates were compared with data of cells treated with 600 nM Cd and 10 µM Cr alone as previously published by Volland et al. (2012). The experimental setup was the same as above. All experiments were carried out in triplets starting with 10 interphase cells.

RESULTS

Effects of Cd, Cr and Pb on *Micrasterias*

40 µM Pb disturbed development of young half-cells in short-term treatments and occasionally led to cell death by bursting. The number of lobes was decreased and the tips were abnormally rounded (Figure 1c) when compared to untreated controls (Figure 1a and b). The polar lobe was often reduced, the lateral lobes were enlarged and the younger half-cells showed deeper indentations than controls. Nevertheless, the basal symmetry of the cell pattern was maintained. Cell vitality was reduced to 76.37% by exposure to 5 µM Pb for 21 days (Table 1). Interestingly, electron microscopic investigations did not indicate any ultrastructural changes of Pb short- and long-term treated cells (Figure 2c and d) when compared to controls (Figure 2a and b). Pb could not be detected by EELS (data not shown). The considerable energy loss of the Pb $M_{4,5}$edge at 2.484 eV and the arising unfavorable signal to noise ratio, however, may prevent detection of low Pb amounts. Photosynthetic oxygen turnover and PS II activity was not changed after 21 days treatment with 5 µM Pb (F_v/F_m = 0.73) in comparison to controls (F_v/F_m = 0.78; Figure 4a and b).

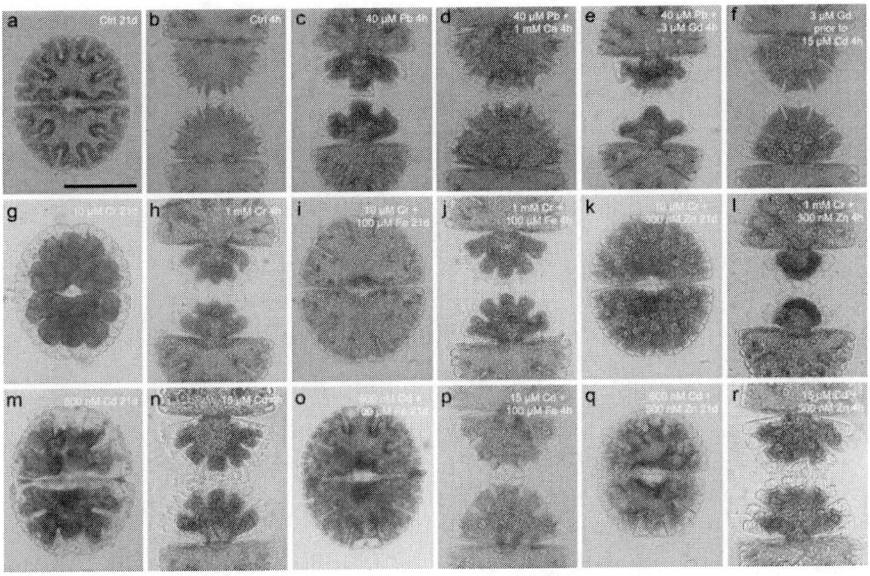

Figure 1. *Micrasterias* control cell in interphase (a) and fully developed approximately 5 h after mitosis (b); *Micrasterias* cells after various metals treatments (c–r): 40 μM Pb 4 h (c), 40 μM Pb + 1 mM Ca 4 h (d), 40 μM Pb + 3 μM Gd 4 h (e), 3 μM Gd (30 min) prior to 15 μM Cd 4 h (f), 10 μM Cr, 21 days (g), 1 mM Cr 4 h (h), 10 μM Cr + 100 μM Fe-EDTA, 21 days (i), 1 mM Cr + 100 μM Fe-EDTA 4 h (j), 10 μM Cr + 300 nM Zn, 21 days (k), 1 mM Cr + 300 nM Zn 4 h (l), 600 nM Cd, 21 days (m), 15 μM Cd 4 h (n), 600 nM Cd + 100 μM Fe-EDTA, 21 days (o), 15 μM Cd + 100 μM Fe-EDTA 4 h (p), 600 nM Cd + 300 nM Zn, 21 days (q), 15 μM Cd + 300 nM Zn 4 h (r); Scale bar 100 μm.

Table 1. *T*-test: Cd and Cr tested against control, Cd treatment tested against Cd + Fe and Cd + Zn, Cr treatment tested against Cr + Fe and Cr + Zn; (paired *T*-test).

Treatment	Mean in %	SD	*T*-test	
Control	97.33	1.16		
600 nM Cd	55.33	8.08	0.0048	**
600 nM Cd + 100 μM Fe	96	2	0.008	**
600 nM Cd + 300 nM Zn	70.66	9.87	0.0145	*
10 μM Cr	75.33	11.55	0.0443	*
10 μM Cr + 100 μM Fe	96	0	0.0451	*
10 μM Cr + 300 nM Zn	90	6	0.0343	*
5 μM Pb	76.37	2.86	0.0051	**

* $p < 0.05\%$.
** $p < 0.01$.

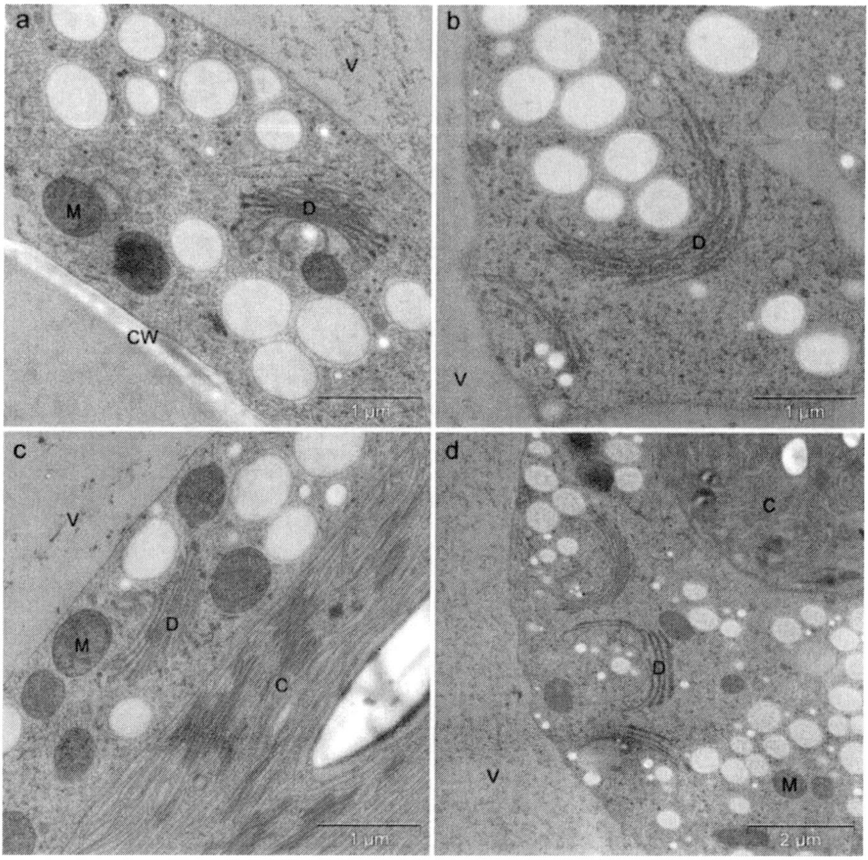

Figure 2. TEM micrographs of *Micrasterias* from control cells (a and b) and lead treated cells (c and d). Control 4 h (a) and 21 d (b) after mitosis. Short-term treatment (4 h) with 40 µM Pb (c) and long-term treatment (21 d) with 5 µM Pb. No ultrastructural differences between controls and treated cells. C chloroplast, CW cell wall, D dictyosome, M mitochondrium, V vacuole.

In concentrations between 5 µM and 150 µM Cd, short-term treatments led to inhibition of cell growth and morphogenesis (Figure 1n), the chloroplast appeared contracted and dramatic changes in cytoplasmic structures were observed (see also Volland et al., 2011; Andosch et al., 2012). Ultrastructural investigations revealed that the cells were strongly vacuolated, dictyosomes were disintegrated and autophagosomes appeared (Andosch et al., 2012). In *Micrasterias* cells treated with 600 nM Cd for 21 days, cell division rates were almost completely inhibited (Figure 5c) and cell vitality dropped to 55.3% of the control level (Table 1). Cells were vacuolated and the chloroplast contracted (Figure 1m). Net-photosynthesis reached negative values, respiration was increased and the electron transport in photosystem II was inhibited, as the flattened fluorescence induction curves ($F_v/F_m = 0.22$) in comparison to the control show ($F_v/F_m = 0.75$; Figure 5a and b).

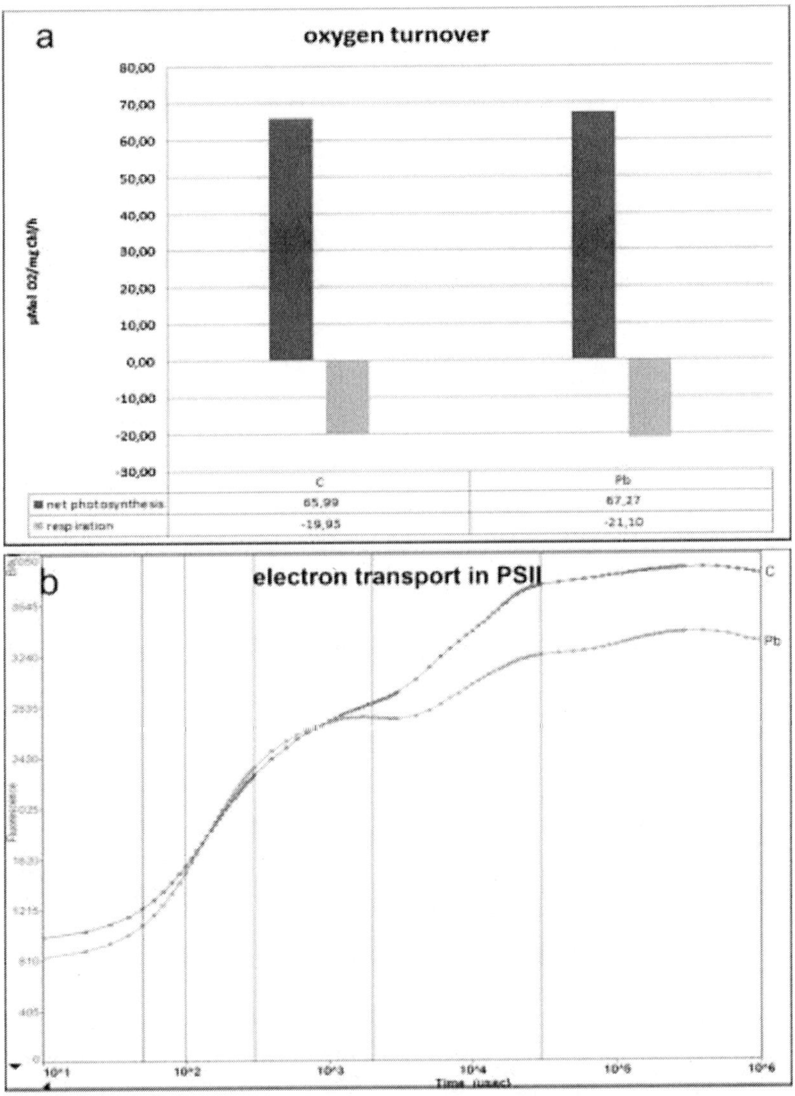

Figure 4. Oxygen turnover (a) and fast electron transport kinetics in PS II (b) after incubation with 5 µM Pb for 21 days.

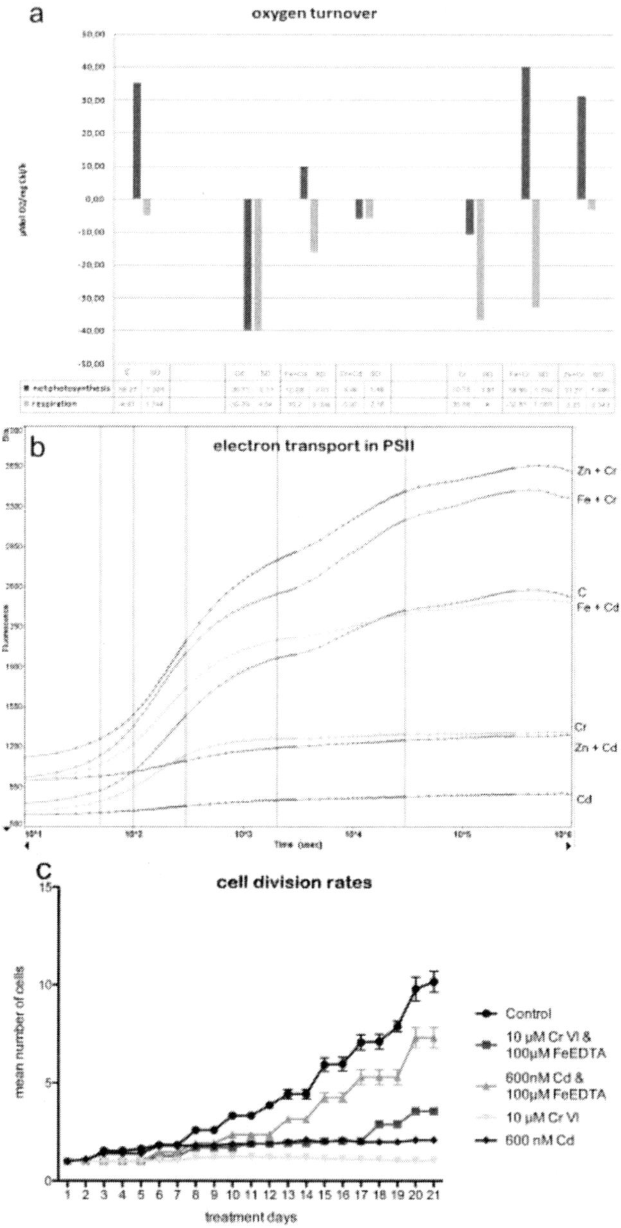

Figure 5. Oxygen turnover (a) and fast electron transport kinetics in PS II (b) after 21 days treatment time with 600 nM Cd and 10 µM Cr, with and without 100 µM Fe-EDTA and 300 nM Zn pre-treatment. Cell division rates over the course of 21 days during Cd and Cr treatment with and without Fe-EDTA pre-treatment (c).

Cr in its hexavalent form was also found to inhibit cell growth and morphogenesis in short-term treated *Micrasterias* cells (Figure 1h; see also Volland et al., 2012). Further, cell division rates were stagnant during long-term treatment (Figure 5c) and the chloroplast appeared dark and contracted while lobe tips were vacuolated (Figure 1g). Net-photosynthesis reached negative levels and the electron transport kinetics is disturbed (F_v/F_m = 0.54; Figure 5a and b). Cell vitality was reduced to 75.33% of that of controls (Table 1). Depositions of Cr in a compound with Fe and O were detected via EELS and ESI in specific, bag-like structures in the inner side of the cell wall during a recent study (Volland et al., 2012). Additionally, ultrastructural alterations were observed in the chloroplast and secretory activity was inhibited.

Zn ameliorates the effects of Cd and Cr

Pre-treatment with 300 nM Zn could not rescue Cd or Cr effects on growth and development in short-term treated cells (Figure 1l and r). In long-term treatments however, Zn pre-treatment affected the general appearance of Cr treated cells positively. Lobe tips were barely vacuolated and the chloroplast was not condensed (Figure 1k). Net-photosynthesis was rescued to a sufficient level for cells to survive and respiration rates which went up during Cr treatment alone were drastically reduced by the pre-incubation with Zn (Figure 5a). On the other hand, pre-treatment with Zn could not ameliorate effects of Cd treatment on general cell appearance (Figure 1q) or on electron transport in PS II (F_v/F_m = 0.31; Figure 5b). Cd treated cells had no functioning electron transport, while 21 days Cr treated cells only showed the first hump of the typical "O-J-I-P" curve (Figure 5b). These findings corroborate the effects previously found by Volland et al. (2012). Cr treated algae cells pre-treated with Zn showed a re-established electron transport chain. The fast kinetics of the electron transport in PS II was rescued although the basic fluorescence (F_0) was elevated (Fm/Fv = 0.71; Figure 5b). Further, Zn had no ameliorating effect on the cell division rates of Cd and Cr treated cells (data not shown), but was able the improve cell vitality from 55.3% to 70.66% in the Cd treatment and from 75.3% to 90% in the Cr treated cell cultures (Table 1).

Fe ameliorates the effects of Cd and Cr

Pre-treatment with 100 μM Fe was able to diminish the inhibiting effects of Cd and Cr on the cell development during short-term treatment and the younger semi-cells were able to develop further than after Cd and Cr treatment alone (Figure 1j and p). The effects were even more distinct after long-term treatment (Figure 1i and o), where the Cd and Cr exposed cells pre-treated with Fe-EDTA did not or barely differ from the appearance of the control cells (Figure 1a). Ultrastructural investigations showed that Fe did not improve the impact of Cr in short-term treatments (data shown in Volland et al., 2012), when compared to the ultrastructure of cells treated with Cr alone. Fe pre-treated long-term Cr exposed cells, however, led to a decreased deposition of electron dense material along the inner side of the cell wall (Figure 3b) in comparison to Cr treatment alone, where pronounced bag-like depositions appeared (data shown in Volland et al., 2012). Interestingly, in these depositions (measurement areas indicated in Figure 3b) only O and Fe were measured at the O–K and the $L_{-2,3}$ edge (Figure 3c and d) and ESI (Figure 3f). Cr could no longer be detected via EELS (Figure 3c) or ESI as indicated by the element distribution map of Cr (Figure 3e). Pre-treatment with Fe was further able to improve net photosynthesis of Cd and Cr treatment, but was not able to diminish respiration down to control levels (Figure 5a). The inhibited electron transport in photosystem II was also rescued by the pre-treatment of Fe in Cd ($F_v/F_m = 0.65$) and Cr ($F_v/F_m = 0.73$) exposed cells (Figure 5b) and cell division rates were improved (Figure 5c). Additionally Fe pre-treatment improved cell vitality from 55.3% to 96% in Cd treated cells and from 75.3% to 96% in Cr treated cell cultures (Table 1).

Ca and Gd ameliorate the effects of Cd and Pb

Dividing *Micrasterias* cells that were treated simultaneously with 40 μM Pb and 1 mM Ca finished their development like controls and no shape alterations were observed (Figure 1d). Only in some cases the lobe tips were slightly and abnormally rounded. A simultaneous exposure to 40 μM Pb and 3 μM Gd did not ameliorate Pb effects. Growth of young half-cells was inhibited and cell deformations were similar to those after treatments with Pb alone (Figure 1e). Treatment with 3 μM Gd for 30 min before exposure to 15 μM $CdSO_4$ for 4 h, however, ameliorated Cd effects (Figure 1f). Development of the young half-cells was almost completed in contrast to 15 μM Cd treated cells (Figure 1n). The chloroplast appeared normal and lobe tips did not seem vacuolated (Figure 1f). These findings are in good agreement with the results of a previous study (Andosch et al., 2012) giving evidence for ameliorating effects of extracellular pre-treatment of Cd exposed *Micrasterias* cells with calcium

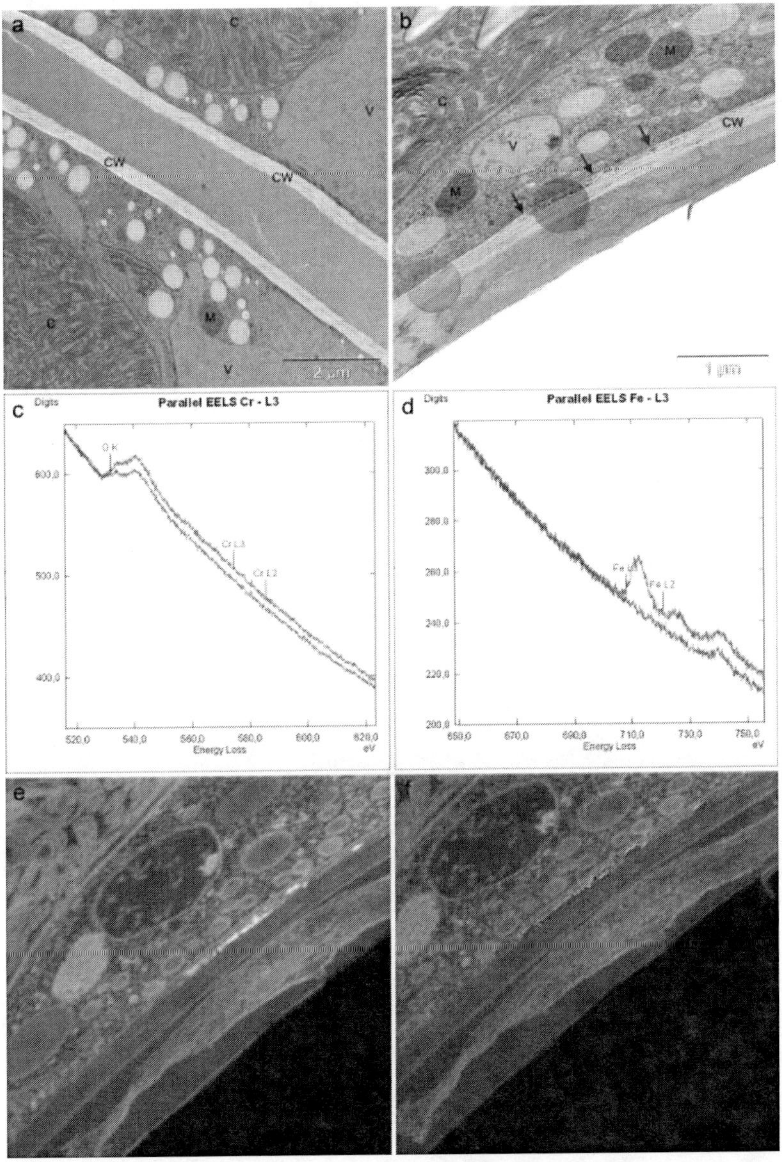

Figure 3. Detail of *Micrasterias* control cell (a) and ultrastructure, EELS and ESI of 10 µM Cr treated cell after 21 days pre-treatment with 100 µM Fe-EDTA (b–f). Arrow points at Cr induced depositions. Areas of EELS measurements indicated: green area without deposition, red area with deposition (b). EELS measurements of Cr-$L_{2,3}$ edge (c), Fe-$L_{2,3}$ edge (d). ESI overlay image indicating Cr distribution (e) and Fe distribution (f) in red. C chloroplast, CW cell wall, M mitochondrium, V vacuole.

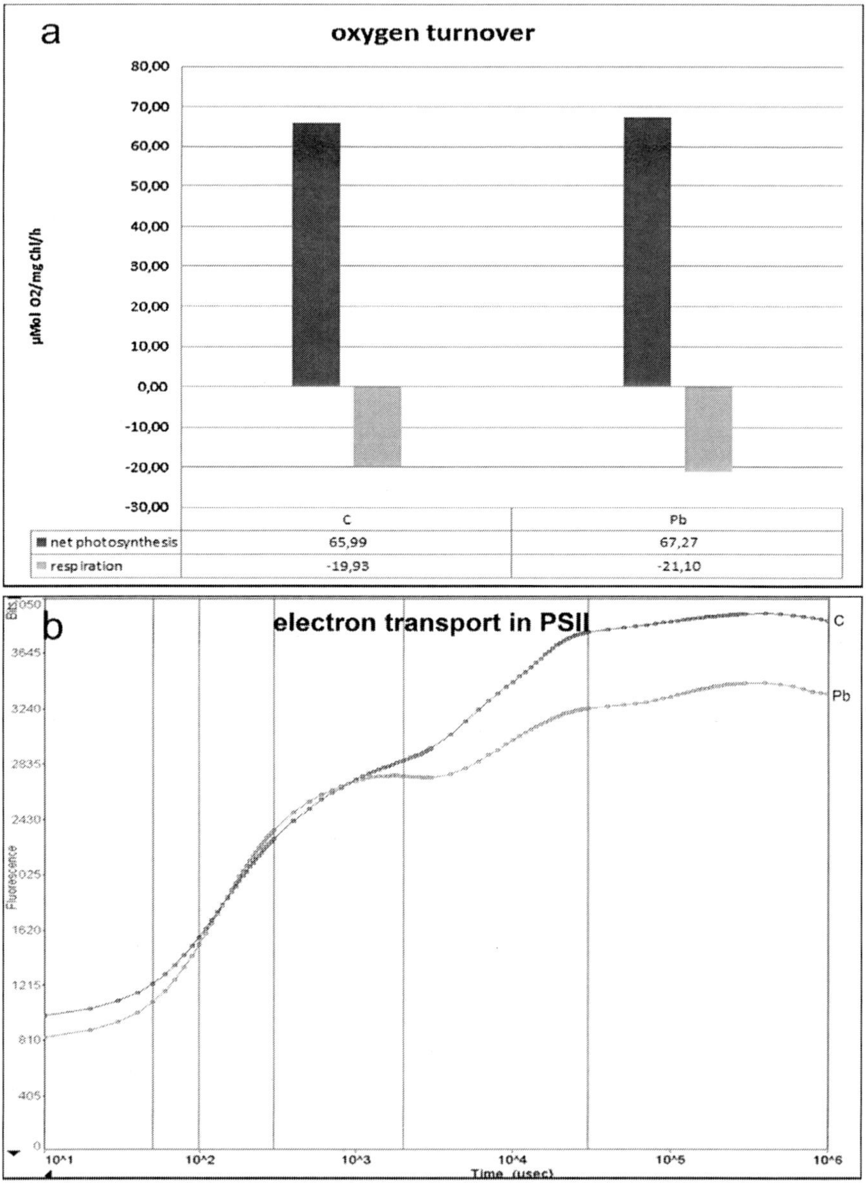

Figure 4. Oxygen turnover (a) and fast electron transport kinetics in PS II (b) after incubation with 5 M Pb for 21 days.

SA, AA and GSH were not able to ameliorate metal effects

In short-term treated Cd cells pre-treated with 300 nM GSH, growth and differentiation was slightly improved when compared to cells treated with 15 μM Cd. AA and SA had no positive effects on growth and development in short-term Cd or Cr treated cells. In long-term metal treated cells SA, AA and GSH did also not exhibit any ameliorating effects. Further cell division rates of metal treated cells pre-treated with SA, AA and GSH did not improve compared to cell division rates from metal only treated cells. Cd and Cr treated cell vitality rates did not change significantly in combined treatment with SA, AA and GSH (data not shown). Additionally, ultrastructural analyses of Cd treated cells pre-treated with SA were performed, but SA could not ameliorate the damaging effects of Cd on ultrastructure in short- or long-term treatments (data not shown).

DISCUSSION

Our results show that depending on the nature of the heavy metal, the divalent ions Ca, Fe, Zn and Gd are able to ameliorate previously reported negative metal impact on cell development, growth, division rates, photosynthesis and ultrastructure of the bog alga *Micrasterias* which grows under extremely nutrient-depleted conditions.

Pre-treatment with Fe was able to diminish the inhibiting effects of Cd and Cr on *Micrasterias* which were reported in previous studies (Volland et al., 2011, 2012; Andosch et al., 2012). Cell morphogenesis, photosynthesis, cell division rates and the overall appearance of the chloroplast in the algae were improved distinctly by Fe after long-term treatments. Zn on the other hand, only improved negative effects on photosynthesis and the general appearance of *Micrasterias* cells after long-term Cr treatment, but did not positively influence any other aspects of Cr, nor Cd toxicity. The powerful ameliorating effects of Zn and Fe seem to mainly derive from their chemical similarity to Cd and Cr (Shanker et al., 2005; Verbruggen et al., 2009). Cd has frequently been found to be taken up via Fe, Zn and Ca transporters in higher plants (DalCorso et al., 2008), whereas Cr uptake is known to compete with Fe, S and P for carrier binding (Shanker et al., 2005). These findings agree with our study concerning the ameliorating effects of Fe on Cd and Cr toxicity in *Micrasterias*. On the contrary, our results suggest that Cd uptake is not mediated through Zn transporters as in higher plants. Instead, Cr seems to additionally enter the algae cell via Zn specific carriers. Increased availability of the essential micronutrients Fe and Zn might not only have led to a decrease of heavy metal uptake, by competing for carrier binding, but might also have protected essential enzymes and proteins by

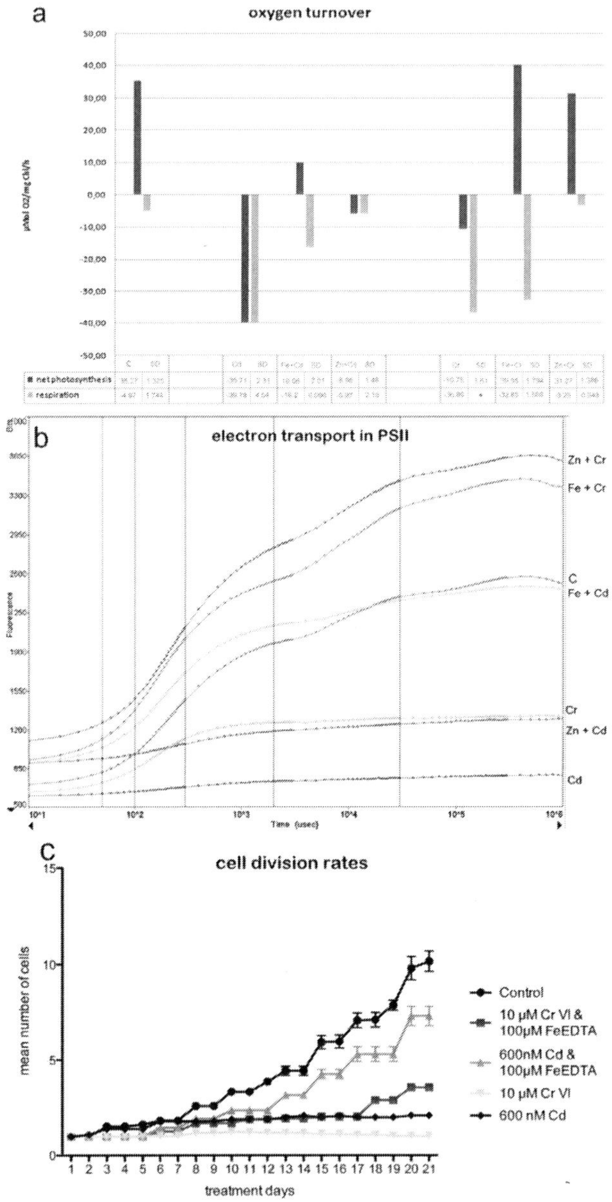

Figure 5. Oxygen turnover (a) and fast electron transport kinetics in PS II (b) after 21 days treatment time with 600 nM Cd and 10 M Cr, with and without 100 M FeEDTA and 300 nM Zn pre-treatment. Cell division rates over the course of 21 days during Cd and Cr treatment with and without Fe-EDTA pre-treatment (c).

their increased abundance within *Micrasterias* cells. Displacement of functionally active ions like Fe and Zn from enzymes and other proteins by heavy metals has been reported as another important reason for metal toxicity (di Toppi and Gabbrielli, 1999; DalCorso et al., 2008).

Zn has been found to completely inhibit Cd uptake in *Euglena gracilis* (Mendoza-Cozatl and Moreno-Sanchez, 2005) but cannot influence Cd effects in *Micrasterias*. This suggests different uptake mechanisms for Cd in these two algae. In contrast, increased concentrations of Fe have been frequently found to lead to reduced Cd uptake and toxicity in higher plants (Liu et al., 2008; Peralta-Videa et al., 2009; Mallick et al., 2010) and were also able to ameliorate the negative impact of Cd and Cr in *Micrasterias*. This points toward Cd and Cr uptake via Fe transporters in both cases. Zn also rescued the effects of Cr on photosynthesis, respiration and the electron transport in PS II. Thus, aside from a likely decreased uptake of heavy metals by excess amounts of Fe and Zn, these micronutrients also appear to have a protective or restoring function on photosynthetic activity in *Micrasterias*. This is possibly due to the fact that some enzymes in photosynthesis are Zn and Fe dependent like for example RuBisCO (Ribulose-1,5-bisphosphate carboxylase oxygenase) which is stabilized by Zn and catalyses the first step of carbon fixation (Stiborová et al., 1987; Curie and Briat, 2003).

The ultrastructure of Cr treated *Micrasterias* cells pre-treated with Fe only revealed ameliorating effects after long-term treatment. The most striking effect found was that dark granular precipitation contained in vesicle-like accumulations on the inner side of the cell wall after Cr treatment alone were reduced or disappeared completely when cells were pre-treated with Fe. Particular bag-like structures forming during Cr treatment alone (Volland et al., 2012) could no longer be observed after Fe pre-treatment. As identified by EELS and ESI, Cr exposed cells contained precipitations of Cr, Fe and O (Volland et al., 2012). In cells that have been exposed to Fe prior to Cr treatment these precipitations contained only Fe and O, whereas Cr could no longer be measured, indicating that less or no Cr was taken up in *Micrasterias* during combined metal treatment. This again confirmed our previous finding that a substantial amount of Cr enters the green algal cell through Fe transporters.

Pb causes shape alterations and cell death in developing *Micrasterias* cells but electron microscopic studies could not identify any structural changes in the cytoplasm, suggesting that Pb does not enter the cells. This is in agreement with the findings of Meindl and Röderer (1990), who discovered that $PbCl_2$ induces cell death by bursting, but does not affect the ultrastructure of *Micrasterias*. However, these results do not correspond to studies in other plants where Pb has been frequently shown to be taken up into the cells and to influence intracellular components (e.g. Eun et al.,

2000; Sacan et al., 2007; Kopittke et al., 2008; Jiang and Liu, 2010). Ca is able to compensate the impact of Pb on short-term treated *Micrasterias* cells, while Gd, a well-known Ca^{2+} channel blocker (Caldwell et al., 1998) has no ameliorative effects. This indicates that Ca^{2+} channels are not involved in the uptake and toxicity mechanisms of Pb in *Micrasterias*, even though they have been described as possible influx sites for Pb ions in other cells (Pourrut et al., 2011). We therefore suggest that disturbed cell shapes *in Micrasterias* after Pb exposure are probably due to an exchange of cell wall bound Ca^{2+} by Pb^{2+}. Low-methyl-esterified pectins, components of the primary wall of *Micrasterias* are linked via Ca^{2+} ions (Eder and Lütz-Meindl, 2008), which can be replaced by di- or trivalent ions exhibiting a higher affinity for pectins, like for example Pb^{2+} (Krzeslowska, 2011). A subsequent change of cell wall plasticity during development could be responsible for the disturbed formation of the cell pattern, resulting in the bursting of cells. This competition of Ca^{2+} and Pb^{2+} for binding sites in the cell wall (Krzeslowska, 2011) could explain why Pb effects are diminished in the presence of high Ca concentrations in *Micrasterias* cells. Ameliorating effects of Ca on Pb toxicity have also been reported in higher plants like barley and *Festuca ovina* (Garland and Wilkins, 1981), in moss protonemata (Krzeslowska et al., 2004), maize, rye, tomato and mustard (Antosiewicz, 2005). Slaveykova and Wilkinson (2002) discovered that Pb uptake of the green alga *Chlorella vulgaris* was decreased in the presence of elevated Ca concentrations. There seems to be a connection between Pb effects and Ca throughout the plant kingdom, whereas the different underlying mechanisms are still not fully understood.

Previous experiments showed that the effects of 150 μM Cd on short-term treated *Micrasterias* cells were reversible upon pre-treatment with 2 mM $CaSO_4$. Especially the detrimental Cd effects on photosynthesis, promotion of autophagy and in parts also the negative degradative effects on dictyosomal ultrastructure were rescued (Andosch et al., 2012). Cd has been reported to have a primary impact on Ca homeostasis, since it is able to displace Ca from binding sites and further shows a strong interference with the movements of K^+ and Ca^{2+} in the cells (di Toppi and Gabbrielli, 1999). In the present study the Ca^{2+} channel blocker Gd was able to almost completely reverse the inhibiting effects of 15 μM Cd on developing *Micrasterias* cells. This indicates that Cd ions are at least partly taken up via Ca^{2+}channels, in contrast to Pb ions (as discussed above). Uptake mechanisms of Cd can differ in plant cells but have been reported to be partly mediated through Ca^{2+}channels (Mendoza-Cozatl and Moreno-Sanchez, 2005). Our study provides first evidence of a decreased uptake of Cd in the presence of Gd in plant cells, as to our knowledge the latter has so far only been employed in animal cells in this respect (Hinkle et al., 1987).

Rescue experiments by extracellular application of SA, AA and GSH were not successful concerning any aspect of Cd or Cr toxicity in *Micrasterias*, only GSH was able to slightly improve cell morphogenesis under Cd stress. To explain why these antioxidants were not able to significantly ameliorate heavy metal effects in *Micrasterias* is difficult. Heavy metals, especially Cr, are known to induce increased ROS production in general (Pandey et al., 2009) and also in *Micrasterias* (Volland et al., 2012) and a strong antioxidative defence and strong redox homeostasis has been linked to metal tolerance (Sharma and Dietz, 2008). GSH and ascorbate accumulate up to mM concentrations in stressed plant cells and play a crucial role in the defence against oxidative damage caused by ROS (Noctor et al., 2012). In the defence against heavy metal effects GSH has another important part as metal ligand and precursor for phytochelatins, commonly found during Cd detoxification in plants (Cobbett, 2000). In *Micrasterias* the phytochelatins PC_{2-4} have been identified after Cd exposure by means of HPLC and UPLC-MS (Volland et al., 2013). Nevertheless AA and GSH had no effect on Cd or Cr toxicity in *Micrasterias* and were probably either altered before entering the algae cells or not taken up at all. SA, though known to alleviate Cd induced growth inhibition in different plants by enhancing the oxidative defence mechanisms (Guo et al., 2009), did not ameliorate any of the heavy metal effects in *Micrasterias.*

In this study we were able to show that adding essential cations like Fe, Zn and Ca to the nutrient solution was able to diminish the effects of the heavy metals like Cd, Cr and Pb on cell development and morphology, ultrastructure, cell division rates and photosynthesis in *Micrasterias.* Our results further revealed possible uptake mechanisms of the investigated metals: Cd seems to be taken up by Ca and Fe transporters, Cr enters the algae cells via Fe and Zn transporters and Pb is not taken up at all by *Micrasterias* but exerts its negative effects via physical changes of the cell wall. Externally added antioxidants SA, AA and GSH were not able to ameliorate heavy metal effects on the green alga *Micrasterias* at all.

ACKNOWLEDGEMENT

We acknowledge the financial support of this study by the Austrian Science Fund (FWF Project 21035-B16 to U. L.-M.).

REFERENCES

1. Affenzeller MJ, Darehshouri A, Andosch A, Lütz C, Lütz-Meindl U. Salt stressinduced cell death in the unicellular green alga Micrasterias denticulata. J Exp Bot 2009;60(3):939–54.
2. Ahmad P, Nabi G, Ashraf M. Cadmium-induced oxidative damage in mustard [Brassica juncea (L.) Czern, & Coss.] plants can be alleviated by salicylic acid. S Afr J Bot 2011;77(1):36–44.
3. Aichinger N, Lütz-Meindl U. Use of energy-filtering transmission electron microscopy for routine ultrastructural analysis of high-pressure-frozen or chemically fixed plant cells. Protoplasma 2004;223(3):155–62.
4. Andosch A, Affenzeller MJ, Lütz C, Lütz-Meindl U. A freshwater green alga under cadmium stress: ameliorating calcium effects on ultrastructure and photosynthesis in the unicellular model Micrasterias. J Plant Physiol 2012;169:1489–500.
5. Antosiewicz DM. Study of calcium-dependent lead-tolerance on plants differing in their level of Ca-deficiency tolerance. Environ Pollut 2005;134(1): 23–34.
6. Belkhadi A, Hediji J, Abbes Z, Nouairi I, Barhoumi Z, Zarrouk M, et al. Effects of exogenous salicylic acid pre-treatment on cadmium toxicity and leaf lipid content in Linum usitatissimum L. Ecotoxicol Environ Saf 2010;73(5):1004–11.
7. Bielen A, Remans T, Vangronsveld J, Cuypers A. The influence of metal stress on the availability and redox state of ascorbate, and possible interference with its cellular functions. Int J Mol Sci 2013;14(3):6382–413.
8. Branzini A, Gonzalez RS, Zubillaga M. Absorption and translocation of copper, zinc and chromium by Sesbania virgata. J Environ Manage 2012;102:50–4.
9. Caldwell RA, Clemo HF, Baumgarten CM. Using gadolinium to identify stretch-activated channels: technical considerations. Am J Physiol 1998;275(2):C619–21.
10. Choi YE, Harada E, Wada M, Tsuboi H, Morita Y, Kusano T, et al. Detoxification of cadmium in tobacco plants: formation and excretion of crystals containing cadmium and calcium through trichoms. Planta 2001;213:45–50.
11. Clemens S. Molecular mechanisms of plant metal tolerance and homeostasis. Planta 2001;212(4):475–86.
12. Cobbett CS. Phytochelatins and their roles in heavy metal detoxification. Plant Physiol 2000;123(3):825–32.
13. Curie C, Briat JF. Ion transport and signaling in plants. Annu Rev Plant Biol 2003;54:183–206.
14. DalCorso G, Farinati S, Maistri S, Furini A. How plants cope with cadmium: staking all on metabolism and gene expression. J Integr Plant Biol 2008;50(10): 1268–80.
15. di Toppi LS, Gabbrielli R. Response to cadmium in higher plants. Environ Exp Bot 1999;41(2):105–30.
16. Eder M, Lütz-Meindl U. Pectin-like carbohydrates in the green alga Micrasterias characterized by cytochemical analysis and energy filtering TEM. J Microsc 2008;231(6):201–14.
17. El-Naggar AH, El-Sheekh MM. Abolishing cadmium toxicity in Chlorella vulgaris by ascorbic acid, calcium, glucose and reduced glutathione. Environ Pollut 1998;101(2):169–74.
18. Eun SO, Youn HS, Lee Y. Lead disturbs microtubule organization in the root meristem of Zea mays. Physiol Plant 2000;110(3):357–65.
19. Garland CJ, Wilkins DA. Effect of calcium on the uptake and toxicity of lead in Hordeum vulgare L. and Festuca ovina L. New Phytol 1981;87(3): 581–93.
20. Guo B, Liang Y, Zhu Y. Does salicylic acid regulate antioxidant defense system, cell death, cadmium uptake and partitioning to acquire cadmium tolerance in rice? J Plant Physiol 2009;166(1):20–31.
21. Hinkle PM, Kinsella PA, Osterhoudt KC. Cadmium uptake and toxicity via voltagesensitive calcium channels. J Biol Chem 1987;262(34):16333–7.

22. Jiang WS, Liu DH. Pb-induced cellular defense system in the root meristematic cells of Allium sativum L. BMC Plant Biol 2010;10:40.
23. Kopittke PM, Asher CJ, Blamey FP, Auchterlonie GJ, Guo YN, Menzies NW. Localization and chemical speciation of Pb in roots of signal grass (Brachiaria decumbens) and Rhodes grass (Chloris gayana). Environ Sci Technol 2008;42(1): 4595–9.
24. Kovacik J, Klejdus B, Hedbavny J, Backor M. Effect of copper and salicylic acid on phenolic metabolites and free amino acids in Scenedesmus quadricauda (Chlorophyceae). Plant Sci 2010;178(3):307–11.
25. Krzeslowska M. The cell wall in plant cell response to trace metals: polysaccharide remodeling and its role in defense strategy. Acta Physiol Plant 2011;33:35–51.
26. Krzeslowska M, Wozny A, Konieczna-Koperska J. Calcium ameliorates effects of lead in protonema of Funaria hygrometrica Hedw. Biol Plant 2004;48(4): 569–74.
27. Liu HJ, Zhang JL, Christie P, Zhang FS. Influence of iron plaque on uptake and accumulation of Cd by rice (Oryza sativa L.) seedlings grown in soil. Sci Total Environ 2008;394(2–3):361–8.
28. Mallick S, Sinam G, Kumar Mishra R, Sinha S. Interactive effects of Cr and Fe treatments on plants growth, nutrition and oxidative status in Zea mays L. Ecotoxicol Environ Saf 2010;73(5):987–95.
29. Meindl U. Micrasterias cells as amodel systemfor researchonmor phogenesis.Microbiol Rev 1993;57(2):415–33.
30. Meindl U, Lütz C. Effects of UV irradiation on cell development and ultrastructure of the green alga Micrasterias. J Photochem Photobiol B 1996;36:285–92.
31. Meindl U, Röderer G. Influence of inorganic and triethyl lead on nuclear migration and ultrastructure of Micrasterias. Ecotoxicol Environ Saf 1990;19(2):192–203.
32. Meindl U, Lancelle S, Hepler PK. Vesicle production and fusion during lobe formation in Micrasterias visualized by high-pressure freeze fixation. Protoplasma 1992;170(3–4):88–97.
33. Mendoza-Cozatl DG, Moreno-Sanchez R. Cd2+ transport and storage in the chloroplast of Euglena gracilis. Biochim Biophys Acta 2005;1706(1–2):88–97.
34. Metraux JP. Recent breakthroughs in the study of salicylic acid biosynthesis. Trends Plant Sci 2002;7(8):332–4.
35. Noctor G, Mhamdi A, Chaouche S, Han Y, Neukermans J, Marquez-Garcia B, et al. Glutathione in plants: an integrated overview. Plant Cell Environ 2012;35(2):454–84.
36. Oertel A, Aichinger N, Hochreiter R, Thalhamer J, Lütz-Meindl U. Analysis of mucilage secretion and excretion in Micrasterias (Chlorophyta) by means of immunoelectron microscopy and digital time lapse video microscopy. J Phycol 2004;40(4):711–20.
37. Panda SK, Choudhury S. Chromium stress in plants. Braz J Plant Physiol 2005;17(1):95–102.
38. Pandey V, Dixit V, Shyam R. Chromium effect on ROS generation and detoxification in pea (Pisum sativum) leaf chloroplasts. Protoplasma 2009;236(1–4):85–95.
39. Peralta-Videa JR, Lopez ML, Narayan M, Saupe G, Gardea-Torresdey J. The biochemistry of environmental heavy metal uptake by plants: implications for the food chain. Int J Biochem Cell Biol 2009;41(8–9):1665–77.
40. Pinto E, Sigaud-Kutner TCS, Leitao MAS, Okamoto OK, Morse D, Colepicolo P. Heavy metal-induced oxidative stress in algae. J Phycol 2003;39(6):1008–18.
41. Porra RJ, Thompson WA, Kreidemann PE. Determination of accurate extinction coefficients and simultaneous equations for assaying chlorophyll a and b extracted with four different solvents: verification of the concentration of chlorophyll standards by absorption spectroscopy. Biochim Biophys Acta 1989;975:384–94.
42. Pourrut B, Shahid M, Dumat C, Winterton P, Pinelli E. Lead uptake, toxicity, and detoxification in plants. Rev Environ Contam Toxicol 2011;213:113–36.

43. Sacan MT, Oztay F, Bolkent S. Exposure of Dunaliella tertiolecta to lead and aluminum: toxicity and effects on ultrastructure. Biol Trace Elem Res 2007;120(1-3):264–72.
44. Schlösser UG. Sammlungen von Algenkulturen. Ber Deutsch Bot Ges 1982;95:181-276.
45. Shah J. The salicylic acid loop in plant defense. Curr Opin Plant Biol 2003;6(4):365-71.
46. Shanker AK, Cervantes C, Loza-Tavera H, Avudainayagam S. Chromium toxicity in plants. Environ Int 2005;31(5):739–53.
47. Shanmugam V, Lo JC, Wu CL, Wang SL, Lai CC, Connolly EL, et al. Differential expression and regulation of iron-regulated metal transporters in Arabidopsis halleri and Arabidopsis thaliana - the role in zinc tolerance. New Phytol 2011;190(1):125–37.
48. Sharma SS, Dietz KJ. The relationship between metal toxicity and cellular redox imbalance. Trends Plant Sci 2008;14(1):43–50.
49. Slaveykova VI, Wilkinson KJ. Physicochemical aspects of lead bioaccumulation by Chlorella vulgaris. Environ Sci Technol 2002;36(5):969–75.
50. Stiborová M, Ditrichová M, Brenzinová A. Effect of heavy metal ions on growth and biochemical characteristics of photosynthesis of barley and maize seedlings. Biol Plant 1987;29(6):453–67.
51. Strasser R, Srivasatava J, Govindjee A. Polyphasic chlorophyll a fluorescence transient in plants and cyanobacteria. Photochem Photobiol 1995;61(1):32–42.
52. Suzuki N. Alleviation by calcium of cadmium-induced root growth inhibition in Arabidopsis seedlings. Plant Biotechnol 2005;22:19–25.
53. Verbruggen N, Hermans C, Schat H. Mechanisms to cope with arsenic or cadmium excess in plants. Curr Opin Plant Biol 2009;12(3):364–72.
54. Volland S, Andosch A, Milla M, Stöger B, Lütz C, Lütz-Meindl U. Intracellular metal compartmentalization in the green algal model system Micrasterias denticulata (Streptophyta) measured by transmission electron microscopy-coupled electron energy loss spectroscopy. J Phycol 2011;47(3):565–79.
55. Volland S, Lütz C, Michalke B, Lütz-Meindl U. Intracellular chromium localization and cell physiological response in the unicellular alga Micrasterias. Aquat Toxicol 2012;109:59–69.
56. Volland S, Schaumlöffel D, Wesenberg D, Krauss G-J, Lütz-Meindl U. Identification of phytochelatins in the cadmium-stressed conjugating green alga Micrasterias denticulata. Chemosphere 2013;91(4):448–54.
57. Wan G, Najeeb U, Jilani G, Naeem M, Zhou W. Caclium invigorates the cadmiumstressed Brassica napus L. plants by strengthening their photosynthetic system. Environ Sci Pollut Res Int 2011;18:1478–86.
58. Wodniok S, Brinkmann H, Glöckner G, Heidel AJ, Philippe H, Melkonian M, et al. Origin of land plants: do conjugating green algae hold the key? BMC Evol Biol 2011;11:104. Yadav SK. Heavy metals toxicity in plants: an overview on the role of glutathione and phytochelatins in heavy metal stress tolerance of plants. S Afr J Bot 2010;76(2):167–79. Zechmann B, Mauch F, Sticher L, Müller M. Subcellular immunocytochemical analysis detects the highest concentrations of glutathione in mitochondria and not in plastids. J Exp Bot 2008;59(14):4017–27. Zheng LJ, Peer T, Seybold V, Lütz-Meindl U. Pb-induced ultrastructural alterations and subcellular localization of Pb in two species of Lespedeza by TEM-coupled electron energy loss spectroscopy. Environ Exp Bot 2012;77:196–206.

CHAPTER 14

Protocol Optimization for Enhanced Production of Pigments in *Spirulina*

Devendra Kumar[1], Neeraj Kumar[2], Sunil Pabbi[1], Suresh Walia[3], Dolly Wattal Dhar[1]

1.Centre for Conservation and Utilisation of Blue Green AlgaeIndian Agricultural Research InstituteNew DelhiIndia

2.Department of MicrobiologyKurukshetra UniversityKurukshetraIndia

3.Division of Agricultural ChemicalsIndian Agricultural Research InstituteNew DelhiIndia

ABSTRACT

Spirulina has attracted special attention due to its importance as human foodstuff and natural colours with specific functional properties. These functional properties have been attributed to phycobilins, carotenoids, phenolics and unsaturated fatty acids. Present study was conducted under controlled phytotron conditions to identify the efficient strains of *Spirulina* in terms of pigment synthesis and to optimize their enhanced production. Methodology for enhanced production was standardized by varying specific environmental parameters (light intensity, temperature, carbon dioxide concentration, pH and NaCl level). Different strains of *Spirulina* depicted variability and environmental parameters showed distinct influence on pigments. Growth and pigment production was recorded to be most efficient under optimized conditions of light intensity (70 μmol m^{-2} s^{-1}), temperature (30 °C), CO_2 concentration (550 ppm and 750 ppm), pH (10.5) and NaCl level (2 g L^{-1}).

Keywords: Mass production Pigments *Spirulina*

Spirulina, a cyanobacterial genus has been reported to be the complete organic food source and contains high proteins with well balanced amino acids. It is also rich in carbohydrates, vitamins, minerals, phenolics, pigments (chlorophyll, carotenoids and phycobilins) and poly-unsaturated fatty acids (Miranda et al. 1998; Anupama 2000). It is successfully used in aquaculture and poultry industries (Belay et al. 1993; Estrada et al. 2001). This genus is reported to be photosynthetic, filamentous in an open left-hand helix, with predominance from aquatic habitats having high levels of carbonates, bicarbonates and alkaline pH (up to 11). The large scale production of *Spirulina* depends on nutrient availability, temperature, light, pH and CO_2 concentration. In view of this, a study was conducted to optimize the protocol for enhanced production of pigments in *Spirulina*. Ten strains procured from the culture collection of Centre for Conservation and Utilisation of Blue-Green Algae (CCUBGA), IARI, New Delhi were grown in chemically defined Z-medium, which is rich in carbonate and thus, has successively served as a common culture medium (Zarrouk 1966). *Spirulina* strains were grown under diffused light intensity of 55 μmol m^{-2} s^{-1} at 28 °C ± 2 °C temperature with 16:8 h light: dark cycle. Initial biomass of 0.1 g L^{-1} (Costa et al. 2000) was taken as inoculum in 100 mL medium in conical flasks. The influence of different pH (9.5, 10.5, 11.5, 12.5, 13.5) and NaCl concentrations (0, 0.5, 1.0, 1.5, 2.0 g L^{-1}) was studied on the pigment production under culture room facility. Studies were also conducted under varying temperature (25, 30 and 35 °C), light intensity (55, 70 and 85 μmol m^{-2} s^{-1}) and CO_2 concentration (350, 550 and 750 ppm) in growth chambers of National Phytotron Facility, IARI. Homogenized suspension was drawn at 15th day of incubation for the estimation of chlorophyll (Lichtenthaler 1987), carotenoids (Jensen 1978) and phycobiliproteins (Bennett and Bogorad 1993).

Most of the strains exhibited uniform suspension in the medium and some showed planktonic growth behaviour. Difference in the thallus colour observed is reported to be under the control of chlorophyll to phycobilin ratio. Pigments can also be subjected to oxidation by molecular oxygen, which is known to be a potent inhibitor of microbial processes including photosynthesis (Paerl and Pinckney 1996). The environmental variables can influence growth and pigment production which in turn may change the composition of the biomass by affecting the metabolism. Varied light intensity exhibited differential influence on the accumulation of chlorophyll, carotenoids and phycobilins. Results indicated that a light intensity of 55 μmol m^{-2} s^{-1} was optimum for maximum chlorophyll and that of 85 μmol m^{-2} s^{-1} was inhibitory in most cases, except in case of

CCC538. Variation in the temperature influenced the pigment profile with 35 °C being the optimum for enhanced chlorophyll. Danesi et al. (2001) reported highest biomass production at 30 °C temperature. High temperatures are the key factors for large scale *Spirulina* cultures outdoors with the optimal temperature reported in range of 35–38 °C (Zarrouk 1966; Hu 2004). The temperature below 20 °C and above 40 °C retarded the growth considerably (Borowitzka and Borowitzka 1988). The highest biomass at 30 °C than at 35 °C may be due to higher partial pressure of CO_2 in the medium leading to enhanced concentration of bicarbonates with consequent increase in the rate of photosynthesis (Colla et al. 2004). Further, a longer dark cycle respiratory activity in which cells use reserve material for respiration accordingly can decrease the cell weight (Vonshak et al. 1982). In general, chlorophyll content enhanced with increase in carbon dioxide concentration and a concentration of 750 ppm was optimum. Effect of pH was distinct with 10.5 pH being the optimum for enhanced chlorophyll production as has also been reported by Richmond and Grobbelaar (1986). *Spirulina* requires relatively high pH between 9.5 and 9.8 (Hu 2004), which inhibit the contamination by most algae in the culture. Therefore, high amounts of sodium bicarbonate is essential in the culture medium to sustain high pH. High alkalinity is mandatory for the growth and bicarbonate is used to maintain the high pH (Vonshak and Tomaselli 2000). The significant pigment production may be the result of an increase in pH due to the formation of a $CO_2/H_2CO_3/HCO^{3-}/CO_3^{-2}$ system, which may function as useful buffer system for maintaining the alkaline pH. This is important for the optimum growth and helps to prevent the carbon depletion (Richmond 1990; Vonshak 1997). Sodium chloride concentration of 2 g L^{-1} depicted highest chlorophyll, however, several species of *Spirulina* exhibit inhibition of growth in hostile environments of saline lakes (Henrikson 1994). Exposure of *S. platensis* to an enhanced sodium concentration resulted in a 30 % increase in intracellular accumulation of Na^+ accompanied by small changes in dry mass and chlorophyll content (Kavita and Mohanty 2000) (Table 1). Carotenoids and phycobiliproteins were highest at 70 µmol m^{-2} s^{-1} light intensity and 10.5 pH was optimum for higher carotenoids and phycobiliproteins. In conventional cultivation, light absorption is proportional to the phycocyanin and chlorophyll content of the cells and a slight reduction in phycocyanin may be accompanied by an increase in chlorophyll (Cornet et al. 1992). Carotenoids were maximum at 35 °C temperature and enhanced CO_2 resulted in their increase with 750 ppm to be an optimum concentration. Sodium chloride concentration of 1.5 g L^{-1} showed highest carotenoids. Phycobiliproteins enhanced significantly at 35 °C in some cultures, whereas decline was observed in others. Enhancement in CO_2 concentration increased the phycobiliproteins, and sodium chloride concentration of 2 g L^{-1} was most suitable for phycobiliprotein production (Tables 2, 3). Study clearly showed that the pigments from *Spirulina*, which have the potential applications in biotechnology can be enhanced by manipulating cultural conditions and environmental variables. Standardized protocol for enhanced production of pigments involved an optimized light intensity: 70 μmol m^{-2} s^{-1} for chlorophyll and carotenoids, and 85 μmol m^{-2} s^{-1} for phycobiliproteins, temperature: 35 °C for chlorophyll and phycobiliproteins, and 25 °C for carotenoids, CO_2 concentration: 500 ppm and 750 ppm, pH: 10.5 and NaCl concentration: 2 g L^{-1}.

Table 1. Influence of environmental variables on chlorophyll (mg g^{-1} dry weight) content of *Spirulina* strains

Environmental variables	*Spirulina* strains[a]										SE (m)±	CD (P = 0.05)
	CCC477	CCC478	CCC479	CCC480	CCC481	CCC482	CCC483	CCC538	CCC539	CCC540		
Light (μmol m^{-2} s^{-1})												
55	1.550	1.934	1.795	2.611	1.445	1.781	3.188	1.725	2.148	2.553	0.20	0.41
70	1.192	0.924	0.611	1.359	1.245	2.341	2.783	2.156	1.808	2.358	0.23	0.47
85	1.014	1.250	1.143	1.304	1.227	1.620	1.298	2.134	1.548	1.991	0.14	0.29
Temperature (°C)												
25	0.408	0.587	0.907	1.513	1.147	1.626	1.293	1.650	0.919	1.693	0.16	0.33
30	0.640	0.859	1.103	1.133	1.294	0.932	1.416	1.388	1.157	3.252	1.29	2.63
35	1.416	0.992	1.542	1.848	1.444	2.238	1.916	2.992	1.875	1.804	0.24	0.50
CO_2 concentration (ppm)												
350	2.547	2.648	1.257	2.008	0.973	0.665	2.324	1.314	1.235	0.858	0.27	0.56
550	0.745	0.979	1.225	1.052	1.339	2.176	2.576	2.150	2.074	2.160	0.24	0.50
750	3.163	4.555	4.035	4.320	2.952	5.961	1.140	3.861	3.056	4.466	0.40	0.82
pH												
9.5	1.632	1.207	1.134	2.062	1.365	0.762	1.019	1.352	1.617	0.953	0.27	0.56
10.5	1.752	2.783	1.579	4.412	2.641	2.466	2.206	2.593	1.527	2.141	0.32	0.66
11.5	1.648	1.787	0.959	1.334	1.552	0.731	0.530	0.459	0.486	0.611	0.07	0.15
12.5	1.369	1.902	0.447	0.965	0.771	0.268	0.579	0.359	0.529	0.193	0.11	0.24
13.5	0.318	1.859	0.993	0.421	1.285	0.442	0.114	0.153	0.187	0.198	0.22	0.45
NaCl concentration (g L^{-1})												
0	0.620	0.872	1.243	0.563	0.661	0.531	0.234	0.638	0.566	0.877	0.02	0.05
0.5	0.981	0.457	0.936	0.782	1.835	0.657	0.373	0.668	0.764	2.065	0.04	0.08
1	1.371	1.749	0.357	0.702	0.349	1.860	0.468	0.749	0.463	1.662	0.01	0.02
1.5	0.830	1.348	0.574	0.878	0.654	1.842	0.665	0.645	0.848	1.325	0.02	0.05
2	2.252	1.247	1.562	1.137	1.374	0.481	0.648	0.647	0.704	1.638	0.02	0.04

[a] CCC477, CCC478, CCC479, CCC480 *Spirulina platensis*, CCC481 *Spirulina maxima*, CCC482 *Spirulina lonar*, CCC483 *Spirulina platensis* (mutant), CCC538, CCC539 *Arthrospira* sp., CCC540 *Spirulina* sp.

Table 2. Influence of environmental variables on carotenoids (mg g^{-1} dry weight) content of *Spirulina* strains

Environmental variables	*Spirulina* strains[a]										SE (m)±	CD ($P = 0.05$)
	CCC477	CCC478	CCC479	CCC480	CCC481	CCC482	CCC483	CCC538	CCC539	CCC540		
Light (μmol m^{-2} s^{-1})												
55	0.243	0.306	0.257	0.287	0.141	0.193	0.268	0.133	0.222	0.266	0.01	0.03
70	0.147	0.144	0.094	0.228	0.151	0.307	0.307	0.233	0.204	0.271	0.02	0.05
84	0.250	0.246	0.202	0.179	0.106	0.231	0.193	0.170	0.202	0.169	0.02	0.04
Temperature (°C)												
25	0.123	0.073	0.157	0.301	0.190	0.235	0.231	0.184	0.252	0.228	0.02	0.05
30	0.149	0.285	0.184	0.186	0.143	0.176	0.146	0.173	0.179	0.208	0.02	0.05
35	0.206	0.224	0.280	0.148	0.166	0.184	0.129	0.181	0.149	0.158	0.02	0.05
CO_2 concentration (ppm)												
350	0.133	0.244	0.180	0.290	0.188	0.216	0.235	0.158	0.181	0.208	0.01	0.03
550	0.159	0.154	0.241	0.393	0.244	0.222	0.274	0.244	0.214	0.203	0.02	0.04
750	0.269	0.208	0.179	0.229	0.263	0.203	0.360	0.312	0.224	0.304	0.03	0.06
pH												
9.5	0.144	0.200	0.358	0.140	0.109	0.114	0.117	0.180	0.183	0.121	0.02	0.04
10.5	0.233	0.292	0.331	0.283	0.203	0.185	0.306	0.269	0.217	0.182	0.05	0.11
11.5	0.180	0.259	0.261	0.073	0.069	0.111	0.107	0.069	0.122	0.112	0.02	0.04
12.5	0.169	0.285	0.166	0.048	0.090	0.030	0.080	0.045	0.124	0.063	0.02	0.05
13.5	0.055	0.152	0.157	0.042	0.013	0.019	0.005	0.003	0.037	0.013	0.01	0.03
NaCl concentration (g L^{-1})												
0	0.079	0.074	0.045	0.094	0.034	0.141	0.044	0.040	0.100	0.103	0.08	0.02
0.5	0.137	0.136	0.057	0.081	0.057	0.230	0.093	0.045	0.096	0.151	0.09	0.03
1	0.034	0.073	0.083	0.048	0.080	0.105	0.036	0.074	0.077	0.056	0.07	0.01
1.5	0.030	0.133	0.114	0.024	0.123	0.105	0.029	0.125	0.120	0.017	0.05	0.01
2	0.042	0.010	0.020	0.034	0.021	0.014	0.019	0.018	0.035	0.030	0.03	0.08

[a] CCC477, CCC478, CCC479, CCC480- *Spirulina platensis*, CCC481-*Spirulina maxima*, CCC482-*Spirulina lonar*, CCC483-*Spirulina platensis* (mutant), CCC538, CCC539-*Arthrospira* sp., CCC540-*Spirulina* sp.

Table 3. Influence of environmental variables on phycobiliproteins (mg g^{-1} dry weight) content of *Spirulina* strains

Environmental variables	*Spirulina* strains[a]										SE(m)±	CD ($P = 0.05$)
	CCC477	CCC478	CCC479	CCC480	CCC481	CCC482	CCC483	CCC538	CCC539	CCC540		
Light (μmol m^{-2} s^{-1})												
55	1.341	1.160	4.059	3.723	1.558	2.091	1.817	1.111	2.277	1.603	0.09	0.19
70	0.354	0.569	2.325	2.194	1.749	4.650	6.235	4.091	3.574	4.922	0.36	0.75
84	0.785	0.614	2.798	3.030	1.703	4.035	3.300	5.134	5.540	3.622	0.15	0.30
Temperature (°C)												
25	1.163	1.051	5.175	3.151	2.199	1.663	1.278	1.431	0.599	0.731	0.13	0.28
30	0.881	2.333	3.063	3.713	2.660	4.803	4.497	3.283	3.247	2.708	0.38	0.78
35	6.651	3.321	11.746	1.766	8.199	2.346	4.489	3.722	0.955	2.627	0.38	0.78
CO_2 concentration (ppm)												
350	7.266	7.524	4.731	8.557	2.239	1.367	5.178	3.431	2.938	1.848	0.10	0.20
550	3.036	4.255	10.505	2.930	9.699	4.954	4.639	4.866	7.979	4.483	0.09	0.03
750	4.562	6.918	6.805	8.089	7.582	8.455	1.848	6.275	2.477	8.400	0.09	0.02
pH												
9.5	5.023	2.908	2.244	7.183	11.243	0.995	4.462	4.418	9.796	2.485	0.14	0.24
10.5	8.603	3.922	6.318	20.266	5.985	8.995	12.197	15.536	5.126	9.219	0.33	0.66
11.5	2.789	1.818	2.675	2.265	1.760	2.278	4.156	2.705	2.169	3.200	0.23	0.45
12.5	3.635	1.118	1.932	1.575	1.752	1.150	2.915	1.520	1.538	0.739	0.22	0.56
13.5	1.274	2.113	0.754	0.619	1.043	0.872	0.822	0.331	0.806	0.664	0.19	0.34
NaCl concentration (g L^{-1})												
0	3.192	3.796	2.889	2.614	5.782	4.390	3.640	2.946	2.592	2.836	0.07	0.01
0.5	3.697	5.951	6.206	4.583	5.976	6.425	6.306	5.206	4.545	5.937	0.13	0.29
1	5.122	4.622	5.625	4.236	5.713	6.765	4.913	6.931	5.929	5.390	0.15	0.34
1.5	6.349	5.613	5.277	6.512	8.514	6.636	5.475	6.310	6.600	6.798	0.19	0.34
2	6.492	6.611	11.486	5.739	6.764	4.259	5.512	10.323	11.792	4.629	0.09	0.20

[a] CCC477, CCC478, CCC479, CCC480- *Spirulina platensis*, CCC481-*Spirulina maxima*, CCC482-*Spirulina lonar*, CCC483-*Spirulina platensis* (mutant), CCC538, CCC539-*Arthrospira* sp., CCC540-*Spirulina* sp.

ACKNOWLEDGMENTS

The authors are thankful to Director IARI, New Delhi for allowing the present study at the Institute and to NAIP, ICAR for financial support.

REFERENCES

1. Anupama, P. R. (2000). Value-added food: Single cell protein. *Biotechnology Advances, 18,* 459–479.
2. Belay, A., Ota, Y., Miyakawa, K., & Shimamatsu, H. (1993). Current knowledge on potential health benefits of *Spirulina. Journal of Applied Phycology, 5,* 235–241.
3. Bennett, A., & Bogorad, L. (1993). Complementary chromatic adaptation in a filamentous blue green alga. *Journal of Cell Biology, 58,* 419–435.
4. Borowitzka, M. A., & Borowitzka, L. J. (1988). *Microalgal biotechnology.* Cambridge: Cambridge University Press
5. Colla, L. M., Reinehr, C. O., Reichert, C., & Costa, J. A. V. (2004). Production of biomass and nutraceutical compounds by *Spirulina platensis* under different temperature and nitrogen regimes. *Bioresource Technology, 98,* 1489–1493.
6. Cornet, J. F., Dussa, C. G., Cluzel, P., & Dubertret, G. A. (1992). Structured model for simulation of cyanobacterium *Spirulina platensis* in photobioreactors: Identification of kinetic parameters under light and mineral limitations. *Biotechnology and Bioengineering, 40,* 826–834.
7. Costa, J. A. V., Linde, G. A., Atala, D. I. P., Mibieli, G. M., & Kruger, R. T. (2000). Modelling of growth conditions for cyanobacterium *Spirulina platensis* in microcosms. *World Journal of Microbiology & Biotechnology, 16,* 15–18.
8. Danesi, E. D. G., Rangel, C. O., Pelzer, L. H., Carvalho, J. C. M., Sat, S. & Moraes, I. O. (2001). Production of *Spirulina platensis* under different temperatures and urea feeding regimes for chlorophyll attainment. *Proceedings: Eighth International Congress in Engineering and Food* (Vol. 2 In: 1978–1982).
9. Estrada, J. E., Bescos, P., & Villar Del Fresno, A. M. (2001). Antioxidant activity of different fractions of *Spirulina platensis* protein extract. *Il Farmaco, 56,* 497–500.
10. Henrikson, R. (1994). *Microalga Spirulina-Superalimento del future.* Bercelona: Ediciones Urano S A. ISBN 84-7953-047-2.
11. Hu, Q. (2004). Industrial production of microalgal cell mass and secondary products-major industrial species: *Arthrospira (Spirulina) platensis.* In A. Richmond (Ed.), *Handbook of microalgal culture: Biotechnology and applied phycology* (pp. 264–272). Oxford: Blackwell Science Ltd.
12. Jensen, A. (1978). Chlorophylls and carotenoids. In J. A. Hellebust & J. S. Caraig (Eds.), *Phycological methods: Physiological and biochemical methods* (pp. 59–70). Cambridge: Cambridge University Press.

13. Kavita, V., & Mohanty, P.. (2000). Alterations in the structure of phycobilisomes of the cyanobacterium *Spirulina platensis* in response to enhanced Na^{+} level. *World Journal of Microbiology & Biotechnology, 16*, 795–798.

14. Lichtenthaler, H. K. (1987). Chlorophyll and carotenoids: pigments of photosynthetic biomembranes. In L. Paker & R. Douce (Eds.), *Methods in enzymology*, (Vol. 148, pp. 350–382). Boca Raton: CRC Press.

15. Miranda, M. S., Cintra, R. G., Barros, S. B. M., & Filho, J. M. (1998). Antioxidant activity of the microalga *Spirulina maxima*. *Brazilian Journal of Medical and Biological Research, 31*, 1075–1079.

16. Paerl, H. W., & Pinckney, J. L. (1996). A minireview of microbial consortia: Their role in aquatic production and biogeochemical cycling. *Microbial Ecology, 31*, 225–247.

17. Richmond, A. (1990). *Handbook of microalgal mass culture*. Boca Raton: CRC Press. ISBN 0-8493-3240-0.

18. Richmond, A., & Grobbelaar, J. U. (1986). Factors affecting the output rate of *Spirulina platensis* with reference to mass cultivation. *Biomass, 10*, 253–264.

19. Vonshak, A. (1997). *Spirulina*: Growth, physiology and biochemistry. In A. Vonshak (Ed.), *Spirulina platensis (Arthrospira)*: *Physiology, cell-biology and biotechnology* (pp. 43–65). London: Taylor & Francis.

20. Vonshak, A., Abeliovich, A., Boussiba, S., Arad, S., & Richmond, A. (1982). Production of *Spirulina* biomass: Effects of environmental factors and population density. *Biomass, 2*, 175–185.

21. Vonshak, A., & Tomaselli, L. (2000). *Arthrospira* (*Spirulina)*: Systematics and ecophysiology. In A. Whitton & E. Potts (Eds.), *The ecology of cyanobacteria* (pp. 505–522). The Netherlands: Kluwer Academic Publishers.

22. Zarrouk, C. (1966). Contribution a l' etude d' une cyanobacterie: Influence de divers facteurs physiques et chimiques et la photosynthese de *Spirulina maxima* (Setchell et Gardener) Geitler. Ph.D. Thesis, University of paris, France.

CHAPTER 15

Plant Responses to Climate Change: The Case Study of Betulaceae and Poaceae Pollen Seasons (Northern Italy, Vignola, Emilia-Romagna)

Anna Maria Mercuri [1], Paola Torri [1], Rita Fornaciari [1,2,*] and Assunta Florenzano [1]

[1]*Laboratorio di Palinologia e Paleobotanica, Dipartimento di Scienze della Vita, Università degli Studi di Modena e Reggio Emilia, Viale Caduti in Guerra, 127, 41121 Modena, Italy*

[2]*Plant Physiology Lab, Dipartimento di Scienze della Vita, Università degli Studi di Modena e Reggio Emilia, Via Kennedy, 17/I, 42124 Reggio Emilia, Italy*

ABSTRACT

Aerobiological data have especially demonstrated that there is correlation between climate warming and the pollination season of plants. This paper focuses on airborne pollen monitoring of Betulaceae and Poaceae, two of the main plant groups with anemophilous pollen and allergenic proprieties in Northern Italy. The aim is to investigate plant responses to temperature variations by considering long-term pollen series. The 15-year aerobiological analysis is reported from the monitoring station of Vignola (located near Modena, in the Emilia-Romagna region) that had operated in the years 1990–2004 with a Hirst spore trap. The Yearly Pollen Index calculated for these two botanical families has shown contrasting trends in pollen production and release. These trends were well identifiable but fairly variable, depending on both meteorological variables and anthropogenic causes. Based on recent reference literature, we considered that some oscillations in pollen concentration could have been a main effect of temperature variability reflecting global warming. The duration of pollen seasons of Betulaceae and Poaceae, depending on the different species included in each family, has not unequivocally been determined. Phenological responses were particularly evident in *Alnus* and especially in *Corylus* as a general moving up of the end of pollination. The study shows that these trees can be affected by global warming more than other, more tolerant, plants. The research can be a contribution to the understanding of phenological plant responses to climate change and suggests that alder and hazelnut trees have to be taken into high consideration as sensible markers of plant responses to climate change.

Keywords: pollen; aerobiology; global warming; wind-pollinated plants; long-term series; *Betula*; *Alnus*; *Corylus*

1. INTRODUCTION

Plants adopt many strategies to face climate changes. There are strategies including shifting of vegetation at association scale, or strategies raising the rate of physiological processes such as photosynthesis at the community and individual scales. In fragmented landscapes such as urban habitats, current rapid climate change has the potential to overcome the capacity for adaptation in plant species, also altering genetic composition and biodiversity [1].

Phenology shows evident responses to long-term variations of temperatures (e.g., [2]). The male gametophytic phase has proven to be one of the most temperature-vulnerable stages, often highly sensitive to hot or cold stresses on a short-term basis. Pollen germination and pollen tube growth show acceleration under increasing temperatures (in peach: [3]) or, contrarily, can be negatively affected by heat [4]. Among the dramatic climate changes, the recent trend towards global warming is overall accepted, and an increase of about 0.6 degrees on mean temperatures since the late 19th century has been acknowledged [5]. In general, pollen development and fertilization may be the most sensitive reproductive stage to temperature increase and, therefore, the current global warming is recognized as having clear effects on pollen production and fertility.

Aerobiological data have especially demonstrated the correlation between climate warming and pollination seasons [6,7,8,9,10,11]. Following this approach, we report on the trend of airborne pollen data from two of the most important botanical families at mid latitudes, Betulaceae and Poaceae. In European countries, they represent large part of the current plant cover as woody (Birch family) and herbaceous (grasses) species. Both the Birch family (*Alnus*, *Betula*, *Carpinus*, *Corylus*, *Ostrya*) and grasses (wild and cultivated, including cereals) are wind-pollinated and produce among the vast amounts of pollen in the spring and summer airflows in Europe [12]. Despite cleistogamous species obviously making a light contribution to airborne pollen rain, grasses are responsible for the most important allergy symptoms registered in the world (e.g., [13]).

This paper focuses on temperature as a major factor influencing the pollen season with respect to precipitation [14,15]. Precipitation, wind speed, and wind direction have an important role in the spreading and amount (concentration) of pollen in the air. However, temperatures higher than normal are able to improve pollen quantity and diffusion, frequently resulting in longer seasons of pollination in many plant species ([16,17,18,19]; see [20] for grasses). In this paper, we focus on this aspect and observe the trend of a long series of airborne pollen monitoring by considering the possible correspondence between pollen concentrations and variation of temperatures.

The paper reports on data collected from one of the first airborne monitoring station located in Italy (at Vignola, MO2). The monitoring station started running in 1989. It has produced only fragmented data in recent years, and it is not working anymore. Therefore, we selected the long interval 1990–2004 during which data had been collected on a continuous basis and for the whole year. The 15-year set of data represents a long series offering the possibility to compare pollen trends and temperatures over a relatively long timespan. Other examples of long series published in the last couple of years are: 16-year period (1994–2009) to predict olive crop production [21]; 15-year period (1993–2007, coupled with another 1992–2011) to model and predict the dynamic of *Vitis* flowering in the field [22]; an 11-year study (2003–2013) to quantify the effects of environmental factors including pollen on asthma hospitalization [23]; a 30-year series (1981–2010) is useful to compare historical data with atmospheric dispersion model for airborne pathogen concentrations [24].

2. RESULTS

The trends of the Yearly Pollen Index (YPI) of airborne pollen spread of the two families from 1990–2004 show comparable and variable concentrations, with general contrasting patterns (Figure 1). Below, the general data are reported aiming at introducing the main features of the concentration variability that has been observed in the pollen rain produced by these anemophilous pollen grains in the studied area. They may be paradigmatic of pollen concentration trends under changing weather conditions over 15 years.

a. Betulaceae

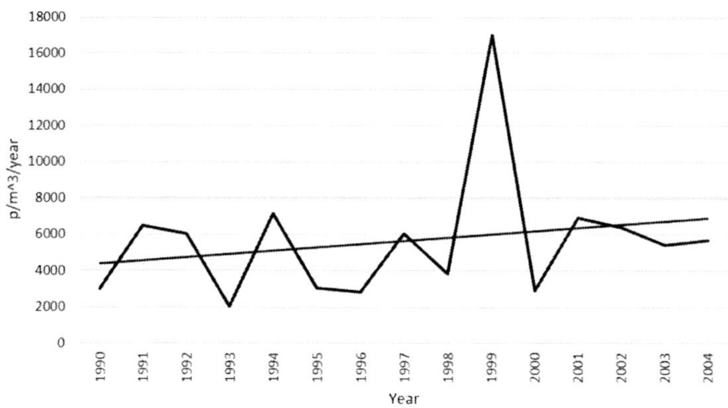

b. Poaceae

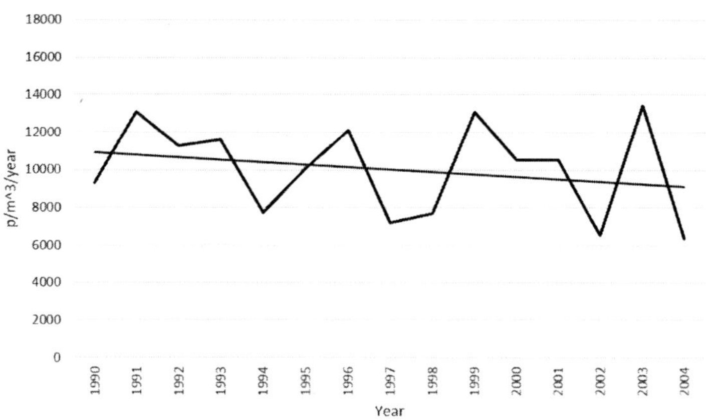

Figure 1. Yearly Pollen Index (YPI) of Betulaceae (**a**) and Poaceae (**b**) from 1990 to 2004.

2.1. Betulaceae

The pollen season is 163 days per year on average. The longest season occurred in 2001 (192 days, mainly due to *Alnus*) whilst the shortest season occurred in 2000 (140 days; Figure 2a). The highest YPI (Figure 1a) occurred in 1999 (17,012 p/m^3): this value was obtained from the very high values of *Ostrya* pollen concentration (12,653 p/m^3) followed by significant values of *Alnus* (1869 p/m^3) and *Corylus* (1656 p/m^3).

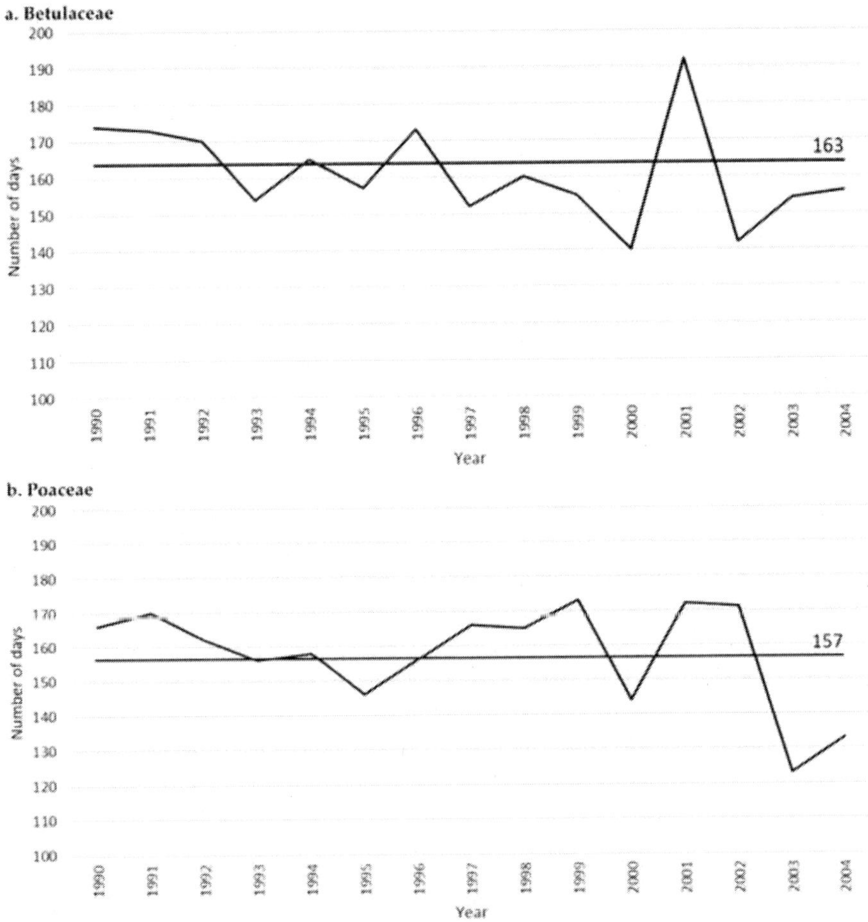

Figure 2. Duration of pollen season of Betulaceae (**a**) and Poaceae (**b**) from 1990 to 2004.

The months composing the pollen season were from January to late July (Figure 3a). Depending on the different plant genera/species included in the family, the pollen seasons started in January (until May for *Corylus*, July for *Alnus*) and February (until June for *Betula* and *Ostrya*, May for *Carpinus*).

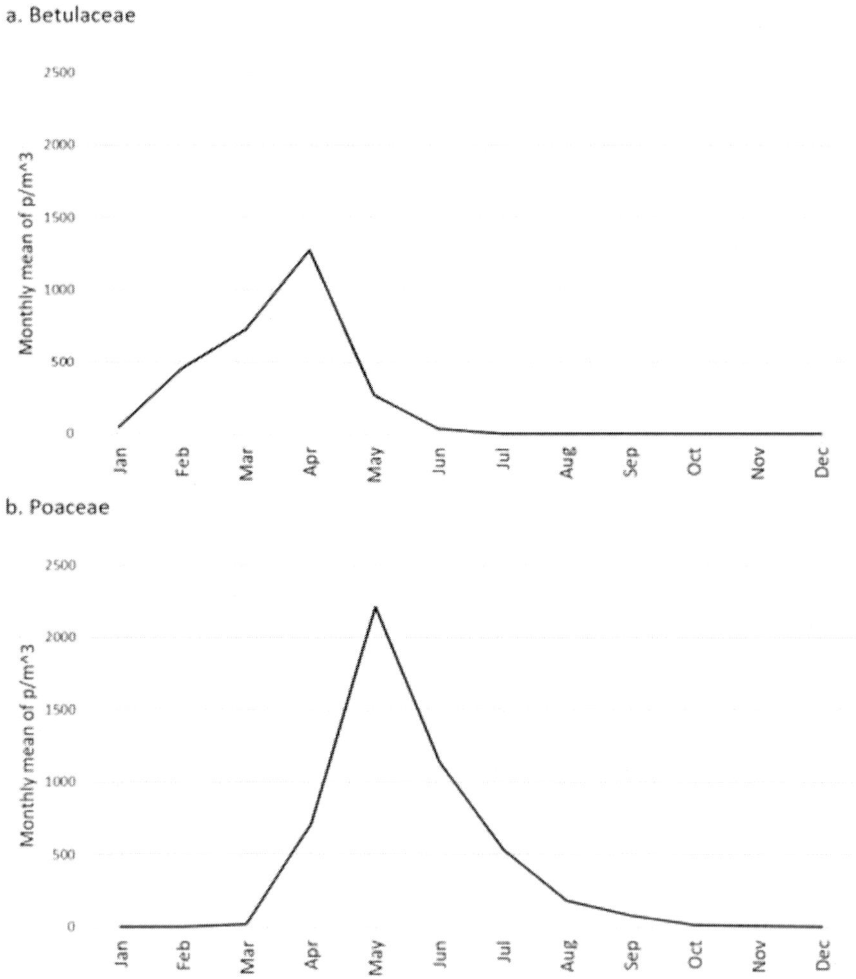

Figure 3. Months composing the pollen season of Betulaceae (**a**) and Poaceae (**b**): total monthly mean of each year from 1990 to 2004.

The daily peaks of pollen concentration were observed in the years 1994 (1100 p/m^3 in March), 1999 (1460 p/m^3 in April), and in 2001 (800 p/m^3 in April) whilst a very low maximum value (<150 p/m^3) was observed in 1993 and 2000 (Figure 4). In the other years, the daily peaks were about 200–600 p/m^3 and, in general, were reached in April. In all years, the pollen shows a decreasing trend in summer with some re-increase of concentration in June due to the blooming of *Alnus viridis* (especially visible in 1994 (see also [25])).

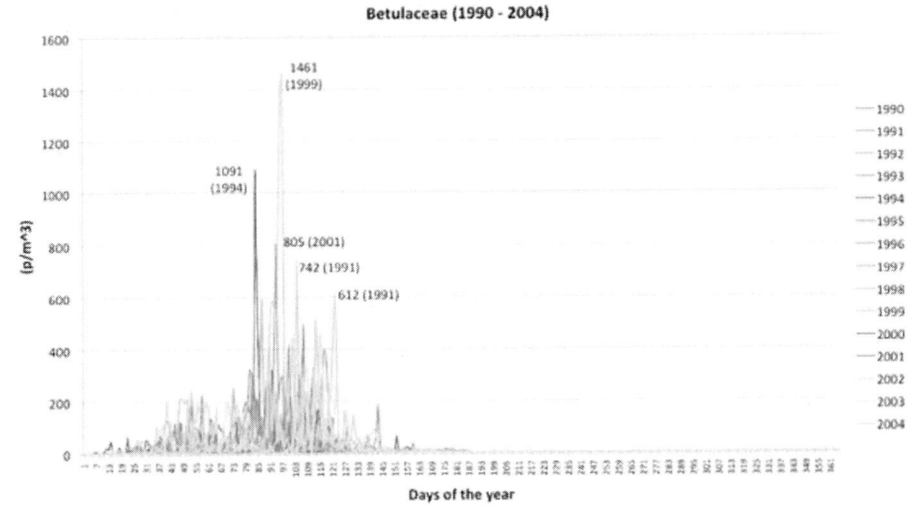

Figure 4. Daily peaks of pollen concentration of Betulaceae: a comparison from 1990 to 2004.

The end of the pollen season of this family has been actually recorded at this time in the considered period.

2.2. Poaceae

The pollen season consists of >100 days per year (157 days on average). The longest season occurred in 1999 (173 days) whilst the shortest season occurred in 2003 (123 days; Figure 2b). Accordingly, the highest YPI (Figure 1b) occurred in the years with the longest pollen season (1991, besides 1999).

The months composing the pollen season were from late March to early September (Figure 3b). In 1994, most of the pollen concentration was especially concentrated at the early period of the season (late March), and in general pollination started in this period during other three years

(1997, 2001, 2002). In the other years, the first half of April saw the start of the pollen season.

The daily peaks of pollen concentration (>1200 p/m^3) were observed in the years 2000 (in April) and 2003 (in May) whilst low maximum values (<300 p/m^3) were observed in 1994, 1997, 2002, and 2004 (Figure 5). In the other years, the daily peaks were 600 p/m^3 and, in general, were reached in May. The years 1991 and 1994 are exceptions as the daily maximum value was reached in June. In all years, the pollen shows a decreasing trend in summer with some re-increase of concentration at the end of July (in 1995 and 1999), or at the end of August (in 1990).

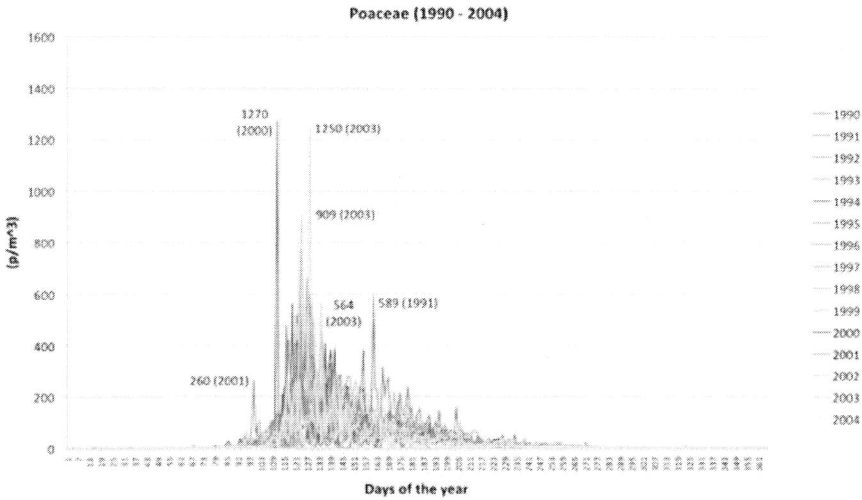

Figure 5. Daily peaks of pollen concentrations of Poaceae: a comparison from 1990 to 2004.

The end of the pollen season has been actually recorded at this time in a few years (1995, 2000, 2001, 2003) whilst it occurred in September in many years of the considered period.

3. DISCUSSION

Aerobiological data can explain unpredictable annual variability of concentrations over the years but cannot be fully responsible for the increasing trend of Betulaceae (Figure 1a) versus the decreasing trend of Poaceae (Figure 1b). In general, the monthly trends and concentrations follow patterns that are typical and repeated after a seasonal basis. Thus, the Betulaceae pollen season is January–July with acme in April whilst the Poaceae pollen season is March–September with acme in May (Figure 3). Data are in agreement with the general information that is usually obtained from the European pollen monitoring network [26].

Aerobiological data from Betulaceae and Poaceae show that pollen seasons depend on the phenology of the different species included in each family. On one hand, human interference with flowering cycles or plant distribution and growing are common as cutting of trees, mowing of herbs, and garden or park maintenance in urban habitats. They can be responsible for unpredictable anomalies in airborne pollen concentrations. On the other hand, a high number of species is included in these families, obviously influencing trends and the variability of pollen production. The genera and species have variable blooming seasons. For example, among Betulaceae, there are species of *Alnus* blooming in April and others in July–August. In the study region, during the spring period in March–April, spore traps can capture alder airborne pollen from *Alnus cordata* (Loisel.) Duby, Italian Alder; *A. glutinosa* (L.) Gaertn., European Alder; and *A. incana* (L.) Moench, Gray Alder. The latter species blooms until May. Data from the monitoring station of Vignola, here discussed, also systematically showed alder pollen in the air in July and August; this is clearly related to another species, *A. viridis* (Chaix) DC that does not live in Emilia Romagna. The Green Alder is known to be distributed in Northern Italy, on mountains from the West Piedmont to the East Friuli. This means that its pollen has arrived, and still arrives, from the northern Alps as a long-distance transport with summer air masses [25,27].

Poaceae is a large group including species with different flowering periods, and very different pollen production [28], which may result in more than one peak during the grass pollen season. Moreover, there are grasses with long periods of pollination or blooming more than once in the year: e.g., *Poa pratensis* L., smooth or common meadow-grass, from May to July; *Lolium perenne* L., perennial ryegrass, and *Cynodon dactylon* (L.) Pers., Bermuda grass, from May to September. According to their nature of herbaceous plants, frequently owing a short-life cycle and growth limits due to the anthropogenic control, Poaceae show more evident (casual) oscillations in the YPI than Betulaceae. Interestingly, the maxima daily concentrations (>1200 p/m^3) of Poaceae were observed in the years 2000 and 2003 (Figure 5) that are among the warmest years of the last century in the considered period (see especially summer temperatures in Figure 6). In particular, according to a survey by the National Environmental Agency, mean temperatures of the same years had an increase up to +1.38 °C in Italy. Moreover, the combination of the highest YPI and the shortest pollen season of Poaceae occurred in 2003.

By comparing the analyses from several towns located in Spain [29], some similarities are evident: oscillations of pollen concentration do not follow a clear trend, pollen season never exceeds 200 days, and the maximum value of concentration was reached in 2003.

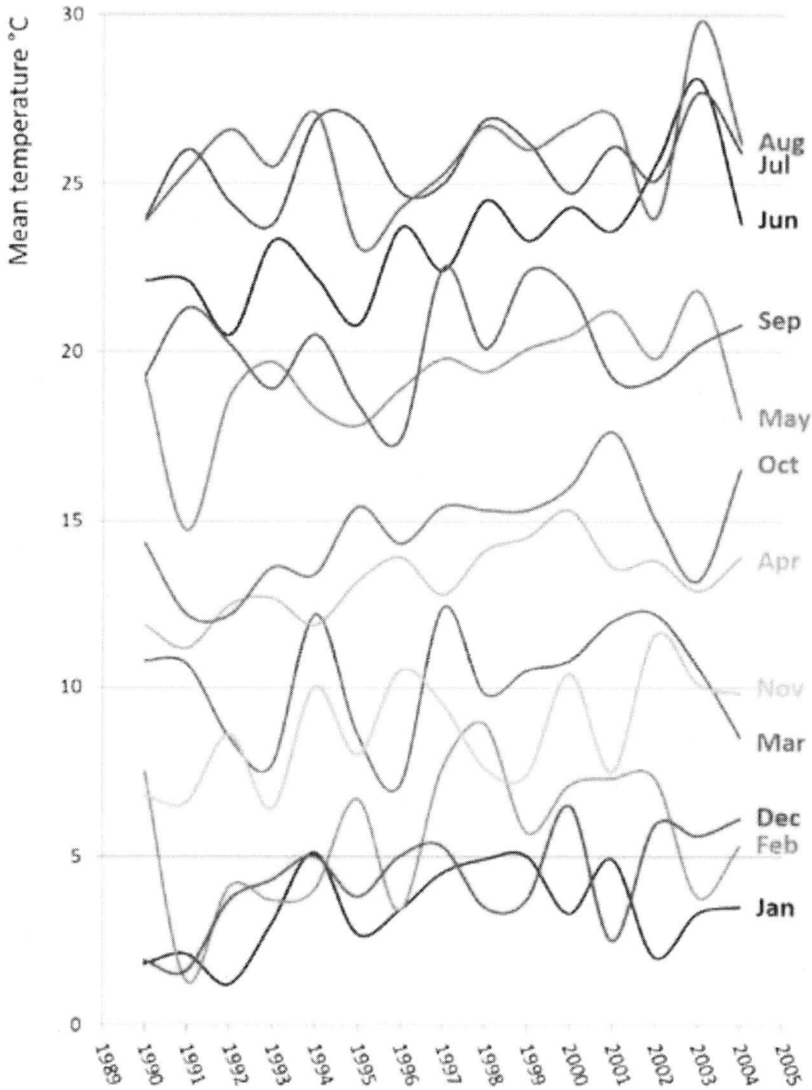

Figure 6. Monthly mean temperature during the considered period (1990–2004).

In general, these data support the idea that Poaceae replies to temperature increase with a higher production of pollen, eventually resulting in high concentration for a short period.

Poaceae data show that the duration of the pollen season influences the YPI. For example, in the years 1991 and 1999 there were high values of grass pollen concentration and long pollen seasons (Figure 2b). In 1999, an anomalous (high) value of the YPI of Betulaceae was also registered, especially due to *Ostrya*, and partly also due to *Alnus* and *Corylus*.

In general, the pollen season of Betulaceae ranges between 140 and 200 days in different years. The Birch family tends to decrease the number of days with pollen in the air but important distinctions were observed among the genera. *Ostrya* and *Carpinus* had a quite steady duration of the pollen season, while the number of days with pollen tends to increase in *Betula* and to decrease in *Alnus* and *Corylus* (Figure 7). For example, there is a slight increasing trend of *Betula* from 56 to 58 days with pollen in the air, passing from the 1990–1995 interval to the 1996–2004 interval. The duration of the pollen season of *Alnus* decreases from 160 in 1990–1995 to 150 days in 1996–2004. *Corylus* pollen season ranges as mean from 97 days in 1990–1995 to 75 days in the years 1996–2004. Interestingly, there are signs that a similar behavior characterized the way the pollen seasons changed. All the genera show a tendency to maintain almost steady pollination in the early days of the pollination season all over the years. Conversely, the end days show important differences: they pass from a delay of about 2 days for *Betula* (corresponding to an increase of 3.5% of the number of days with *Betula* pollen in the air), to a reduction of about 10 days in *Alnus* (decrease of 6.5%), and of about 22 days in *Corylus* (decrease of 25.6%; Figure 7).

Despite the obvious observation that airborne pollen concentration for all species decreases with precipitation, and pollen concentration is low/insignificant in a day with rain or the day after (e.g., [30]), most research points to the main significance of temperature for pollination. An 18 year long-term study (1989–2006) on the characterization of the male flowering and pollen flow of birch in Finland-northern Europe was carried out to understand the proximal causes contributing to pollination, and the annual variability and spatial synchronization of male catkin numbers of silver birch (*Betula pendula* Roth.) and downy birch (*Betula pubescens* Ehrn.). The research demonstrated that there is no correlation between catkin numbers, annual airborne pollen sums, and the amount of precipitation during the pollination season [31]. Indeed, studies report on the correlation between temperatures and pollen seasons where different tree species give different responses to climate change. The focus has commonly been directed to the start of pollination.

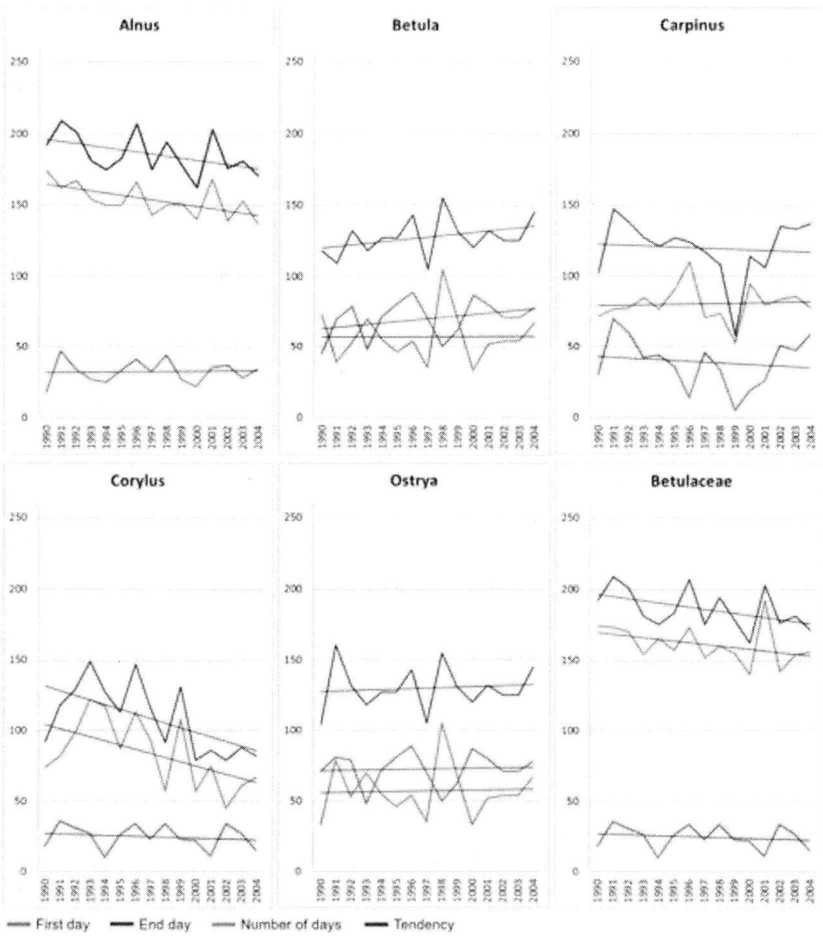

Figure 7. Pollen season of the five genera in the Birch family and in total Betulaceae, from 1990 to 2004.

Studying the warming conditions of the last years, from monitoring stations of Northern Europe it emerges that *Alnus*, *Betula*, and *Corylus* generally have an early pollination start. As *Betula* lives at low temperatures in mid to high latitude countries, data from Central Europe suggest that an increase in temperature could limit the physiological performance of birch species, including the production of pollen [32]. European data, however, reveal that *Alnus*, *Betula*, and *Corylus* trees are more associated with high latitudes and low temperatures than others, and their annual concentrations have a negative correlation with temperature that could reflect the limited presence of these species at warmer sites [33]. In Italy, moving to different local geobotanical features, in the last years the pollination start tends to become delayed in winter-blooming plants (*Alnus*, *Corylus*) while the most important tendency to an early pollination occurs in spring-blooming trees (*Betula*) [6,19]. Our data add information about the behavior of the last day of pollination periods. Pollen monitoring suggests that there is a tendency of *Alnus* and *Corylus* to behave as winter-blooming trees thus anticipating the end (earlier last day) of their pollen seasons. Contrarily, the spring-blooming *Betula* trees tend to delay the end of the pollen season. Comparing these results, we can conclude that the pollen season of alder and hazelnut trees tends to be shorter, meanwhile the season of birch becomes longer. This seems to demonstrate that warm climate, as the current environmental conditions are, may be generally unfavorable to the winter-blooming trees that have a later pollination after exceptionally warm autumn temperatures [27].

4. MATERIALS AND METHODS

The pollen monitoring station was located in Vignola (MO2; 125 m a.s.l.; 44°29′ N 11°00′ E), a small town at the center of Emilia-Romagna, in Northern Italy. The region includes the Po plain, the Tuscan-Emilian Apennines to the south, and the Adriatic Sea coast to the east (Figure 8). The Po plain has a climate gradient ranging from the Mediterranean warm-temperate to the cold-temperate climate of the Apennines. Based on the mean temperature of the coldest month (January; from −4 °C to +4 °C), Emilia-Romagna falls in the temperate (C) and warm-temperate (D) thermic zones by Pignatti [34].

The airborne pollen monitoring station was part of both the Italian and the European A.I.A.—Aeroallergen monitoring networks. The station was in continuous operation from 1990 to 2004. A seven-day Lanzoni VPPS 2000 spore trap (a Hirst volumetric sampler; Figure 8) was located at approximately 12 m on the roof of the Local Health Agency. Sampling and slide mounting followed standard methods of the aerobiological networks [35,36,37]. The daily slides were analysed with a light microscope at 250× magnification (400–1000× for details), and pollen grains were counted along four longitudinal transects. Pollen identification was done through current pollen atlases and keys and the reference pollen collection. Pollen data were elaborated using the softwares GEPO and "Aerobiologia 2.1" [38].

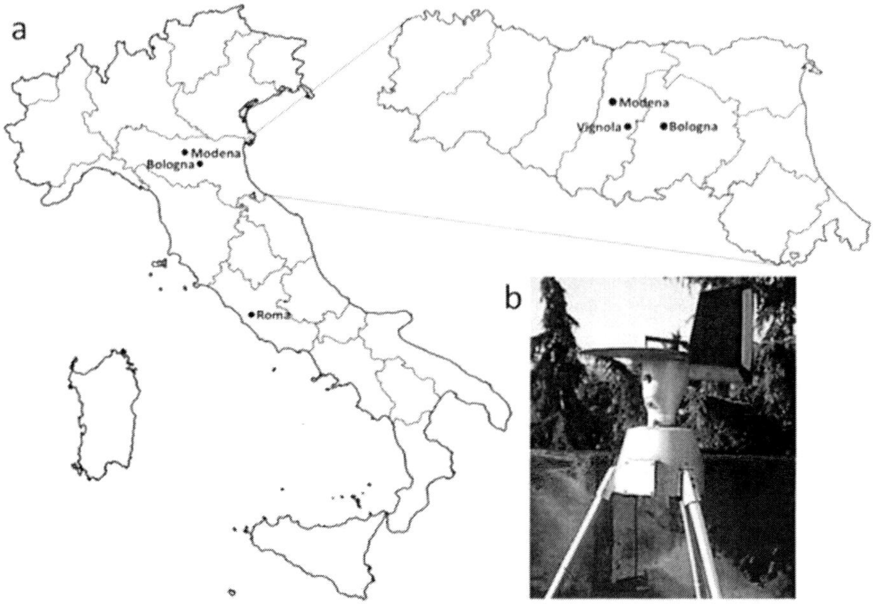

Figure 8. (**a**) Location map of Vignola, Emilia-Romagna, Northern Italy; (**b**) Lanzoni VPPS 2000 spore trap.

The Yearly Pollen Index (YPI) is the total amount of one pollen type in one year and it is given by the sum of the daily mean concentrations. The pollination period corresponds to the days between start and end days of pollen in the air, and refers to the continuous presence of the studied pollen in the air. The start of pollination was established as the first time that the pollen grains of the studied taxon were observed in slides from three subsequent days [9,39,40]. The pollination period considered for each specific pollen taxon include >90% of its yearly total pollen.

Data were imported into Microsoft Excel 2013 and used for building the graphs. The linear trend was calculated by the coefficient of determination (R^2) using linear analysis tools.

Temperature and precipitation data were available from one meteorological station of Modena (e.g., [41]) and from the Regional Agency for the Protection of the Environment (A.R.P.A.).

5. CONCLUSIONS

The 15-year aerobiological data analysis reported from the monitoring station of Vignola shows that trends in pollen production and release are easily identifiable but fairly variable, depending on meteorological variables besides anthropogenic causes. Despite data collection being interrupted about 12 years ago, the interest of these results is still effective as they provide an opportunity to observe phenological responses to temperature changes that had occurred in the considered years, under increasing mean temperature values.

The pollen season of Betulaceae, like most of the tree families at middle latitudes, covers the first part of the year, from winter to early summer. The pollen season of Poaceae starts later, principally covering the spring and summer seasons. In our data, the highest YPI and the shortest pollen season of Poaceae occurred in 2003 suggesting that, in general, these phenological events match the presence of average warm temperatures. Poaceae perform a higher pollen production (i.e., high pollen concentration in the air for a short period) as a response to temperature increase.

Differently from what has been observed for pollen of *Taxus* [9]—which represents a woody species highly sensible to temperature increase—by considering the general trend of the families, we can conclude that the duration of the pollen season of Betulaceae and Poaceae was not unequivocally altered by warming climate. Deepening the role of different trees in the Birch family, significant differences emerged by distinguishing the winter and spring blooming species: they actually are sensible to the temperatures of the season when microsporogenis starts.

Most evident consequences to climate change by reducing the release of pollen in the air are those in *Alnus*, and especially in *Corylus*. This study shows that winter-blooming tree species can be phenologically affected by global warming as this phenomenon has caused the increase of autumn temperatures. These trees react by end of the season earlier, and in general by shortening their pollen season. Surprisingly, this occurs in two widespread tree/shrub meso-hygrophilous genera – *Alnus* and *Corylus* – that therefore result less warmth-tolerance than other plants such as birch. This is partly in contrast with Central European models built by gridding meteorological/pollen data and by using them as predictor variables to build a model of high *Alnus*, *Corylus*, and *Betula* pollen concentration levels for spatially continuous areas of 11 cities in Poland [42]. In that research, the maximum range of the *Corylus* pollen season was 6–150 days of the year, 14–145 days for *Alnus*, and 35–164 days for *Betula* (calculated using the 99% method).

The model predicts high pollen concentrations for the three genera, only suggesting a proportionally reduced increase of *Corylus*. Actually, the examination of data from pollen monitoring stations located in the same nation [43] suggests that hazel and alder pollination and the occurrence of airborne pollen vary greatly and are significantly influenced by meteorological conditions. The research introduces several variables, such as presence of locations exposed or sheltered to sun and wind, that have important influence on the pollen season of hazels and alders: for example, a delay in pollination start was observed in quite sunny but very windy sites.

Our data refer to mean data from a long series of aerobiological monitoring. Results support that the trends of pollen concentrations have some relationship with variations in temperatures. Probably these results especially reflect local evidence but they also mirror regional responses of plants to climate change. The phenology of *Alnus* and *Corylus* seems particularly affected by current global warming, and more than other Betulaceae trees and grasses. These trees have to be particularly taken into account as possible markers of plant responses to climate change.

ACKNOWLEDGMENTS

The aeropalynological analyses are the fruit of the work of a large number of people who have worked in the lab as researchers and students for several years. Only to mention the last contributions, two students of bachelor thesis in Biology have collected and elaborated the pollen data: the focus on Betulaceae was done by Raffaella Ranieri, while that on Poaceae was done by Eleonora Clò, both students at the University of Modena and Reggio Emilia.

AUTHOR CONTRIBUTIONS

Anna Maria Mercuri and Paola Torri designed the research and made pollen analyses. Rita Fornaciari and Assunta Florenzano performed data elaboration. Anna Maria Mercuri wrote the text, and all authors contributed to the final manuscript.

CONFLICTS OF INTEREST

The authors declare no conflict of interest.

REFERENCES

1. Jump, A.; Penuelas, J. Running to stand still: Adaptation and the response of plants to rapid climate change. Ecol. Lett. 2005, 8, 1010–1020.
2. Hedhly, A. Sensitivity of flowering plant gametophytes to temperature fluctuations. Environ. Exp. Bot. 2011, 74, 9–16.
3. Hedhly, A.; Hormaza, J.I.; Herrero, M. The effect of temperature on pollen germination, pollen tube growth, and stigmatic receptivity in peach. Plant Biol. 2005, 7, 476–483.
4. Zinn, K.E.; Tunc-Ozdemir, M.; Harper, J.F. Temperature stress and plant sexual reproduction: Uncovering the weakest links. J. Exp. Bot. 2010, 61, 1959–1968.
5. NOAA Climate of 2005—Annual Report. National Climatic Data Center, 2006. Available online: http://www.ncdc.noaa.gov/oa/climate/research /2005/ann /global.html#top (accessed on 24 October 2016).
6. Frenguelli, G. Interaction between climatic changes and allergenic plants. Monaldi Arch. Chest Dis. 2002, 57, 141–143.
7. Ariano, R.; Canonica, G.W.; Passalacqua, G. Possible role of climate changes in variations in pollen seasons and allergic sensitizations during 27 years. Ann. Allergy Asthma Immunol. 2010, 104, 215–222.
8. Ziska, L.; Knowlton, K.; Rogers, C.; Dalan, D.; Tierney, N.; Elder, M.A.; Filley, W.; Shropshire, J.; Ford, L.B.; Hedberg, C.; et al. Recent warming by latitude associated with increased length of ragweed pollen season in central North America. PNAS 2011, 108, 4248–4251.
9. Mercuri, A.M.; Torri, P.; Casini, E.; Olmi, L. Climate warming and the decline of Taxus airborne pollen in urban pollen rain (Emilia Romagna, Northern Italy). Plant Biol. 2013, 15, 70–82.
10. Smith, M.; Jäger, S.; Berger, U.; Šikoparija, B.; Hallsdottir, M.; Sauliene, I.; Bergmann, K.-C.; Pashley, C.H.; de Weger, L.; Majkowska-Wojciechowska, B.; et al. Geographic and temporal variations in pollen exposure across Europe. Allergy 2014, 69, 913–923.
11. Galán, C.; Alcázar, P.; Oteros, J.; García-Mozo, H.; Aira, M.J.; Belmonte, J.; de la Guardia, C.D.; Fernández-González, D.; Gutierrez-Bustillo, M.; Moreno-Grau, S.; et al. Airborne pollen trends in the Iberian Peninsula. Sci. Total Environ.2016, 550, 53–59.
12. Sofiev, M.; Siljamo, P.; Ranta, H.; Linkosalo, T.; Jaeger, S.; Rasmussen, A.; Rantio-Lehtimaki, A.; Severova, E.; Kukkonen, J. A numerical model of birch pollen emission and dispersion in the atmosphere. Description of the emission module. Int. J. Biometeorol. 2013, 57, 45–58.
13. Kmentaa, M.; Bastla, K.; Kramerc, M.F.; Hewingsc, S.J.; Mwangec, J.; Zetterb, R.; Bergera, U. The grass pollen season 2014 in Vienna: A pilot study combining phenology, aerobiology and symptom data. Sci. Total Environ. 2016, 566–567, 1614–1620.
14. Van Vliet, A.J.H.; Overeem, A.; de Groot, R.S.; Jacobs, A.F.G.; Spieksma, F.T.M. The influence of temperature and climate change on the timing of pollen release in the Netherlands. Int. J. Climatol. 2002, 22, 1757–1767.
15. Tormo-Molina, R.; Gonzalo-Garijo, M.A.; Silva-Palacios, I.; Muñoz-Rodríguez, A.F. General trends in airborne pollen production and pollination periods at a

Mediterranean site (Badajoz, Southwest Spain). J. Investig. Allergol. Clin. Immunol. 2010, 20, 567–574.
16. Frenguelli, G. L'influenza dei cambiamenti climatici sulla presenza dei pollini in atmosfera. Aria Ambient. Salut. 2004, 4, 23–25. (In Italian)
17. Bortenschlager, S.; Bortenschlager, I. Altering airborne pollen concentrations due to the Global Warming. A comparative analysis of airborne pollen records from Innsbruck and Obergurgl (Austria) for the period 1980–2001. Grana 2005, 44, 172–180.
18. Gehrig, R. The influence of the hot and dry summer 2003 on the pollen season in Switzerland. Aerobiologia 2006, 22, 27–34.
19. Tedeschini, E.; Rodríguez-Rajo, F.J.; Caramiello, R.; Jato, V.; Frenguelli, G. The influence of climate changes in Platanus spp. pollination in Spain and Italy. Grana 2006, 45, 222–229.
20. Ianovici, N. Relation between Poaceae pollen concentrations and meteorological factors during 2000–2010 in Timisoara, Romania. Acta Agrobot. 2015, 68, 373–381.
21. Aguilera, F.; Ruiz-Valenzuela, L. Forecasting olive crop yields based on long-term aerobiological data series and bioclimatic conditions for the southern Iberian. Span. J. Agric. Res. 2014, 12, 215–224.
22. Cunha, M.; Ribeiro, H.; Costa, P.; Abreu, I. A comparative study of vineyard phenology and pollen metrics extracted from airborne pollen time series. Aerobiologia 2015, 31, 45–56.
23. Chen, K.; Clonek, G.; Hansen, A.; Peng Bi, L. The effects of air pollution on asthma hospital admissions in Adelaide, South Australia, 2003–2013: Time-series and case-crossover analyses. Clin. Exp. Allergy 2016, 46, 1416–1430.
24. Van Leuken, J.P.G.; Swart, A.N.; Droogers, P.; van Pul, A.; Heederik, D.; Havelaar, A.H. Climate change effects on airborne pathogenic bioaerosol concentrations: A scenario analysis. Aerobiologia 2016, 32, 607–617.
25. Accorsi, C.A.; Bandini Mazzanti, M.; Lombroso, L.; Santangelo, R.; Torri, P. La stazione meteorologica-aeropollinica di Modena (Nord-Italia): Il caso di Alnus viridis. In Studi in Ricordo di Daria Bertolani Marchetti. Deputazione di Storia Patria per le Antiche Province Modenesi; Accorsi, C.A., Bandini Mazzanti, M., Labate, D., Trevisan Grandi, G., Eds.; Aedes Muratoriana: Modena, Italy, 1998; Volume 150, pp. 205–212.
26. D'Amato, G.; Cecchi, L.; Bonini, S.; Nunes, C.; Annesi-Maesano, I.; Behrendt, H.; Liccardi, G.; Popov, T.; van Cauwenberge, P. Allergenic pollen and pollen allergy in Europe. Allergy 2007, 62, 976–990.
27. Mercuri, A.M. Applied palynology as a trans-disciplinary science: The contribution of aerobiology data to forensic and palaeoenvironmental issues. Aerobiologia 2015, 31, 323–339.
28. Aboulaich, N.; Bouziane, H.; Kadiri, M.; Trigo, M.M.; Riadi, H.; Kazzaz, M.; Merzouki, A. Pollen production in anemophilous species of the Poaceae family in Tetouan (NW Morocco). Aerobiologia 2009, 25, 27–38.
29. García-Mozo, H.; Galan, C.; Alcazar, P.; de la Guardia, C.D.; Nieto-Lugilde, D.; Recio, M.; Hidalgo, P.; Gonzalez-Minero, F.; Ruiz, L.; Domínguez-Vilches, E. Trends in grass pollen season in southern Spain. Aerobiologia 2010, 26, 157–169.
30. Peternel, R.; Srnec, L.; Čulig, J.; Zaninović, K.; Mitić, B.; Vukušić, I. Atmospheric pollen season in Zagreb (Croatia) and its relationship with temperature and precipitation. Int. J. Biometeorol. 2004, 48, 186–191.
31. Ranta, H.; Hokkanen, T.; Linkosalo, T.; Laukkanen, L.; Bondestam, K.; Oksanen, A. Male flowering of birch: Spatial synchronization, year-to-year variation and relation of catkin numbers and airborne pollen counts. For. Ecol. Manag.2008, 255, 643–650.
32. Emberlin, J.; Detandt, M.; Gehrig, R.; Jaeger, S.; Nolard, N.; Rantio-Lehtimäki, A. Responses in the start of Betula (birch) pollen seasons to recent changes in spring temperatures across Europe. Int. J. Biometeorol. 2002, 46, 159–170.
33. Ziello, C.; Sparks, T.H.; Estrella, N.; Belmonte, J.; Bergmann, K.C.; Bucher, E.; Brighetti, M.A.; Damialis, A.; Detandt, M.; Galán, C.; et al. Changes to airborne pollen counts across Europe. PLoS ONE 2012, 7, e34076.

34. Pignatti, G. Forest vegetation in view of some scenarios of climate change in Italy. Forest@ 2011, 8, 1–12.
35. Frenguelli, G.; Mandrioli, P. Airborne pollen monitoring in Italy. Rev. Palaeobot. Palynol. 1990, 64, 41–43.
36. Mandrioli, P. The Italian aeroallergen network: Report 1990. Aerobiologia 1990, 6, 2–59.
37. Galán, C.; Smith, M.; Thibaudon, M.; Frenguelli, G.; Oteros, J.; Gehrig, R.; Berger, U.; Clot, B.; Brandao, R.; EAS QC Working Group. Pollen monitoring: Minimum requirements and reproducibility of analysis. Aerobiologia 2014, 30, 385–395.
38. Mercuri, A.M.; Accorsi, C.A.; Bandini Mazzanti, M.; Ferrarini, S.; Venturi, L. Aerobiologia 2.0: A software program for processing Aerpollen Data. J. Environ. Pathol. Toxicol. Oncol. 1997, 16, 191–194.
39. Clot, B. Forecast of the Poaceae pollination in Zurich and Basle (Switzerland). Aerobiologia 1998, 14, 267–268.
40. Jato, V.; Rodríguez-Rajo, F.J.; Alcázar, P.; de Nuntiis, P.; Galán, C.; Mandrioli, P. May the definition of pollen season influence aerobiological results? Aerobiologia 2006, 22, 13–25.
41. Lombroso, L.; Quattrocchi, S. Il Clima e le Misure Meteo-Climatiche a Modena; Memorie Dell'accademia di Scienze, Lettere ed Arti: Modena, Italy, 2002; pp. 391–401.
42. Nowosad, J. Spatiotemporal models for predicting high pollen concentration level of Corylus, Alnus and Betula. Int. J. Biometeorol. 2016, 60, 843–855.
43. Puc, M.; Kasprzyk, I. The patterns of Corylus and Alnus pollen seasons and pollination periods in two Polish cities located in different climatic regions. Aerobiologia 2013, 29, 495–511.

CHAPTER 16

Thermal Requirement of Indian Mustard (*Brassica Juncea*) at Different Phonological Stages under Late Sown Condition

Manoj Pratap Singh[1], Lallu1, N. B. Singh[2]

[1].*Department of Crop PhysiologyC.S.A. University of Agriculture and TechnologyKanpurIndia*

[2].*Section of Rabi CerealC.S.A. University of Agriculture and TechnologyKanpurIndia*

ABSTRACT

Indian mustard [*Brassica juncea* (L.) Czern & Coss.] is a long day plant, which requires fairly cool climatic condition during growth and development for obtaining better seed yield. Various workers have correlated crop growth and development with energy requirement parameters, such as growing degree days (GDD), photo-thermal unit (PTU), helios thermal unit (HTU), photo-thermal index (PTI) and heat use efficiency (HUE). Therefore, GDD requirement for different phenological stages of 22 newly developed Indian mustard varieties was studies during winter (*rabi*) seasons of 2009–10 and 2010–11 at student instructional farm of C.S. Azad University of Agriculture and Technology, Kanpur, (Utter Pradesh). Study revealed that RH-8814, NRCDR-02 and BPR-549-9 recorded higher GDD (1703.0, 1662.9 and 1648.0), PTU (19129.8, 18694.2 and 18379.8), HTU (11397.7, 11072.2 and 10876.0), PTI (13.25, 13.14 and 13.08) and HUE (4.11, 3.84 and 3.71) at physiological maturity, while higher HUE was recorded (9.62, 8.99 and 8.91 kg ha^{-1} degrees-day) at days after sowing (DAS) to 50 % flowering. On the basis of study mustard genotypes RH-8814, NRCDR-02 and BPR-549-9 were identified as most heat-tolerant, as they maintained higher values for energy related parameters. Seed yield was highly positively correlated with GDD (r = 0.61, 0.65 and 0.75), PTU (r = 0.66, 0.64 and 0.74), HTU(r = 0.79, 0.68 and 0.73) at the above these three phenological stages, while negatively correlated with PTI at anthesis and at maturity. Hence, these parents could be used in crossing programme for achieving further improvement.

Keywords: Growing degree days Heat use efficiency Helios thermal unit Photo-thermal index Photo-thermal unit

INTRODUCTION

Indian mustard [*Brassica juncea* (L.) Czern & Coss.] is a crop of tropical as well as temperate zones requiring somewhat cool and dry weather for satisfactory growth and development. High temperature at flowering stage causes reduction in seed yield as it may lead to pollen sterility. Although mustard is a long day plant requiring 16 h of light period in 24 h cycle, it can made to flowering if it is provided with a cycle of 8 h of light period with 4 h of dark period (short night). Mustard can be made to flower in about 50 days under 16/8 h light/dark period.

Growing degree days (GDD), photo-thermal unit (PTU), helios thermal unit (HTU), photo-thermal index (PTI) and heat use efficiency (HUE) have frequently been used as a weather based parameters for assessing crop phenology. Therefore, all growth and development stages of crop may be estimated more accurately on the basis of GDD rather than calendar method (Warthington and Hatchinson 2005). Mustard crop require different amount of GDD, PTU, HTU, PTI and HUE for growth and development stages. The GDD is used to quantify effect of temperature and described the timing of different biological process (McMaster and Wilhelm 1997; Qiao-yan et al. 2012). The present investigation was carried out to quantify relationship of GDD, PTU, HTU, PTI and HUE with phenological development of crop.

MATERIALS AND METHODS

Field experiments were conducted during the winter (*Rabi*) seasons of 2009–10 and 2010–11 at student instructional farm of C.S. Azad University of Agriculture & Technology, Kanpur, Utter Pradesh, India, with the objective to study the relationship between GDD, PTU, HTU, PTI and HUE at different phenological stages with seed yield, and thereby to evaluate twenty two Indian mustard genotypes for high temperature tolerance under late sown condition. The experiment site situated at a height 125.9 m above mean sea level at 26°39′N latitude and 80°18′E longitude had a semi-arid type of climate. The soil of the experimental field was sandy loam in texture having 0.45 % organic carbon, 267.0 kg ha^{-1} available N, 18.2 kg ha^{-1} available P and 168.5 kg ha^{-1} available K with pH 7.4. Average temperature during crop grown season was 20.5 ± 1 °C. Experiment were laid out in randomize block design with three replications. The weather conditions (Table 1) (climatic features) in respect of rainfall, temperature, relative humidity and sun shine hours of the relevant years were obtained from the Agronomy Department of the C.S. Azad university of Agriculture and Technology, Kanpur.

All energy requirement parameters were recorded for three durations, viz., sowing to anthesis, sowing to 50 % flowering and sowing to physiological maturity.

Growing degree days was computed with 5 °C as base temperature on the basis of daily mean temperature with the help of following formula:

$$GDD = \frac{\sum(Tem.\max - Tem.\min)}{2} - Tb$$

Where, *Tem.max* and *Tem.min* are daily maximum and minimum temperature, Tb is the base temperature (5 °C).

Photo-thermal unit (PTU) (degree-days hours) was calculated on the basis of GDD and day length with the formula given below:

PTU=GDD×DaylengthPTU=GDD×Daylength

Helios thermal unit (HTU) (degree-days hours) was calculated on the basis of GDD and sunshine hours by the following formula:

HTU=GDD×Duration of sunshine hoursHTU=GDD×Duration of sunshine hours

Table 1. Weather data of standard meteorological week's from November to April, 2009–10 and 2010–11 at C.S. Azad university of Agriculture and Technology, Kanpur (*T* Temperature, *RH* relative humidity, *RF* rainfall, *BSS* bright sun shine h)

Meteorological weeks	2009–10						2010–11					
	T (°C)		RH (%)		RF day^{-1} (mm)	BSS (h)	T (°C)		RH (%)		RF day^{-1} (mm)	BSS (h)
	Max.	Min.	Max.	Min.			Max.	Min.	Max.	Min.		
45	30.66	13.63	90.14	38.86	00.0	7.5	29.96	17.66	90.14	48.29	0.0	3.0
46	27.27	16.90	87.00	58.43	1.5	3.9	26.40	17.06	91.29	73.57	8.1	3.5
47	24.84	12.56	82.71	47.57	00.0	3.8	26.23	16.74	95.57	69.14	2.5	3.5
48	26.17	8.94	85.86	31.86	00.0	3.4	24.17	9.29	86.43	66.00	1.5	7.8
49	25.54	8.90	91.43	41.86	00.0	5.6	24.63	11.40	83.86	50.86	0.0	6.1
50	25.69	10.74	90.00	41.43	00.0	3.5	23.54	7.73	85.29	35.29	0.0	5.6
51	24.14	10.57	89.29	43.43	00.0	3.7	23.60	7.31	84.86	35.29	0.0	5.9
52	22.10	6.70	89.57	41.29	00.0	5.6	23.06	9.64	91.71	48.29	0.2	2.4
53	17.36	8.03	94.29	69.57	1.4	2.4	–	–	–	–	–	–
1	16.04	7.56	95.57	81.71	00.0	0.5	11.73	4.23	95.71	74.57	0.0	0.2
2	14.09	8.54	95.57	80.00	0.5	0.5	20.51	5.29	89.43	48.71	0.0	3.4
3	15.94	6.63	93.57	71.57	00.0	1.6	22.80	6.80	84.00	32.71	0.0	7.0
4	24.11	8.03	91.57	46.14	00.0	6.1	23.14	8.06	86.43	49.86	0.0	6.2
5	25.14	9.29	90.57	39.29	00.0	7.0	25.23	9.71	90.57	44.14	0.0	6.2
6	23.10	13.71	95.14	69.29	1.6	3.3	26.56	11.91	85.57	38.71	0.0	8.8
7	25.31	10.97	85.00	54.86	00.0	7.0	23.40	11.69	92.00	55.00	0.9	4.6
8	28.94	13.90	86.14	43.86	00.0	8.6	25.17	11.61	89.43	45.71	0.3	6.6
9	31.86	16.26	80.14	36.57	00.0	8.0	28.09	13.39	77.14	38.00	0.0	8.1
10	30.83	14.19	76.71	29.71	00.0	9.7	29.51	14.91	73.57	36.29	0.0	9.8
11	34.23	17.37	80.14	28.00	00.0	8.7	33.51	16.63	77.57	43.14	0.0	8.5
12	39.14	19.10	74.14	20.71	00.0	8.3	34.86	16.71	72.57	47.29	0.0	9.5
13	38.46	19.63	57.71	24.14	00.0	8.2	33.77	19.17	82.43	58.29	0.0	8.1
14	40.14	20.96	32.14	17.43	00.0	8.3	36.23	16.89	55.71	27.71	0.1	8.5
15	42.54	22.46	35.43	13.14	00.0	9.7	37.37	19.63	50.00	21.29	0.4	6.7
16	41.46	28.20	38.43	15.29	00.0	5.83	37.23	20.80	59.86	22.57	1.1	8.8
17	41.44	34.20	60.00	22.40	00.0	7.64	37.80	23.60	56.00	28.29	0.0	8.0

Photo-thermal index (PTI) (degree-days day^{-1}) was calculated using the following equation:

PTI=GDD/GrowthdaysPTI=GDD/Growthdays

Heat use efficiency (HUE) (kg ha^{-1} degrees-day) was calculated with the help of seed yield (kg^{-1}ha) per GDD with the help of following equation:

HUE=Seed yield(kg ha−1)/GDDHUE=Seed yield(kg ha−1)/GDD

RESULTS AND DISCUSSION

Growing degree days (GDD) (degree-days) calculated for durations of sowing to anthesis, sowing to 50 % flowering and sowing to physiological maturity showed significant differences among tested mustard genotypes under late sown situation (Table 2). Pooled data analysis of both the years indicated higher values of GDD at anthesis, 50 % flowering and at physiological maturity in RH-8814 (595.1, 726.6 and 1703.0) followed by NRCDR-02 (573.6, 713.5 and 1669.2) and in BPR-549-9 (572.0, 685.8 and 1648.0), while the lowest GDD values were recorded in NPJ-113 (445.2, 543.5 and 1351.5) and in NPJ-117 (450.4, 559.0 and 1391.8) under late sown condition. Results on GDD at various phonological stages are in conformity of results reported by Renganayaki and Krishnasamy (2013).

Table 2. Growing degree days, photo-thermal unit and helios thermal unit at different growth stages in Indian mustard under late sown condition (pooled 2 years)

Genotypes	Traits								
	GDD (degree-days)			PTU (degree-days hour)			HTU (degree-days hour)		
	Days to anthesis	Days to 50 % flowering	Days to Physiological maturity	Days to anthesis	Days to 50 % flowering	Days to physiological maturity	Days to anthesis	Days to 50 % flowering	Days to physiological maturity
DRMR-802	540.3	662.2	1,569.2	5,555.7	6,897.5	17,497.8	2,338.1	3,036.2	10,276.2
PBR-331	527.7	632.1	1,533.6	5,417.7	6,565.5	16,946.4	2,266.9	2,834.6	9,928.4
NDC-601	533.2	650.1	1,512.1	5,479.2	6,625.2	16,822.8	2,233.6	2,935.5	9,848.8
RH-0,216	504.9	594.0	1,453.9	5,166.9	6,146.4	15,990.9	2,205.4	2,615.9	9,218.0
NPJ-113	445.2	543.5	1,351.5	4,535.2	5,591.5	14,761.5	2,084.3	2,407.2	8,281.7
RB-50	527.1	637.8	1,510.7	5,410.5	6,628.2	16,672.5	2,225.9	2,865.4	9,737.5
NPJ-117	450.4	559.0	1,391.8	4,587.4	5,762.0	15,245.1	2,084.3	2,485.1	8,656.3
RH-0,447	496.4	615.8	1,432.4	5,215.3	6,386.2	15,732.3	2,212.5	2,738.7	9,014.6
DLM-2	522.4	650.1	1,567.6	5,358.8	6,625.2	17,354.4	2,289.8	2,935.5	10,202.1
RH-0,305	523.1	605.8	1,402.2	5,224.6	6,276.2	15,609.9	2,181.6	2,681.2	8,934.0
NRCDR-701	524.6	649.1	1,521.7	5,453.4	6,753.4	16,803.9	2,353.3	2,949.1	9,825.1
RH-0,116	524.4	662.4	1,535.1	5,381.4	6,899.1	16,965.0	2,266.9	3,026.8	9,948.3
NRCDR-02	573.6	713.5	1,669.2	5,992.4	7,460.9	18,694.2	2,556.5	3,344.5	11,072.2
RB-55	543.8	643.6	1,589.6	5,594.8	6,692.3	17,618.4	2,342.4	2,897.5	10,344.6
RH-8,814	595.1	726.6	1,703.0	6,228.4	7,605.6	19,129.8	2,643.3	3,448.7	11,397.7
DRMR-537-40	536.2	650.1	1,569.2	5,511.2	6,625.2	17,373.6	2,303.3	2,935.5	10,205.8
RGN-773	548.4	674.3	1,635.9	5,644.8	7,029.7	18,234.0	2,371.0	3,130.2	10,783.0
PBR-330	546.4	640.0	1,600.5	5,624.4	6,652.4	17,779.2	2,345.8	2,904.1	10,462.5
RGN-197	522.0	627.3	1,565.8	5,355.0	6,513.0	17,332.8	2,266.9	2,801.6	10,171.0
BPR-349-9	557.0	655.7	1,600.5	5,739.4	6,686.3	17,779.2	2,461.2	2,987.1	10,462.5
BPR-549-9	572.0	685.8	1,648.0	5,904.4	7,156.8	18,379.8	2,492.7	3,209.7	10,876.0
RH-555A	529.0	655.5	1,577.6	5,431.4	6,822.9	17,474.9	2,303.3	3,000.9	10,270.5
Mean	529.2	642.4	1,542.7	5,446.0	6,654.6	17,099.9	2,310.4	2,916.9	9,742.2
SE (d) ±	10.2	8.6	3.9	21.5	96.4	156.1	27.4	48.5	62.5
CD at 5 %	20.6	17.3	8.0	43.3	194.3	315.1	55.2	97.7	126.1

Photo-thermal unit (PTU) (degree-days hours) increased linearly with increase in plant age up to physiological maturity of crop in both the years of study. PTU recorded at anthesis, 50 % flowering and at physiological maturity ranged from 4961.7 to 6488.2, 5851.3 to 7619.0 degree-days hours. Based on pooled analysis the highest value of PTU was observed in RH-8814 (6228.4, 7605.6 and 19129.8), followed by NRCDR-02 (5992.4, 7460.9 and 18694.2) and in BPR-549-9 (5904.4, 7156.8 and 18379.8), while the lowest PTU were recorded in NPJ-113 (4535.2, 5591.5 and 14761.5) and NPJ-117 (4587.4, 5762.0 and 15245.1) at anthesis, 50 % flowering and physiological maturity under late sown condition. Khushu et al. (2008) reported that the photo-thermal units (PTU) were highest in early sowing (D_1), followed by late sowings (D_2 and D_3).

Helios thermal unit (HTU) (degree-days hours) recorded at anthesis, 50 % flowering and at physiological maturity (Table 2) increased linearly with increase in duration of crop during both the years of investigation, and showed significant variability among 22 mustard genotypes. Based on pooled analysis, at anthesis, 50 % flowering and at physiological maturity, the higher values of HTU were recorded in RH-8814 (2643.3, 3448.7 and 11397.7), followed by NRCDR-02 (2556.5, 3344.5 and 11072.2) and BPR-549-9 (2492.7, 32.09.7 and 10876.0), while the lower values of HTU were recorded in NPJ-113 (2084.3, 2407.2 and 8281.7), followed by NPJ-117 (2084.3, 2485.1 and 8656.3) under late sown condition. These result are in conformity of the results reported on phenological development in mustard crop by Srivastava and Balkrishna (2003).

Results of PTI (degree-days day^{-1}) at anthesis revealed highest values in genotypes NPJ-113 (10.60), followed by NPJ-117 (10.47) and RH-447 (10.03) (Table 3). However, at 50 % flowering and physiological maturity the higher values for PTI were observed in genotypes RH-8814 (10.02 and 13.25), NRCDR-02 (9.98 and 13.14) and BPR-549-9 (9.87 and 13.08). Similar results on PTI were supported by Prasanta Neog Chakravarty (2005).

Heat use efficiency (HUE) (kg ha^{-1} degrees-day) recorded at anthesis, 50 % flowering and physiological maturity (Table 3 showed higher values for genotypes RH-8814 (1.17, 9.62 and 4.11), NRCDR-02 (1.12, 8.99 and 3.84) and BPR-549-9 (1.07, 8.91 and 3.74) at anthesis, 50 % flowering and physiological maturity, respectively. Results on HUE are in conformity of the results reported by Prasanta Neog Chakravarty (2005).

Seed yield plant^{-1} recorded at harvest (Table 3) showed significant genotypic variation among 22 genotypes under late sown condition during 2009–10 and 2010–11. Crop faces high temperature under late sown condition at seed development phase. Pooled data of two crop seasons (Table 4) revealed that seed yield per plant was highest in RH-8814 (43.7 g), NRCDR-02 (40.10 g), and BPR-549-9 (38.20 g). These finding are in conformity with the results reported by Singh et al. (2001) and (2013). Studies of correlation matrix of seed yield with GDD ($r = 0.61$, 0.65 and 0.75), PTU ($r = 0.66$, 0.64 and 0.74), HTU ($r = 0.79$, 0.68 and 0.73) were highly significant at anthesis, 50 % flowering and at physiological maturity (Table 4). From the data it is evident that GDD, PTU, HTU, PTI and HUE can be recommended as indices for selection of genotypes for high temperature tolerance. Further, genotypes RH-8814, NRCDR-02 and BPR-549-9 having highest GDD, PTU, HTU, PTI and HUE as well as seed yield can be recommended for breeding programme for developing high yielding genotypes for late sown-high temperature condition.

Table 3. Photo-thermal index, heat use efficiency and seed yield plant^{-1} in Indian mustard under late sown condition (Pooled 2 years)

Genotypes	Traits						
	Photo-thermal index (degree-days day^{-1})			Heat use efficiency (kg ha^{-1} degrees-day)			Seed yield (g plant^{-1})
	Days to anthesis	Days to 50 % flowering	Days to physiological maturity	Days to anthesis	Days to 50 % flowering	Days to physiological maturity	
RH-0305	9.87	9.69	12.30	0.62	5.34	2.30	20.20
RB-50	9.85	9.74	12.59	0.72	5.92	2.50	23.60
RH-0447	10.03	9.70	12.30	0.83	6.65	2.86	25.60
RGN-773	9.79	9.84	13.04	0.85	6.95	2.87	29.30
RH-0116	9.89	9.81	12.69	0.89	7.03	3.03	29.10
RH-555A	9.89	9.78	12.83	0.91	7.35	3.05	30.10
BPR-349-9	9.86	9.79	12.91	0.93	7.88	3.23	32.30
PBR-330	9.76	9.77	12.91	0.95	8.10	3.24	32.40
DRMR-802	9.82	9.81	12.81	0.94	7.71	3.25	31.90
DLM-2	9.95	9.78	12.80	0.98	7.90	3.28	32.10
NPJ-117	10.47	9.89	12.16	1.02	8.19	3.29	28.60
RH-0216	10.00	9.74	12.37	0.96	8.13	3.32	30.20
NDC-601	9.78	9.78	12.60	0.95	7.75	3.33	31.50
RGN-197	9.67	9.73	12.78	1.01	8.42	3.37	33.00
PBR-331	9.86	9.72	12.67	0.99	8.25	3.40	32.60
NPJ-113	10.60	9.88	12.01	1.04	8.54	3.43	29.00
DRMR-537-40	9.84	9.78	12.81	1.03	8.49	3.52	34.50
NRCDR-701	9.81	9.76	12.63	1.04	8.38	3.57	34.00
RB-55	9.80	9.75	12.87	1.06	8.92	3.61	35.90
BPR-549-9	9.78	9.87	13.08	1.07	8.91	3.71	38.20
NRCDR-02	9.64	9.98	13.14	1.12	8.99	3.84	40.10
RH-8814	9.60	10.02	13.25	1.17	9.62	4.11	43.70
Mean	9.87	9.81	12.72	0.96	7.90	3.29	31.72
SE (d) ±	0. 11	0.11	0.11	0.09	0.10	0.09	0.16
CD at 5 %	0.22	0.22	0.22	0.18	0.21	0.18	0.32

Table 4. Correlation of growing degree days, photo-thermal unit, helios thermal unit, photo-thermal index and heat use efficiency with seed yield of mustard genotypes under late sown condition

Traits	2	3	4	5	6	7	8	9	10	11	12	13	14	15	16
Seed yield (1)	0.61**	0.65**	0.75**	0.66**	0.64**	0.74**	0.79**	0.68**	0.73**	–0.01	0.05	–0.17	0.15	0.15	0.18*
GDD at anthesis (2)	–	0.94**	0.91**	0.99**	0.93**	0.93**	0.92**	0.92**	0.93**	–0.08	0.01	–0.07	0.24**	0.24**	0.27**
GDD at 50 % flowering (3)		–	0.92**	0.96**	0.99**	0.93**	0.91**	0.99**	0.94**	–0.03	0.01	–0.06	0.21*	0.23**	0.26**
GDD at maturity (4)			–	0.92**	0.92**	0.99**	0.15	0.18*	0.26**	–0.14	0.07	–0.03	0.31**	0.33**	0.37**
PTU at anthesis (5)				–	0.95**	0.94**	0.94**	0.94**	0.94**	–0.08	0.01	–0.08	0.24**	0.24**	0.24**
PTU at 50 % flowering (6)					–	0.93**	0.91**	0.99**	0.93**	–0.07	0.00	–0.03	0.21*	0.22*	0.26**
PTU at maturity (7)						–	0.92**	0.94**	0.99**	–0.14	0.05	–0.03	0.30**	0.31**	0.36**
HTU at anthesis (8)							–	0.93**	0.92**	–0.11	0.10	–0.01	0.30**	0.30**	0.34**
HTU at 50 % flowering (9)								–	0.94**	–0.05	0.03	–0.13	0.23**	0.24**	0.28**
HTU at maturity (10)									–	–0.13	0.05	–0.05	0.29**	0.31**	0.35**
PTI at anthesis (11)										–	0.01	–0.82	–0.06	–0.09	–0.18
PTI at 50 % flowering (12)											–	0.40**	0.63**	0.58**	0.63**
PTI at maturity (13)												–	0.39**	0.40**	0.46**
HUE at anthesis (14)													–	0.99**	0.98**
HUE at 50 % flowering (15)														–	0.98**
HUE at maturity (16)															1.00

*, ** significant 0.05 and 0.01 levels

OPEN ACCESS

REFERENCES

1. Khushu, M.K. Naseer-u-Rahman Mahender Singh Arvind Prakash Tiku, A.K. Bali, A.S. (2008). Thermal time indices for some mustard genotypes in the Jammu region, *Journal of Agrometeorology 10*(2): 224–227.
2. McMaster, G. S., & Wilhelm, W. W. (1997). Growing degree days: One equation, two interpretations. *Agricultural and Forest Meteorology, 87*, 291–300.
3. Prasanta, N. C. (2005). Thermal indices in Brassica grown under a semi-arid environment. *Annals of Agricultural Research, 26*(2), 291–296
4. Qiao-yan, L. I., Jun, Y. I. N., Wan-dai, Liu., Mei, Zhou., Lei, L. I., Ji-shan, Niu., et al. (2012). Determination of optimum growing degree days (GDD) range before winter for wheat cultivars with different growth characteristics in North China Plain. *Journal of Integrative Agriculture,11*, 405–415.
5. Renganayaki, P. R., & Krishnasamy, V. (2013). Correlation equation for synchronized flowering in sunflower (*Helianthus annuus* L.) hybrid KBSH-1 parental lines based on meteorological parameters. *African Journal of Agricultural Research, 8*(17), 1738–1742.
6. Singh, A., Avtar, R., Singh, D., Sangwan, O., & Balyan, P. (2013). Genetic variability, character association and path analysis for seed yield and component traits under two environments in Indian mustard. *journal of Oilseed Brassica, 4*(1), 43–48.
7. Singh, R., Patidar, M., & Singh, B. (2001). Response of Indian Mustard (*Brassica juncea*) cultivars to different sowing time. *Indian Journal of Agronomy, 46*, 292–295.
8. Srivastava, J. P., & Balakrishna, (2003). Environmental parameters influence phonological development of mustard in relation to yield. *Indian Journal of Plant Physiology, 8*(4), 349–353.
9. Warthinhton, C. M., & Hatchinson, C. M. (2005). Accumulated degree days as a model to determine key development stages and evacuate yield and quality of potato in Northeast Florida. *Proceeding of State Horticulture Society, 118*, 98–101.

CHAPTER 17

Senescence, Stress, and Reactive Oxygen Species

Ivan Jajic [1], Tadeusz Sarna [2] and Kazimierz Strzalka [1,3,*]

[1]*Department of Plant Physiology and Biochemistry, Faculty of Biochemistry, Biophysics and Biotechnology, Jagiellonian University in Krakow, Gronostajowa 7, Krakow 30-387, Poland*

[2]*Department of Biophysics, Faculty of Biochemistry, Biophysics and Biotechnology, Jagiellonian University in Krakow, Gronostajowa 7, Krakow 30-387, Poland*

[3]*Malopolska Centre of Biotechnology, Jagiellonian University in Krakow, Gronostajowa 7, Krakow 30-387, Poland*

ABSTRACT

Generation of reactive oxygen species (ROS) is one of the earliest responses of plant cells to various biotic and abiotic stresses. ROS are capable of inducing cellular damage by oxidation of proteins, inactivation of enzymes, alterations in the gene expression, and decomposition of biomembranes. On the other hand, they also have a signaling role and changes in production of ROS can act as signals that change the transcription of genes that favor the acclimation of plants to abiotic stresses. Among the ROS, it is believed that H_2O_2 causes the largest changes in the levels of gene expression in plants. A wide range of plant responses has been found to be triggered by H_2O_2 such as acclimation to drought, photooxidative stress, and induction of senescence. Our knowledge on signaling roles of singlet oxygen ($^1O_2$) has been limited by its short lifetime, but recent experiments with a *flu* mutant demonstrated that singlet oxygen does not act primarily as a toxin but rather as a signal that activates several stress-response pathways. In this review we summarize the latest progress on the signaling roles of ROS during senescence and abiotic stresses and we give a short overview of the methods that can be used for their assessment.

Keywords: leaf senescence; abiotic stresses; reactive oxygen species; hydrogen peroxide; singlet oxygen; superoxide anion

1. INTRODUCTION

Senescence in plants is a complex deterioration process that can lead to the death of whole organisms or a single organ. It is regulated by autonomous (internal) factors (age, reproductive development, and phytohormone levels) and by environmental signals, including photoperiod, stresses such as drought, ozone, nutrient deficiency, wounding, and shading [1]. The generation of reactive oxygen species (ROS) is one of the earliest responses of plant cells under abiotic stresses and senescence [2,3]. In plants, ROS are formed as byproducts of aerobic energy metabolism and of plants being exposed to various biotic and abiotic stresses [4,5,6]. Under normal conditions, the production of ROS in cells is maintained at low levels by antioxidant enzymes. This balance can be disrupted by a depletion of antioxidants or the excess accumulation of ROS, leading to oxidative stress, and consequently to damage to cellular macromolecules and membranes and an increase in lipid peroxidation [7,8]. ROS-induced oxidative stress limits agricultural yields worldwide [9]; in the United States alone, it is estimated that the deleterious effects of abiotic stresses on agricultural production are responsible for losses amounting to billions of dollars annually [10]. Plants have evolved different mechanisms to protect themselves from adverse environmental conditions, such as the process of acclimation, which involves less ROS production coupled with an efficient antioxidant defense [11,12] and the activation of different signaling pathways [13,14].

Initially, ROS were exclusively considered toxic metabolic products that can damage cellular components, but now it is clear that ROS play a dual role in plants both as toxic compounds and as key regulators of many biological processes [15,16,17]. The important role of ROS in signaling has been demonstrated in many studies. It has been shown that ROS modulate the activity of key signaling compounds such as MAP kinases [18], provide protection against pathogen invasion [19,20], stimulate abiotic stress tolerance [21], and have an important role during early responses to wounding [22]. Despite the importance of ROS, our knowledge of the mechanism of their action is still limited. In this review we will try to summarize the latest progress on the roles of reactive oxygen species during senescence and abiotic stresses.

2. ROS DETECTION

The main problem with an accurate determination of the role of ROS in senescence and abiotic stresses is the simultaneous generation of ROS [23,24] and the limited number of non-invasive and specific methods that can be used for their determination. For example, in plants suffering from moderate light stress singlet oxygen ($^1O_2$), superoxide anion ($O^{\bullet}{}_2^-$) and hydrogen peroxide (H_2O_2) are released simultaneously [23], making it difficult to establish their individual roles. This is further complicated by significant differences in the production of ROS when senescence is induced artificially and when the plant ages naturally [25]. The effects of H_2O_2 on gene expression have also been reported to be different when the H_2O_2 was applied exogenously and when it was induced in response to high light [26]. Finally, several studies revealed that multiple stressors, as usually encountered by plants in nature, could substantially change the expression patterns of genes determined in a single factor analysis [27,28,29,30]. There is a need for an experimental model that can take into account all the factors that can influence the outcome of research into the role of ROS.

There are a great number of user-friendly assays that can be used for the measurement of ROS production in plant tissues, including the fluorometric determination of H_2O_2 with Amplex Red [31] and the colorimetric determination of $O^{\bullet}{}_2^-$ with XTT [32]. Although easy to use, these methods lack specificity. This is because Amplex Red is somewhat unstable—it can be autooxidized and produce $O^{\bullet}{}_2^-$ and H_2O_2 [33], and it can react not only with H_2O_2 but also with other redox active compounds [34], while XTT can be reduced by short-chain sugars [35]. Higher specificity can be achieved by using the spin trapping technique by electron paramagnetic resonance (EPR) spectroscopy. Spin trapping involves the reaction between a nitrone or nitroso compound and a free radical to form a stable spin adduct [36]. The spin adduct usually yields a distinctive EPR spectrum characteristic of the particular free radical that is trapped. Under normal conditions, the flux of ROS generated in cells is maintained at low levels by the action of antioxidant enzymes and low molecular weight antioxidants, making the detection of ROS difficult. In order to successfully detect ROS, an imbalance between the production of ROS and their decay needs to be created. This can be achieved by the illumination of samples with visible light in the presence of an appropriate spin trap or with the addition of the spin trap immediately after illumination [37,38,39,40]. There are numerous publications with detailed descriptions of how to use EPR spectroscopy for the detection of hydrogen peroxide [39], the superoxide anion [40], the hydroxyl radical [41], and singlet oxygen [38] in plant tissues. One of the limitations of spin trapping is that the method does not provide information on the specific sites of ROS production in tissues due to solvent incompatibility with living tissue and high concentrations of spin traps needed [42], Also, it typically underestimates actual *in situ* ROS concentrations because only a small fraction of the radicals produced is usually trapped [43]. Non-invasive, *in vivo* measurement of ROS can be achieved using fluorescent probes in combination with confocal laser scanning microscopy (CLSM) [44] or fluorescence microscopy [45,46,47]. One advantage of CLSM methodology is the possibility of studying the intracellular location of ROS using simultaneously specific fluorescent probes for different organelles. 2′,7′-Dichlorofluorescein diacetate (DCF-DA) can be used for monitoring H_2O_2 in plant tissues [48]; however, it lacks specificity since it was demonstrated that it can also react with other peroxides [49]. Dihydroethidium (DHE) is a more specific probe that can be used for detection of $O^{\bullet}{}_2^-$ in different plant tissues [50,51]. The determination of singlet oxygen with a singlet oxygen sensor green (SOSG) reagent showed high specificity to $^1O_2$ without the interference of hydroxyl radicals or superoxide [46]. Unfortunately, even in the absence of external $^1O_2$ photosensitisers, the indicator can be converted to its green fluorescent form upon exposure to either UV or visible radiation. This could obviously lead to a wrong estimation of $^1O_2$ levels [52]. Real-time monitoring of reactive oxygen species in living cells and tissues can be achieved with a genetically encoded redox probe such as HyPer and reduction-oxidation sensitive green fluorescent protein (roGFPs) [53]. These probes offer high specificity and can be used for determination of ROS in various subcellular compartments. HyPer is highly sensitive to hydrogen peroxide, is insensitive to other oxidants, and it does not cause artifactual ROS generation, thus having great potential in plant biology [54], while roGFP can be used for determination of H_2O_2, but also for determination of thiol redox state depending on its form [53]. Disadvantages of these probes include the necessity of pH control and possible antioxidant activity, which need to be taken into consideration [53]. In Table 1 we summarize the methods that can be used for measurement of ROS production in plant samples.

Table 1. Overview of the methods for determination of ROS in plan samples.

Method/Probe	Advantage	Disadvantage	Used for
Amplex Red	Easy to use and fast	Can be autooxidized, reacts with other redox reactive compounds	H_2O_2
XTT	Easy to use and fast	Non-specific, can be reduced by short-chain sugars	$O^{\bullet}{}_2^-$
EPR spin trapping	Highly specific, can be used for determination of numerous ROS	Expensive, laborious, method does not provide information on the specific sites of ROS production in tissues	H_2O_2, $O^{\bullet}{}_2^-$, $^1O_2$, OH^-
DCF – DA	*In vivo* determination of intracellular ROS production	Non-specific, reacts with other peroxides, expensive equipment	H_2O_2
DHE	*In vivo* determination of intracellular ROS production	Expensive equipment	$O^{\bullet}{}_2^-$
SOSG	*In vivo* determination of intracellular ROS production	Wrong estimation of $^1O_2$ upon exposure to visible or UV light	$^1O_2$
HyPer and roGFP	Real time monitoring of ROS in living cells and tissues	Necessity for pH control and possible antioxidant activity	H_2O_2

3. SUPEROXIDE ANION

In plants $O^{\bullet}{}_2^-$ is generated in different cell compartments, including chloroplasts, peroxisomes, apoplast, the mitochondrial electron transport chain, and the plasma membrane [55,56,57]. The primary source of the superoxide anion in chloroplasts are Mehler reactions, during which O_2 is reduced by electrons from the photosynthetic electron transport chain [58]. Generated $O^{\bullet}{}_2^-$ is then converted to hydrogen peroxide (H_2O_2), mostly by the action of CuZn-superoxide dismutase (SOD) [59]. Thus, SOD determines the lifetime of $O^{\bullet}{}_2^-$ in cells and the probability of its involvement in biochemical processes. Furthermore, $O^{\bullet}{}_2^-$ is a moderately reactive, short-lived ROS with a half-life of approximately 2–4 μs [17], and it cannot cross the chloroplast membrane [60]. For the reasons listed above, a signaling role of $O^{\bullet}{}_2^-$ during senescence and abiotic stresses seems unconvincing. However, chloroplasts are not the only sites of $O^{\bullet}{}_2^-$ production. In peroxisomes, $O^{\bullet}{}_2^-$ is being generated by two different sources: in peroxisomal matrix via action of enzyme xanthine oxidase [61,62] and by electron transport chain (ETC) in peroxisomal membrane [63]. Peroxisomes can be considered as an important source of signaling molecules since they have capacity to rapidly produce and scavenge H_2O_2 and $O^{\bullet}{}_2^-$ due to the presence of many antioxidants in these organelles. Another important source of $O^{\bullet}{}_2{}^-$ in plant cells are NADPH oxidases (NOX), in plants commonly known as respiratory burst oxidase homologs (Rbohs), which catalyze the production of $O^{\bullet}{}_2{}^-$ [64,65]. Plant Rbohs have been intensively studied recently since they play key roles in many physiological processes, such as ROS signaling and stress responses [66,67]. Finally, $O^{\bullet}{}_2{}^-$ is also produced in cytosol by action of xanthine dehydrogenase and the aldehyde oxidase [68,69]. Numerous studies have reported an increase in the production of $O^{\bullet}{}_2{}^-$during natural and artificially induced senescence [70,71]; however, attributing a specific signaling role to this increase is extremely difficult since the increase in most cases is accompanied by the production of other ROS and the quick conversion of $O^{\bullet}{}_2{}^-$ to H_2O_2. High production of ROS is damaging to the cell due to oxidative modifications of key cellular components and may ultimately lead to plant cell death [72]. In a recent study [73] it was shown that under the high temperature treatments, large amounts of $O^{\bullet}{}_2{}^-$ and H_2O_2 were generated and accumulated in cucumber leaves, leading to premature senescence, which is indicated by the changes in protein, lipid peroxidation (LPO), and chlorophyll content. Nevertheless, a signaling role of $O^{\bullet}{}_2{}^-$ was demonstrated in *Arabidopsis thaliana* plants exposed to methyl viologen, a superoxide anion propagator, under light. The generation of $O^{\bullet}{}_2{}^-$ in the absence of H_2O_2 accumulation revealed a subset of nuclear encoded genes that are likely to be specific for an $O^{\bullet}{}_2{}^-$-mediated signaling pathway [74]. Data analysis identified a strong upregulation of genes belonging to categories functioning in abiotic stress responses, among them WRKY6, which has previously been reported to play a role during senescence and in defense-related processes [75]. Recently, we investigated the production of $O^{\bullet}{}_2{}^-$ during the development and senescence of secondary barley leaves by using EPR--a spin trapping method with DMPO as a spin trap [76]. It was shown that the production of $O^{\bullet}{}_2{}^-$ increases during the development of barley, reaching its highest level right after the onset of senescence. Thereafter, the production of $O^{\bullet}{}_2{}^-$started to decline till the end of the senescence process. This was accompanied with an increase in membrane fluidity during the same period [77], which could be a factor facilitating the increase in the generation of ROS. An increase in $O^{\bullet}{}_2{}^-$was also observed in the interveinal area of senescing tobacco leaves, as well as in the minor veins of mature and senescent leaves, while it was absent in the major veins [78]. It is hypothesized that spatial differences in the superoxide anion are important for the non-uniform downregulation of photosynthesis-associated genes. A further role of $O^{\bullet}{}_2{}^-$ as a signaling molecule was demonstrated during the early wound response in an experiment with *Medicago* leaves where ROS production was inhibited with diphenyleneiodonium (DPI). The rapid (≤3 min) DPI inhibition of phase I $O^{\bullet}{}_2{}^-$ production suppressed the differential regulation of 7 out of 19 wound responsive proteins, showing that early, wound-related

$O^{\bullet}{}_2^-$production (phase I) provides an essential signal for wound-related changes in the leaf apoplast proteome [22]. Increased production of $O^{\bullet}{}_2^-$ was observed in plant responses to cadmium stress in pea (*Pisum sativum* L.) [48]. Exposure to Cd leads to an oxidative stress as a result of disturbance in antioxidant defense and a decrease in NO level. It was demonstrated that NO can mitigate the deleterious effect of Cd on lupine roots [79] and that it has a possible antioxidant effect in its ability to react with $O^{\bullet}{}_2^-$ to prevent oxidative damage [80]. In this way, $O^{\bullet}{}_2^-$ could contribute to plant responses to abiotic stresses.

4. HYDROGEN PEROXIDE

Hydrogen peroxide plays an important role in plants under stress conditions as a signaling molecule that mediates between different physiological processes [81]. It is involved in the regulation of the senescence process [82], protection against pathogen attack [83], the reduction of stress intensity at low light [84], and the alleviation of drought stress [85], and it can influence the expression of hundreds of genes [86]. Hydrogen peroxide is produced in plants via two possible pathways: dismutation of $O^{\bullet}{}_2^-$ with the involvement of SOD [59], and via oxidases such as amino and oxalate oxidases [87]. The level of H_2O_2 is kept under control by a fine-tuned network of enzymatic and low-molecular-weight antioxidants that prevent the excess accumulation of H_2O_2 [88]. Production and scavenging of H_2O_2 in plant cells has been summarized in Figure 1. The balance between SODs and the different H_2O_2-scavenging enzymes in cells is considered to be crucial in determining the steady-state level of H_2O_2 [89]. In comparison with other ROS, H_2O_2 is the most stable and least reactive ROS, and it can easily cross the membrane [81,90], which makes it a good signaling molecule. H_2O_2 plays a versatile role in plants; as a signaling molecule it is involved in the regulation of various abiotic and biotic stresses [81] and, at high concentrations, it has an important role in cell death and during the final stages of senescence, when it contributes to cell degradation [16,17]. The dual role of H_2O_2 was confirmed in a recent study, in which treatment with 600 μM H_2O_2 caused an increase in the vase life of a cut Oriental × Trumpet hybrid lily "Manissa," while concentrations of 800 and 1200 μM resulted in negative effects [91]. Further evidence that the effects of H_2O_2 are dose dependent comes from a study in which wax apple trees were spray-treated with different concentrations of H_2O_2 under field conditions [92]. Spraying wax apple fruits with 5 and 20 mM of H_2O_2 once a week produced better fruit growth and maximized the yield and quality in comparison with the control and with a higher dose of 50 mM of H_2O_2.

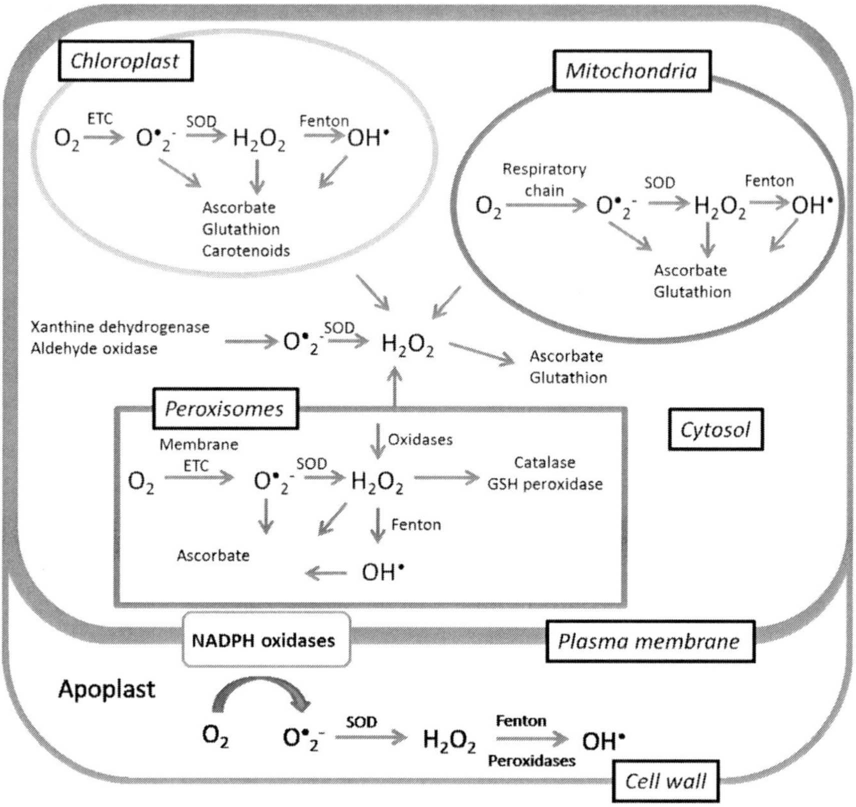

Figure 1. Production and scavenging of ROS in a plant cell. Figure legend: Superoxide dismutase (SOD); Fenton (decomposition of hydrogen peroxide to highly reactive hydroxyl radical in presence of iron); ETC (electron transport chain).

H_2O_2 plays an important role during the senescence process, where it was shown that it could be used as a signal to promote senescence in different plant species, and to be part of a complex regulatory network [93]. It was shown that H_2O_2increases at the point when the plants start to bolt and flower, which is supported by a decrease in ascorbate peroxidase 1 activity at the same time [16]. This probably induces the expression of transcription factors and senescence-associated genes. Indeed, expression analysis showed that H_2O_2 treatment induced 14 out of 36 salt-triggered senescence-associated genes and 15 senescence-associated NAC genes [94], indicating that salt-triggered senescence at least in part involves H_2O_2-mediated signaling through NAC transcription factors. This is supported by studies in which it was demonstrated that

H_2O_2 treatment induces the expression of NAC transcription factors ORS1 [95] JUB1 [96] and ATAF1 [97], which have a significant impact on progression of senescence. Overexpression of ORS1 triggers the expression of senescence-associated genes and accelerates senescence in transgenic plants, whereas its inhibition delays it. Contrary to ORS1, the overexpression of JUB1 strongly delays senescence, dampens intracellular H_2O_2 levels, and enhances tolerance to various abiotic stresses. *ATAF1* overexpression stimulates the progression of senescence by activating senescence promoting transcription factor *ORE1* and repressing chloroplast maintenance transcription factor *GLK1*. The generation of H_2O_2during the development and senescence of barley was investigated in our recent study [76]. The results showed an increase in the production of H_2O_2 with the development of barley, with the highest levels observed right after the induction of senescence and at the very end of the senescence process, which is similar to the pattern observed in the study discussed above [16]. Our research provides further evidence of the important role of H_2O_2 during leaf senescence in two different aspects: as a signal molecule for the induction of senescence, and in the degradation of molecules at later stages of senescence.

Several studies have indicated that H_2O_2 can interplay with other signal molecules that are important for plant development and during senescence such as abscisic acid (ABA) and ethylene [12,98]. It was shown that H_2O_2 could be involved in the signaling of plant growth regulators such as ethephon [98]. The application of ethephon results in an elevation in H_2O_2 levels, which is accompanied by the increased expression of sweet potato catalase. The elimination of H_2O_2 influence by exogenous-reduced glutathione alleviates ethephon-mediated effects. Recently, the interaction between ABA, H_2O_2, and ascorbic acid in Mediterranean shrubs during summer drought was investigated. It was shown that the drought stress ABA–H_2O_2 interaction can induce an increase in ascorbic acid, maintaining and even decreasing the ascorbate oxidative status under summer drought conditions, thereby protecting plants from oxidative damage.

There are numerous studies demonstrating the important role of H_2O_2 in the acquisition of tolerance to different abiotic and biotic stresses [84,85,99,100]. It was recently shown that pre-treatment with H_2O_2 provides protection against heat stress and low light induced oxidative stress by modulating the activity of antioxidant enzymes. The exogenous application of H_2O_2 can induce tolerance to heat stress in seedlings of *Cucumis sativus* cv Lvfeng no. 6 [100]. The pre-treatment of cucumber leaves with H_2O_2 and heat increased antioxidant enzyme activities, decreased lipid peroxidation, and thus protected the ultrastructure of chloroplasts under heat stress. Similarly, it has been shown that exogenous H_2O_2 can have a beneficial effect on low light induced oxidative stress [84]. Low light induces an oxidative stress [101], which increases ROS and causes lipid peroxidation. H_2O_2 pre-treatment of cucumber leaves resulted in decreased levels of $O^{\bullet}{}_2^-$, endogenous H_2O_2, and malonaldehyde by moderating the activities of antioxidant enzymes, thus reducing lipid peroxidation and stress intensity at low light. Pre-treatment with H_2O_2 can also increase drought stress tolerance in soybean leaves by promoting the expression of stress-response genes [85]. Exogenous application of H_2O_2 caused an increase in the mRNA levels of key enzymes for the biosynthesis of oligosaccharides, which are known to help plants tolerate drought stress. This enabled the soybean plant to avoid drought stress through the maintenance of leaf water content and thus to delay foliar wilting. Finally, hydrogen peroxide contributes to defense responses against pathogens. It was demonstrated that H_2O_2 is important for the greater tolerance of kumquat leaves infected with *Xanthomonas axonopodis* than that of grapefruit [19]. Infected kumquat leaves have a high accumulation of H_2O_2, which is promoted by the suppression of ascorbate peroxidase activity and later by the suppression of catalase activity, both involved in maintaining H_2O_2 at low levels. H_2O_2 can then be used as a substrate for the higher activity of Class III peroxidase in the apoplast, which is known to be involved in plant defense against pathogens.

5. SINGLET OXYGEN

Singlet oxygen is the highly reactive, excited state of molecular oxygen that can be formed in a reaction between O_2 and the chlorophyll triplet state [102]. Unlike the formation of H_2O_2 and $O^{\bullet}{}_2^-$, the formation of $^1O_2$ is not accompanied by the transfer of an electron to O_2. Instead, one of the unpaired electrons is promoted to a higher energy orbital [103]. Under normal conditions, $^1O_2$ is generated during photosynthesis by the photo activation of photosensitizers, mainly chlorophylls and their precursors [102]. Singlet oxygen is also generated during senescence [25] and under different abiotic stresses [104,105,106]. Similarly to other ROS, $^1O_2$ has a dual effect. As an oxidizing agent it can react with various biological molecules, causing damage and leading to cell death [107]. It can also play a signaling role by activating the expression of different genes [23,108]. As a result of its high reactivity and short lifetime of 3.1 to 3.9 μs in pure water [109], $^1O_2$ is able to interact with molecules mostly in its nearest environment. The diffusion distance of $^1O_2$ has been calculated to be up to 10 nm in a physiologically relevant situation [110]. On the other hand, it was demonstrated that $^1O_2$ is capable of diffusing a distance of over 270 nm in rat nerve cells [111] and that $^1O_2$ produced in the photosynthetic apparatus of *C. reinhardtii* under high light is capable of leaving the thylakoid membrane and reaching the cytoplasm or even the nucleus [112], which makes its role as a signaling molecule feasible.

Significant progress in the investigation of the role of singlet oxygen in signaling was achieved with the use of a conditional fluorescent (*flu*) mutant of *Arabidopsis* that accumulates the photosensitizer protochlorophyllide in the dark and generates singlet oxygen after transfer to light [107,113]. Following illumination with light, a different set of nuclear genes are activated within the *flu* mutant, and they are different from those induced by $O^{\bullet}{}_2^-$ and/or H_2O_2, suggesting that singlet oxygen does not act primarily as a toxin but rather as a signal that activates several stress-response pathways [107]. Recently, it was reported that $^1O_2$ could be responsible for increased tolerance to photooxidative stress in *Arabidopsis* plants through the action of β-cyclocitral [114]. β-cyclocitral is a β-carotene derivative produced in high light that is able to induce changes in the expression of a large set of genes, which strongly overlap with the network of genes induced by $^1O_2$ [114]. At the same time it has little effect on the expression of H_2O_2 gene markers. β-cyclocitral-induced reprogramming of gene expression is associated with increased tolerance to photooxidative stress, indicating that β-cyclocitral is a stress signal produced in high light that is able to induce defense mechanisms and represents a likely messenger involved in the $^1O_2$signaling pathway in plants [114]. Further evidence that $^1O_2$ participates in acclimation to photooxidative stress comes from a study with an npq1lut2 double mutant [106]. The npq1lut2 mutant specifically accumulates $^1O_2$ due to its selective loss of lutein and zeaxanthin, which participate in the quenching and scavenging of $^3Chl^*$ and $^1O_2$ [115,116]. Following high light illumination, $^1O_2$ accumulates and modifies the expression of a group of genes encoding chloroplast proteins, leading to a significant change in chloroplast composition and functional modifications. High light induced $^1O_2$ responses were also investigated in an *Arabidopsis* cell suspension culture (ACSC) containing functional chloroplast [117]. An experiment with different fluorescent probes showed that the high light treated cultures emitted fluorescence that corresponded with the production of $^1O_2$. This was accompanied by significant changes in the expression of transcripts specifically upregulated by $^1O_2$, which leads us to conclude that $^1O_2$ plays an important role in the initiation of defense responses to high light.

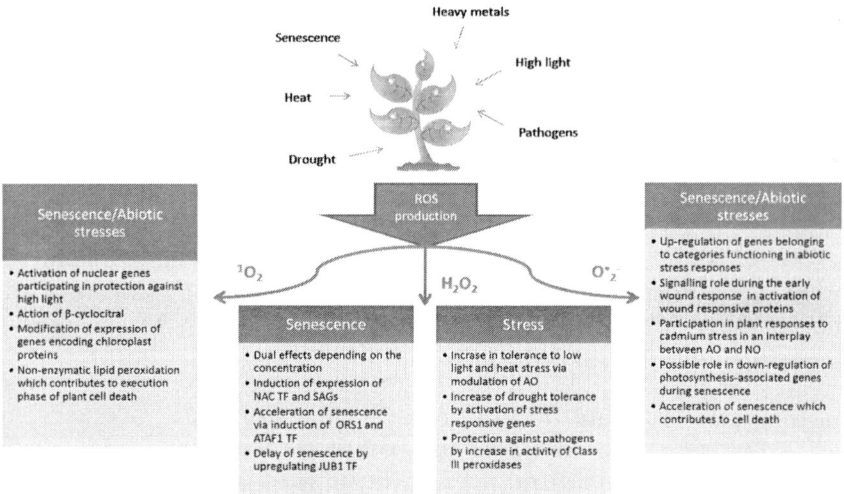

Figure 2. Possible roles of ROS during senescence and abiotic stresses.

When it comes to senescence, our knowledge on the signaling role of the $^1O_2$ is limited by the scarcity of the research relevant to the topic. There is general agreement that ROS production increases during senescence [2,3]. However, the increase in $^1O_2$ is observed simultaneously with that of other ROS, making it difficult to isolate the role of $^1O_2$. It was reported that $^1O_2$ is the main cause of senescence-associated oxidative stress in chloroplasts of sage [118]. However, this was concluded on the basis of the strong degradation of β-carotene and α-tocopherol in drought-stressed plants, which suggests the enhanced formation of singlet oxygen. In addition, in a recent study, a mass generation of singlet oxygen was measured in the early stages of hormone-treated barley but then declined, while in naturally senescing plants there was continuous production of low amounts of singlet oxygen [25]. Together with an increase in $^1O_2$, artificially senescing plants contained oxidative breakdown products of β-carotene such as β-cyclocistral, which is a possible messenger involved in the $^1O_2$ signaling pathway in plants [114], while the same was undetectable in a naturally senescing plant. Similar results were observed in our recent study, where it was shown that during the development and senescence of secondary barley leaves there is a continuous production of low amounts of $^1O_2$ [76]. Another possible role of $^1O_2$ could be its contribution to an increase in lipid peroxidation (LPO), leading to cell death. It is well known that with advancing senescence there is a notable increase in LPO [119,120]. Recently, it has been reported that in optimal growth conditions $^1O_2$ was responsible for more than 80% of the non-enzymatic

LPO in *Arabidopsis* leaf tissues [121]. Lipid peroxidation leads to the generation of free radicals, which can lead to the promotion of senescence [122]. This in turn leads to an increase in lipoxygenase activity, which can further increase LPO and also form $^1O_2$ [123], leading to overproduction of $^1O_2$. Indeed, it was showed that in *Arabidopsis* mutants favoring $^1O_2$ production, photooxidative stress led to a dramatic increase in LPO preceding cell death [121]. Possible roles of ROS during senescence and stress are summarized in Figure 2.

6. CONCLUSIONS

In this review we have given a short overview on the possible role of three classes of ROS during senescence and abiotic stresses. ROS play an important role in different plant processes ranging from plant stress adaptation to defense against pathogen attack. In the ROS family, the signaling roles of H_2O_2 have been most thoroughly studied due to its relative stability and ability to diffuse through membranes. It has been demonstrated that H_2O_2 plays various roles in plant growth, development, and metabolism. It has an important role during the senescence process, where it was shown that it could be used as a signal to promote senescence and during the acquisition of tolerance to different abiotic and biotic stresses. In comparison with H_2O_2, the superoxide anion is less stable and cannot cross the membrane, which makes it less suitable as a signal molecule. Nevertheless, an important signaling role of $O^{\bullet}{}_2^-$ was demonstrated when it was shown that the generation of $O^{\bullet}{}_2^-$ in the absence of H_2O_2 leads to a strong upregulation of the genes that function in abiotic stress responses and during senescence. Finally, in recent years, with the discovery of a *flu* mutant, it was demonstrated that $^1O_2$is capable of activating a set of nuclear genes different from those activated by $O^{\bullet}{}_2^-$ and H_2O_2 and that it plays an important role in plant responses to light.

At present, only the role of H_2O_2 during senescence and abiotic stresses has been extensively studied, while the role of other ROS remains to be further clarified. In recent years there has been significant progress in this area, with development of new techniques and technologies, but still there is no ideal technique that can be applied to a variety of systems and to specific ROS classes. EPR spectrometry can be used to measure specific ROS species but it requires a thorough sample preparation, which prevents the measurement of specific sites of ROS production in tissues. On the other hand, genetically-encoded redox probes can be used to measure ROS production in different cell compartments *in vivo* but they often do not differentiate between different classes of ROS.

Moreover, most of the papers investigate the impact of a group of ROS, while the contribution of individual ROS remains questionable. Significant progress in the future can be made on the signaling roles of $O^{\bullet}{}_2^-$ and $^1O_2$ during senescence. Roles of these two classes during this important process have not yet been sufficiently investigated and many questions wait to be answered. Already it has been shown that $O^{\bullet}{}_2^-$ upregulates some of the genes that are important during senescence such as WRKY6, but not much more is known. Similarly, the role of $^1O_2$during senescence is limited to its contribution to cell death, while several studies reported that production of this species remains unchanged during natural senescence. Whether this is true or not remains to be seen.

ACKNOWLEDGMENTS

This research was supported by the EU in the frame of Marie Curie Actions (MC-ITN-CropLife; Nr 24394).

The Faculty of Biochemistry, Biophysics, and Biotechnology of Jagiellonian University is a partner of the Leading National Research Center (KNOW), supported by the Ministry of Science and Higher Education.

AUTHOR CONTRIBUTIONS

All authors contributed extensively to the work presented in this paper. Ivan Jajic wrote the manuscript; Tadeusz Sarna and Kazimierz Strzalka supervised the work and gave final corrections and valuable input on how to improve it.

CONFLICTS OF INTEREST

The authors declare no conflict of interest.

REFERENCES

1. Gan, S.; Amasino, R.M. Making sense of senescence (molecular genetic regulation and manipulation of leaf senescence). Plant. Physiol. 1997, 113, 313–319.

2. Lee, S.; Seo, P.J.; Lee, H.J.; Park, C.M. A nac transcription factor ntl4 promotes reactive oxygen species production during drought-induced leaf senescence in arabidopsis. Plant J. Cell Mol. Biol. 2012, 70, 831–844.
3. Prochazkova, D.; Sairam, R.K.; Srivastava, G.C.; Singh, D.V. Oxidative stress and antioxidant activity as the basis of senescence in maize leaves. Plant Sci. 2001, 161, 765–771.
4. Selote, D.S.; Khanna-Chopra, R. Drought acclimation confers oxidative stress tolerance by inducing co-ordinated antioxidant defense at cellular and subcellular level in leaves of wheat seedlings. Physiol. Plant. 2006, 127, 494–506.
5. Silva, E.N.; Ferreira-Silva, S.L.; Fontenele, A.D.V.; Ribeiro, R.V.; Viégas, R.A.; Silveira, J.A.G. Photosynthetic changes and protective mechanisms against oxidative damage subjected to isolated and combined drought and heat stresses in Jatropha curcas plants. J. Plant Phsiol. 2010, 167, 1157–1164.
6. Choudhury, S.; Panda, P.; Sahoo, L.; Panda, S.K. Reactive oxygen species signaling in plants under abiotic stress. Plant Signal. Behav. 2013, 8, e23681.
7. Río, L.A.D.; Pastori, G.M.; Palma, J.M.; Sandalio, L.M.; Sevilla, F.; Corpas, F.J.; Jiménez, A.; López-Huertas, E.; Hernández, J.A. The activated oxygen role of peroxisomes in senescence. Plant Physiol. 1998, 116, 1195–1200.
8. Lushchak, V.I. Adaptive response to oxidative stress: Bacteria, fungi, plants and animals. Comp. Biochem. Physiol. Toxicol. Pharmacol. CBP 2011, 153, 175–190.
9. Ashraf, M. Biotechnological approach of improving plant salt tolerance using antioxidants as markers. Biotechnol. Adv. 2009, 27, 84–93.
10. Mittler, R. Abiotic stress, the field environment and stress combination. Trends Plant Sci. 2006, 11, 15–19.
11. Khanna-Chopra, R.; Selote, D.S. Acclimation to drought stress generates oxidative stress tolerance in drought-resistant than -susceptible wheat cultivar under field conditions. Environ. Exp. Bot. 2007, 60, 276–283.
12. Jubany-Mari, T.; Munne-Bosch, S.; Lopez-Carbonell, M.; Alegre, L. Hydrogen peroxide is involved in the acclimation of the mediterranean shrub, cistus albidus l., to summer drought. J. Exp. Bot. 2009, 60, 107–120.
13. Wang, L.; Su, H.; Han, L.; Wang, C.; Sun, Y.; Liu, F. Differential expression profiles of poplar map kinase kinases in response to abiotic stresses and plant hormones, and overexpression of PtMKK4 improves the drought tolerance of poplar. Gene 2014, 545, 141–148.
14. Suzuki, N.; Koussevitzky, S.; Mittler, R.; Miller, G. Ros and redox signalling in the response of plants to abiotic stress. Plant Cell Environ. 2012, 35, 259–270.
15. Foyer, C.H.; Noctor, G. Oxidant and antioxidant signalling in plants: A re-evaluation of the concept of oxidative stress in a physiological context. Plant Cell Environ. 2005, 28, 1056–1071.
16. Zimmermann, P.; Heinlein, C.; Orendi, G.; Zentgraf, U. Senescence-specific regulation of catalases in Arabidopsis thaliana (L.) heynh. Plant Cell Environ. 2006, 29, 1049–1060.
17. Dat, J.; Vandenabeele, S.; Vranová, E.; van Montagu, M.; Inzé, D.; van Breusegem, F. Dual action of the active oxygen species during plant stress responses. Cell. Mol. Life Sci. 2000, 57, 779–795.
18. Rentel, M.C.; Lecourieux, D.; Ouaked, F.; Usher, S.L.; Petersen, L.; Okamoto, H.; Knight, H.; Peck, S.C.; Grierson, C.S.; Hirt, H.; et al. Oxi1 kinase is necessary for oxidative burst-mediated signalling in arabidopsis. Nature 2004, 427, 858–861.
19. Kumar, N.; Ebel, R.C.; Roberts, P.D. H2O2 degradation is suppressed in kumquat leaves infected with xanthomonas axonopodis pv. Citri. Sci. Hortic. 2011, 130, 241–247.
20. Dias, C.V.; Mendes, J.S.; dos Santos, A.C.; Pirovani, C.P.; da Silva Gesteira, A.; Micheli, F.; Gramacho, K.P.; Hammerstone, J.; Mazzafera, P.; de Mattos Cascardo, J.C. Hydrogen peroxide formation in cacao tissues infected by the hemibiotrophic fungus Moniliophthora perniciosa. Plant Physiol. Biochem. 2011, 49, 917–922.

21. Golldack, D.; Li, C.; Mohan, H.; Probst, N. Tolerance to drought and salt stress in plants: Unraveling the signaling networks. Front. Plant Sci. 2014, 5, 151.
22. Soares, N.C.; Wojtkowska, J.; Jackson, P.A. A proteomic analysis of the wound response in medicago leaves reveals the early activation of a ros-sensitive signal pathway. J. Proteomics 2011, 74, 1411–1420.
23. Laloi, C.; Stachowiak, M.; Pers-Kamczyc, E.; Warzych, E.; Murgia, I.; Apel, K. Cross-talk between singlet oxygen- and hydrogen peroxide-dependent signaling of stress responses in Arabidopsis thaliana. Proc. Natl. Acad. Sci. USA 2007, 104, 672–677.
24. Hideg, E.; Barta, C.; Kalai, T.; Vass, I.; Hideg, K.; Asada, K. Detection of singlet oxygen and superoxide with fluorescent sensors in leaves under stress by photoinhibition or uv radiation. Plant Cell Physiol. 2002, 43, 1154–1164.
25. Springer, A.; Acker, G.; Bartsch, S.; Bauerschmitt, H.; Reinbothe, S.; Reinbothe, C. Differences in gene expression between natural and artificially induced leaf senescence in barley. J. Plant Pysiol. 2015, 176, 180–191.
26. Golemiec, E.; Tokarz, K.; Wielanek, M.; Niewiadomska, E. A dissection of the effects of ethylene, h2o2 and high irradiance on antioxidants and several genes associated with stress and senescence in tobacco leaves. J. Plant Pysiol.2014, 171, 269–275.
27. Rizhsky, L.; Liang, H.; Shuman, J.; Shulaev, V.; Davletova, S.; Mittler, R. When defense pathways collide. The response of arabidopsis to a combination of drought and heat stress. Plant Physiol. 2004, 134, 1683–1696.
28. Rizhsky, L.; Liang, H.; Mittler, R. The combined effect of drought stress and heat shock on gene expression in tobacco. Plant Physiol. 2002, 130, 1143–1151.
29. Prasch, C.M.; Sonnewald, U. Simultaneous application of heat, drought and virus to Arabidopsis thaliana plants reveals significant shifts in signaling networks. Plant Physiol. 2013.
30. Atkinson, N.J.; Lilley, C.J.; Urwin, P.E. Identification of genes involved in the response of arabidopsis to simultaneous biotic and abiotic stresses. Plant Physiol. 2013, 162, 2028–2041.
31. Zhou, M.; Diwu, Z.; Panchuk-Voloshina, N.; Haugland, R.P. A stable nonfluorescent derivative of resorufin for the fluorometric determination of trace hydrogen peroxide: Applications in detecting the activity of phagocyte nadph oxidase and other oxidases. Anal. Biochem. 1997, 253, 162–168.
32. Frahry, G.; Schopfer, P. Nadh-stimulated, cyanide-resistant superoxide production in maize coleoptiles analyzed with a tetrazolium-based assay. Planta 2001, 212, 175–183.
33. Dikalov, S.; Griendling, K.K.; Harrison, D.G. Measurement of reactive oxygen species in cardiovascular studies. Hypertension 2007, 49, 717–727.
34. Gomes, A.; Fernandes, E.; Lima, J.L.F.C. Fluorescence probes used for detection of reactive oxygen species. J. Biochem. Biophys. Methods 2005, 65, 45–80.
35. Benov, L.; Fridovich, I. Is reduction of the sulfonated tetrazolium 2,3-bis (2-methoxy-4-nitro-5-sulfophenyl)-2-tetrazolium 5-carboxanilide a reliable measure of intracellular superoxide production? Anal. Biochem. 2002, 310, 186–190.
36. Swartz, H.M.; Khan, N.; Khramtsov, V.V. Use of electron paramagnetic resonance spectroscopy to evaluate the redox state in vivo. Antioxid. Redox Signal. 2007, 9, 1757–1771.
37. Hideg, E.; Kalai, T.; Hideg, K. Direct detection of free radicals and reactive oxygen species in thylakoids. Methods Mol. Biol. 2011, 684, 187–200.
38. Hideg, É.; Spetea, C.; Vass, I. Singlet oxygen production in thylakoid membranes during photoinhibition as detected by epr spectroscopy. Photosynth. Res. 1994, 39, 191–199.
39. Mubarakshina, M.M.; Ivanov, B.N.; Naydov, I.A.; Hillier, W.; Badger, M.R.; Krieger-Liszkay, A. Production and diffusion of chloroplastic h2o2 and its implication to signalling. J. Exp. Bot. 2010, 61, 3577–3587.
40. Krieger-Liszkay, A.; Kós, P.B.; Hideg, É. Superoxide anion radicals generated by methylviologen in photosystem i damage photosystem ii. Physiol. Plant. 2011, 142, 17–25.

41. Heyno, E.; Mary, V.; Schopfer, P.; Krieger-Liszkay, A. Oxygen activation at the plasma membrane: Relation between superoxide and hydroxyl radical production by isolated membranes. Planta 2011, 234, 35–45.
42. Steffen-Heins, A.; Steffens, B. EPR spectroscopy and its use in planta--A promising technique to disentangle the origin of specific ros. Front. Environ. Sci. 2015, 3.
43. Leister, D. Photosynthesis research protocols. Ann. Bot. 2013, 112, vi–vii.
44. Sandalio, L.M.; Rodríguez-Serrano, M.; Romero-Puertas, M.C.; del Río, L.A. Imaging of reactive oxygen species and nitric oxide in vivo in plant tissues. Methods Enzymol. 2008, 440, 397–409.
45. Fryer, M.J.; Oxborough, K.; Mullineaux, P.M.; Baker, N.R. Imaging of photo-oxidative stress responses in leaves. J. Exp. Bot. 2002, 53, 1249–1254.
46. Flors, C.; Fryer, M.J.; Waring, J.; Reeder, B.; Bechtold, U.; Mullineaux, P.M.; Nonell, S.; Wilson, M.T.; Baker, N.R. Imaging the production of singlet oxygen in vivo using a new fluorescent sensor, singlet oxygen sensor green. J. Exp. Bot. 2006, 57, 1725–1734.
47. Costa, A.; Drago, I.; Behera, S.; Zottini, M.; Pizzo, P.; Schroeder, J.I.; Pozzan, T.; Lo Schiavo, F. H2O2 in plant peroxisomes: An in vivo analysis uncovers a ca(2+)-dependent scavenging system. Plant J. Cell Mol. Biol. 2010, 62, 760–772.
48. Rodriguez-Serrano, M.; Romero-Puertas, M.C.; Zabalza, A.; Corpas, F.J.; Gomez, M.; Del Rio, L.A.; Sandalio, L.M. Cadmium effect on oxidative metabolism of pea (Pisum sativum L.) roots. Imaging of reactive oxygen species and nitric oxide accumulation in vivo. Plant Cell Environ. 2006, 29, 1532–1544.
49. Tarpey, M.M.; Wink, D.A.; Grisham, M.B. Methods for detection of reactive metabolites of oxygen and nitrogen: In vitro and in vivo considerations. Am. J. Physiol. Regul. Integr. Comp. Physiol. 2004, 286, R431–R444.
50. Corpas, F.J.; Fernandez-Ocana, A.; Carreras, A.; Valderrama, R.; Luque, F.; Esteban, F.J.; Rodriguez-Serrano, M.; Chaki, M.; Pedrajas, J.R.; Sandalio, L.M.; et al. The expression of different superoxide dismutase forms is cell-type dependent in olive (Olea europaea L.) leaves. Plant Cell Physiol. 2006, 47, 984–994.
51. Yamamoto, Y.; Kobayashi, Y.; Devi, S.R.; Rikiishi, S.; Matsumoto, H. Aluminum toxicity is associated with mitochondrial dysfunction and the production of reactive oxygen species in plant cells. Plant Physiol. 2002, 128, 63–72.
52. Ragas, X.; Jimenez-Banzo, A.; Sanchez-Garcia, D.; Batllori, X.; Nonell, S. Singlet oxygen photosensitisation by the fluorescent probe singlet oxygen sensor green [registered sign]. Chem. Commun. 2009, 28, 2920–2922.
53. Lukyanov, K.A.; Belousov, V.V. Genetically encoded fluorescent redox sensors. Biochim. Biophys. Acta 2014, 1840, 745–756.
54. Hernandez-Barrera, A.; Quinto, C.; Johnson, E.A.; Wu, H.M.; Cheung, A.Y.; Cardenas, L. Using hyper as a molecular probe to visualize hydrogen peroxide in living plant cells: A method with virtually unlimited potential in plant biology. Methods Enzymol. 2013, 527, 275–290.
55. Rhoads, D.M.; Umbach, A.L.; Subbaiah, C.C.; Siedow, J.N. Mitochondrial reactive oxygen species. Contribution to oxidative stress and interorganellar signaling. Plant Physiol. 2006, 141, 357–366.
56. Sandalio, L.; Rodríguez-Serrano, M.; Romero-Puertas, M.; del Río, L.A. Role of peroxisomes as a source of reactive oxygen species (ROS) signaling molecules. Subcell. Biochem. 2013, 69, 231–255.
57. Rosenwasser, S.; Rot, I.; Sollner, E.; Meyer, A.J.; Smith, Y.; Leviatan, N.; Fluhr, R.; Friedman, H. Organelles contribute differentially to reactive oxygen species-related events during extended darkness. Plant Physiol. 2011, 156, 185–201.
58. Allen, J.F.; Hall, D.O. Superoxide reduction as a mechanism of ascorbate-stimulated oxygen uptake by isolated chloroplasts. Biochem. Biophys. Res. Commun. 1973, 52, 856–862.
59. Asada, K. Production and scavenging of reactive oxygen species in chloroplasts and their functions. Plant Physiol.2006, 141, 391–396.

60. Takahashi, M.A.; Asada, K. Superoxide anion permeability of phospholipid membranes and chloroplast thylakoids. Arch. Biochem. Biophys. 1983, 226, 558–566.
61. Sandalio, L.M.; Fernandez, V.M.; Ruperez, F.L.; Del Rio, L.A. Superoxide free radicals are produced in glyoxysomes. Plant Physiol. 1988, 87, 1–4.
62. Corpas, F.J.; Palma, J.M.; Sandalio, L.M.; Valderrama, R.; Barroso, J.B.; Del Rio, L.A. Peroxisomal xanthine oxidoreductase: Characterization of the enzyme from pea (Pisum sativum L.) leaves. J. Plant Pysiol. 2008, 165, 1319–1330.
63. Del Río, L.A.; Donaldson, R.P. Production of superoxide radicals in glyoxysomal membranes from castor bean endosperm. J. Plant Pysiol. 1995, 146, 283–287.
64. Kaur, G.; Sharma, A.; Guruprasad, K.; Pati, P.K. Versatile roles of plant nadph oxidases and emerging concepts. Biotechnol. Adv. 2014, 32, 551–563.
65. Sagi, M.; Fluhr, R. Production of reactive oxygen species by plant nadph oxidases. Plant Physiol. 2006, 141, 336–340.
66. Torres, M.A.; Dangl, J.L. Functions of the respiratory burst oxidase in biotic interactions, abiotic stress and development. Curr. Opin. Plant Biol. 2005, 8, 397–403.
67. Kwak, J.M.; Mori, I.C.; Pei, Z.M.; Leonhardt, N.; Torres, M.A.; Dangl, J.L.; Bloom, R.E.; Bodde, S.; Jones, J.D.; Schroeder, J.I. Nadph oxidase atrbohd and atrbohf genes function in ROS-dependent aba signaling in arabidopsis. EMBO J. 2003, 22, 2623–2633.
68. Zarepour, M.; Kaspari, K.; Stagge, S.; Rethmeier, R.; Mendel, R.; Bittner, F. Xanthine dehydrogenase atxdh1 from Arabidopsis thaliana is a potent producer of superoxide anions via its nadh oxidase activity. Plant. Mol. Biol. 2010, 72, 301–310.
69. Yesbergenova, Z.; Yang, G.; Oron, E.; Soffer, D.; Fluhr, R.; Sagi, M. The plant mo-hydroxylases aldehyde oxidase and xanthine dehydrogenase have distinct reactive oxygen species signatures and are induced by drought and abscisic acid. Plant. J. 2005, 42, 862–876.
70. McRae, D.G.; Thompson, J.E. Senescence-dependent changes in superoxide anion production by illuminated chloroplasts from bean leaves. Planta 1983, 158, 185–193.
71. Pastori, G.M.; Del Rio, L.A. Natural senescence of pea leaves (an activated oxygen-mediated function for peroxisomes). Plant Physiol. 1997, 113, 411–418.
72. Van Breusegem, F.; Dat, J.F. Reactive oxygen species in plant cell death. Plant Physiol. 2006, 141, 384–390.
73. Zhao, X.; Nishimura, Y.; Fukumoto, Y.; Li, J. Effect of high temperature on active oxygen species, senescence and photosynthetic properties in cucumber leaves. Environ. Exp. Bot. 2011, 70, 212–216.
74. Scarpeci, T.E.; Zanor, M.I.; Carrillo, N.; Mueller-Roeber, B.; Valle, E.M. Generation of superoxide anion in chloroplasts of Arabidopsis thaliana during active photosynthesis: A focus on rapidly induced genes. Plant. Mol. Biol. 2008, 66, 361–378.
75. Robatzek, S.; Somssich, I.E. A new member of the arabidopsis wrky transcription factor family, atwrky6, is associated with both senescence- and defence-related processes. Plant J. Cell Mol. Biol. 2001, 28, 123–133.
76. Jajic, I.; Sarna, T.; Szewczyk, G.; Strzalka, K. Changes in production of reactive oxygen species level in illuminated thylakoids isolated during development and senescence of barley leaf senescence. J. Plant Physiol. 2015. submitted.
77. Jajić, I.; Wiśniewska-Becker, A.; Sarna, T.; Jemioła-Rzemińska, M.; Strzałka, K. EPR spin labeling measurements of thylakoid membrane fluidity during barley leaf senescence. J. Plant Pysiol. 2014, 171, 1046–1053.
78. Niewiadomska, E.; Polzien, L.; Desel, C.; Rozpadek, P.; Miszalski, Z.; Krupinska, K. Spatial patterns of senescence and development-dependent distribution of reactive oxygen species in tobacco (Nicotiana tabacum) leaves. J. Plant Pysiol. 2009, 166, 1057–1068.

79. Kopyra, M.; Gwóźdź, E.A. Nitric oxide stimulates seed germination and counteracts the inhibitory effect of heavy metals and salinity on root growth of lupinus luteus. Plant Physiol. Biochem. 2003, 41, 1011–1017.
80. Romero-Puertas, M.C.; Delledonne, M. Nitric oxide signaling in plant-pathogen interactions. IUBMB Life 2003, 55, 579–583.
81. Quan, L.J.; Zhang, B.; Shi, W.W.; Li, H.Y. Hydrogen peroxide in plants: A versatile molecule of the reactive oxygen species network. J. Integr. Plant Biol. 2008, 50, 2–18.
82. Peng, L.T.; Jiang, Y.M.; Yang, S.Z.; Pan, S.Y. Accelerated senescence of fresh-cut chinese water chestnut tissues in relation to hydrogen peroxide accumulation. J. Plant Phsiol. Mol. Biol. 2005, 31, 527–532. (In Chinese).
83. Kumar, N.; Ebel, R.C.; Roberts, P.D. Antioxidant metabolism of grapefruit infected with xanthomonas axonopodis pv. Citri. Environ. Exp. Bot. 2011, 71, 41–49.
84. Zhang, X.-L.; Jia, X.-F.; Yu, B.; Gao, Y.; Bai, J.-G. Exogenous hydrogen peroxide influences antioxidant enzyme activity and lipid peroxidation in cucumber leaves at low light. Sci. Hortic. 2011, 129, 656–662.
85. Ishibashi, Y.; Yamaguchi, H.; Yuasa, T.; Iwaya-Inoue, M.; Arima, S.; Zheng, S.H. Hydrogen peroxide spraying alleviates drought stress in soybean plants. J. Plant Pysiol. 2011, 168, 1562–1567.
86. Yun, K.-Y.; Park, M.; Mohanty, B.; Herath, V.; Xu, F.; Mauleon, R.; Wijaya, E.; Bajic, V.; Bruskiewich, R.; de los Reyes, B. Transcriptional regulatory network triggered by oxidative signals configures the early response mechanisms of japonica rice to chilling stress. BMC Plant. Biol. 2010, 10, 1–29.
87. Hu, X.; Bidney, D.L.; Yalpani, N.; Duvick, J.P.; Crasta, O.; Folkerts, O.; Lu, G. Overexpression of a gene encoding hydrogen peroxide-generating oxalate oxidase evokes defense responses in sunflower. Plant Physiol. 2003, 133, 170–181.
88. Foyer, C.H.; Noctor, G. Redox homeostasis and antioxidant signaling: A metabolic interface between stress perception and physiological responses. Plant Cell 2005, 17, 1866–1875.
89. Mittler, R.; Vanderauwera, S.; Gollery, M.; van Breusegem, F. Reactive oxygen gene network of plants. Trends Plant Sci. 2004, 9, 490–498.
90. Yang, T.; Poovaiah, B.W. Hydrogen peroxide homeostasis: Activation of plant catalase by calcium/calmodulin. Proc. Natl. Acad. Sci. 2002, 99, 4097–4102.
91. Liao, W.-B.; Zhang, M.-L.; Huang, G.-B.; Yu, J.-H. Hydrogen peroxide in the vase solution increases vase life and keeping quality of cut Oriental × Trumpet hybrid lily "manissa". Sci. Hortic. 2012, 139, 32–38.
92. Khandaker, M.M.; Boyce, A.N.; Osman, N. The influence of hydrogen peroxide on the growth, development and quality of wax apple (Syzygium samarangense, [blume] merrill & LM. Perry var. Jambu madu) fruits. Plant Physiol. Biochem. 2012, 53, 101–110.
93. Bieker, S.; Riester, L.; Stahl, M.; Franzaring, J.; Zentgraf, U. Senescence-specific alteration of hydrogen peroxide levels in Arabidopsis thaliana and oilseed rape spring variety Brassica napus L. Cv. Mozart. J. Integr. Plant Biol. 2012, 54, 540–554.
94. Balazadeh, S.; Wu, A.; Mueller-Roeber, B. Salt-triggered expression of the ANAC092-dependent senescence regulon in Arabidopsis thaliana. Plant Signal. Behav. 2010, 5, 733–735.
95. Balazadeh, S.; Kwasniewski, M.; Caldana, C.; Mehrnia, M.; Zanor, M.I.; Xue, G.-P.; Mueller-Roeber, B. ORS1, an H2O2-responsive nac transcription factor, controls senescence in Arabidopsis thaliana. Mol. Plant. 2011, 4, 346–360.
96. Wu, A.; Allu, A.D.; Garapati, P.; Siddiqui, H.; Dortay, H.; Zanor, M.I.; Asensi-Fabado, M.A.; Munne-Bosch, S.; Antonio, C.; Tohge, T.; et al. JUNGBRUNNEN1, a reactive oxygen species-responsive NAC transcription factor, regulates longevity in Arabidopsis. Plant Cell 2012, 24, 482–506.
97. Garapati, P.; Xue, G.P.; Munne-Bosch, S.; Balazadeh, S. Transcription factor ATAF1 in arabidopsis promotes senescence by direct regulation of key chloroplast maintenance and senescence transcriptional cascades. Plant Physiol. 2015.

98. Chen, H.J.; Wu, S.D.; Huang, G.J.; Shen, C.Y.; Afiyanti, M.; Li, W.J.; Lin, Y.H. Expression of a cloned sweet potato catalase SPCAT1 alleviates ethephon-mediated leaf senescence and H2O2 elevation. J. Plant Pysiol. 2012, 169, 86–97.
99. Prasad, T.K.; Anderson, M.D.; Martin, B.A.; Stewart, C.R. Evidence for chilling-induced oxidative stress in maize seedlings and a regulatory role for hydrogen peroxide. Plant Cell 1994, 6, 65–74.
100. Gao, Y.; Guo, Y.-K.; Lin, S.-H.; Fang, Y.-Y.; Bai, J.-G. Hydrogen peroxide pretreatment alters the activity of antioxidant enzymes and protects chloroplast ultrastructure in heat-stressed cucumber leaves. Sci. Hortic. 2010, 126, 20–26.
101. Sielewiesiuk, J. Why there are photodamages to photosystem II at low light intensities. Acta Physiol. Plant. 2002, 24, 399–406.
102. Krieger-Liszkay, A. Singlet oxygen production in photosynthesis. J. Exp. Bot. 2005, 56, 337–346.
103. Gill, S.S.; Tuteja, N. Reactive oxygen species and antioxidant machinery in abiotic stress tolerance in crop plants. Plant Physiol. Biochem. 2010, 48, 909–930.
104. Pospíšil, P.; Šnyrychová, I.; Nauš, J. Dark production of reactive oxygen species in photosystem II membrane particles at elevated temperature: EPR spin-trapping study. Biochim. Biophys. Acta 2007, 1767, 854–859.
105. Morker, K.H.; Roberts, M.R. Light as both an input and an output of wound-induced reactive oxygen formation in Arabidopsis leaves. Plant Signal. Behav. 2011, 6, 1087–1089.
106. Alboresi, A.; Dall'osto, L.; Aprile, A.; Carillo, P.; Roncaglia, E.; Cattivelli, L.; Bassi, R. Reactive oxygen species and transcript analysis upon excess light treatment in wild-type Arabidopsis thaliana vs. a photosensitive mutant lacking zeaxanthin and lutein. BMC Plant. Biol. 2011, 11, 62.
107. Op den Camp, R.G.L.; Przybyla, D.; Ochsenbein, C.; Laloi, C.; Kim, C.; Danon, A.; Wagner, D.; Hideg, É.; Göbel, C.; Feussner, I.; et al. Rapid induction of distinct stress responses after the release of singlet oxygen in arabidopsis. Plant Cell 2003, 15, 2320–2332.
108. Kim, C.; Meskauskiene, R.; Apel, K.; Laloi, C. No single way to understand singlet oxygen signalling in plants. EMBO Rep. 2008, 9, 435–439.
109. Krasnovsky, A.A., Jr. Singlet molecular oxygen in photobiochemical systems: IR phosphorescence studies. Membr. Cell Biol. 1998, 12, 665–690.
110. Sies, H.; Menck, C.F.M. Singlet oxygen induced DNA damage. Mutat. Res./DNAging 1992, 275, 367–375.
111. Skovsen, E.; Snyder, J.W.; Lambert, J.D.C.; Ogilby, P.R. Lifetime and diffusion of singlet oxygen in a cell. J. Phys. Chem. B 2005, 109, 8570–8573.
112. Fischer, B.B.; Krieger-Liszkay, A.; Hideg, E.; Snyrychova, I.; Wiesendanger, M.; Eggen, R.I. Role of singlet oxygen in chloroplast to nucleus retrograde signaling in chlamydomonas reinhardtii. FEBS Lett. 2007, 581, 5555–5560.
113. Wagner, D.; Przybyla, D.; Op den Camp, R.; Kim, C.; Landgraf, F.; Lee, K.P.; Wursch, M.; Laloi, C.; Nater, M.; Hideg, E.; et al. The genetic basis of singlet oxygen-induced stress responses of Arabidopsis thaliana. Science 2004, 306, 1183–1185.
114. Ramel, F.; Birtic, S.; Ginies, C.; Soubigou-Taconnat, L.; Triantaphylides, C.; Havaux, M. Carotenoid oxidation products are stress signals that mediate gene responses to singlet oxygen in plants. Proc. Natl. Acad. Sci. USA 2012, 109, 5535–5540.
115. Dall'Osto, L.; Lico, C.; Alric, J.; Giuliano, G.; Havaux, M.; Bassi, R. Lutein is needed for efficient chlorophyll triplet quenching in the major LHCII antenna complex of higher plants and effective photoprotection in vivo under strong light. BMC Plant. Biol. 2006, 6, 32.
116. Dall'Osto, L.; Cazzaniga, S.; Havaux, M.; Bassi, R. Enhanced photoprotection by protein-bound vs. free xanthophyll pools: A comparative analysis of chlorophyll b and xanthophyll biosynthesis mutants. Mol. Plant. 2010, 3, 576–593.
117. González-Pérez, S.; Gutiérrez, J.; García-García, F.; Osuna, D.; Dopazo, J.; Lorenzo, Ó.; Revuelta, J.L.; Arellano, J.B. Early transcriptional defense responses in arabidopsis

cell suspension culture under high-light conditions. Plant Physiol. 2011, 156, 1439–1456.

118. Munné-Bosch, S.; Jubany-Marí, T.; Alegre, L. Drought-induced senescence is characterized by a loss of antioxidant defences in chloroplasts. Plant Cell Environ. 2001, 24, 1319–1327.

119. Berger, S.; Weichert, H.; Porzel, A.; Wasternack, C.; Kühn, H.; Feussner, I. Enzymatic and non-enzymatic lipid peroxidation in leaf development. Biochim. Biophys. Acta 2001, 1533, 266–276.

120. Zhang, M.P.; Zhang, C.J.; Yu, G.H.; Jiang, Y.Z.; Strasser, R.J.; Yuan, Z.Y.; Yang, X.S.; Chen, G.X. Changes in chloroplast ultrastructure, fatty acid components of thylakoid membrane and chlorophyll a fluorescence transient in flag leaves of a super-high-yield hybrid rice and its parents during the reproductive stage. J. Plant Pysiol. 2010, 167, 277–285.

121. Triantaphylides, C.; Krischke, M.; Hoeberichts, F.A.; Ksas, B.; Gresser, G.; Havaux, M.; van Breusegem, F.; Mueller, M.J. Singlet oxygen is the major reactive oxygen species involved in photooxidative damage to plants. Plant Physiol.2008, 148, 960–968.

122. Arora, A.; Sairam, R.K.; Srivastava, G.C. Oxidative stress and antioxidative system in plants. Curr. Sci. 2002, 82, 1227–1238.

123. Thompson, J.E.; Legge, R.L.; Barber, R.F. The role of free radicals in senescence and wounding. New Phytol. 1987, 105, 317–344.

CHAPTER 18

Phenotypic Characterization of The Arabidopsis UFM1 (Ubiquitin Fold Modifier) Gene Involved in Seed Development

Cornejo Paula and Masuelli Ricardo Williams

Laboratorio de Biología Molecular, Instituto de Biología Agrícola de Mendoza (IBAM), Consejo Nacional de Investigaciones Científicas y Técnicas, Facultad de Ciencias Agrarias, Universidad Nacional de Cuyo, A. Brown 500 (M5528AHB) Chacras de Coria, Mendoza, Argentina

ABSTRACT

Background and Objective: With the completion of the Arabidopsis genome sequencing, the next challenge is the determination of gene function. Post-translational modifications of proteins by small polypeptide are implicated in plant growth and development. Ubiquitin fold modifier 1 is a member of the ubiquitin like protein family. While, the enzymatic conjugation cascade of ubiquitin fold modifier 1 has been elucidated in recent years, its biological role is still unknown. The present study focuses in elucidate the role of the ufm1 in plants development. Materials and Methods: The researchers performed analyses of the development of wild-type Columbia plants and mutants of the ubiquitin fold modifier 1 gene to identify and interpret phenotypic changes in plants and seeds. Data were statistically analyzed with the Info Stat software. Results: In this study, evidence suggesting that ubiquitin fold modifier 1 is involved in the normal development of the seeds. Conclusion: The ufm1 gene would affect the normal development of the seed, particularly of the embryo, causing high percentage of seed abortion.

INTRODUCTION

Post-translational modifications to proteins are one of the principal components of the organization and regulation of function and cell structure as well as the development of organism, refer to covalent and generally enzymatic modification of proteins during or after its biosynthesis. The protein modification systems produce a functional change in target proteins. The 1st protein modifier system discovered was ubiquitin (Ub). Ubiquitin is a highly conserved 76 **AMINO ACID** polypeptide ubiquitously present in eukaryotes, from yeast to human[1-3]. It is covalently ligated to a wide variety of target proteins through the action of ubiquitin-activating enzyme (E1), ubiquitin-conjugating enzyme (E2) and ubiquitin-protein isopeptide ligase (E3)[3-5]. This addition can be reversed by the action of deubiquitylating enzymes (DUBs)[6]. Ubiquitin can participate in a wide variety of cellular processes like protein degradation pathways, histone regulation, transcriptional regulation, RNA metabolism, chromatin remodeling and cell cycle control. Besides ubiquitin, a number of other ubiquitin like proteins (UBLs) have been identified[7-12]. These proteins do not necessarily share a high similarity to ubiquitin, but are structurally related and can be conjugated to target proteins in a similar manner than ubiquitin[2,13-16].

The most recently identified ubiquitin like protein, *ufm1* (ubiquitin fold modifier 1), shares only 14% sequence identity with the ubiquitin proteins family but adopts the typical β-grasp fold[17]. The *ufm1* is synthesized as a precursor protein and is activated by an E1-like protein, *Uba5*, while conjugation is conducted by an E2-Like protein, *ufc1*[18]. Recently, an E3-Like ligase protein termed *ufl1* has been identified[19]. Finally specific *ufsps* can also cleave *ufm1* from its target proteins[20,21]. Cascade of *ufm1* is conserved among nearly all the eukaryotic organisms except yeast[18]. Studies on mammalian cells showed that the only target of *ufm1* identified, until now is the *ufbp1* protein and suggested a limited number of substrates[21,22]. Mutant mice of *Uba5* -/- are embryonic lethal[23] and cultured human cells silenced by *ufm1* and *ufl1* showed accelerated apoptosis[22]. Until now, there are no studies on the role of *ufm1* in plant development. The aim of this study is conduct a functional analysis to understanding the role of *ufm1* gene in plants. The phenotypic analysis described here, in conjunction with **GENE EXPRESSION** profiling, suggest that *ufm1* protein has a fundamental role in development of plant primordial in seed, particularly embryo development.

MATERIALS AND METHODS

Plant material and growth conditions: Plants of *Arabidopsis thaliana* were derived from the Columbia-0 (Col-0) mutant lines *ufm1* (SALK-040955C), locus AT1G77710, obtained from the *Arabidopsis* Biological Resource Center (Columbus, OH), from de SALK T-DNA insertion collection[24]. Surface-sterilized seeds were placed in petri dishes containing 15 mL of agar 0.8% w/v. Petri dishes were placed in a cold room for 3 days at 4°C to synchronize germination. Plates with cold stratified seeds were transferred into a room with day and night temperatures of 22 and 20°C, respectively. The average light intensity at the level of the rosette was ~65 μmol m^{-2} sec^{-1} supplied by fluorescent tubes. Seedlings were transferred to soil when 2 rosette leaves were greater than 1 mm in length. Plants were watered by sub irrigation every 2-3 days, depending on the growth stage. The plants, wild type and mutant were sown in 2 subgroups of 12 individuals each.

Phenotypic characterization: All plants were analyzed for the following phenotypic characteristics: Leaf morphology, rosette radius, flower morphology, length of major axis of silique and size of embryos. The phenotype of the mutant plants were compared to that of control plants during the different stages of growth.

All measurements were performed by analysis of digital images using AxioVision Rel. 4.8 software (Carl Zeiss).

Histology: For whole-mount preparations of embryos, silique were dissected with hypodermic needles. Dissected siliques were fixed on ice with FAA (10:7:2:1 ethanol:distilled water:acetic acid:formaldehyde 37%) for 30 min, hydrated in a graded ethanol series to 50 mM $NaPOH_4$ buffer, pH 7.2. Embryos were excised using fine forceps and surgical blades under a magnifying scope and mounted on a microscope slide in a clearing solution 8:2:1 chloral hydrate:distilled water:glycerol. Samples were cleared for 1 h at 4°C before inspection. Light microscope preparations were examined using a Zeiss Axioplan microscope (Jena, Germany). Pictures were taken with a Cannon PowerShot G11 camera and edited with Adobe Photoshop.

Tetrazolium staining: Seeds were imbibed for 2 h and embryos were excised using fine forceps and surgical blades under a magnifying scope. Embryos were subjected to staining in 0.5% w/v 2, 3, 5-triphenyl tetrazolium chloride solution.

Germination assay: Seed lots were harvested from individual plants grown under identical conditions. Each plant population was sown in 5 petri dishes (20 seeds from each individual plant per petri dish) on agar and incubated in a climate-controlled room. Germination was measured 5 days after imbibition and was expressed as the percentage of plants with emerged roots in each plant population (**GERMINATION PERCENTAGE**).

Quantification of DNA: Excised embryos were stained in the dark with DAPI dihydrochloride (Invitrogen, San Diego, Calif) 300 nM. Images were taken in a fluorescence Zeiss Axioplan microscope (Jena, Germany) with 358 nm excitation and 461 nm emission wavelengths, recorded with Cannon Power Shot G11 camera and processed using Image J.

RNA extraction: Siliques of mutant *ufm1* and wild type plants were dissected to isolate the seeds. Total RNA was extracted from a pool of seeds per plant using the TRIZOL reagent (Invitrogen GIBCO/Life Technologies, Grand Island, NY) according to the protocol by Meng and Feldman[25].

Semiquantitative reverse transcription (RT)-PCR analysis: Equal amounts of RNA from seeds were treated with DNase and reverse transcribed using reverse transcriptase M-MLV (Invitrogen, San Diego, Calif) as indicated by the manufacturer. The cDNA was amplified with specific primers designed based on the sequence of *ufm1* transcript of *Arabidopsis thaliana* (forward primer, CGTTGGTATTATCTCTGGTTTCAC, reverse primer, CAACTCACTTCAATTGTGACGA) and specific primers of the constitutive control gene action (forward primer, CATCAGGAAGGACTTGTACGG, reverse primer, GATGGACCTGACTCGTCATAC). As template, 40 ng of cDNA were used in a 15 μL reaction volume that contained 20 mmol L^{-1} tris-HCl, pH 8.4, 50 mmol L^{-1} KCl, 1.5 mmol L^{-1} $MgCl_2$, 300 μmol L^{-1} of each dNTP, 0.4 μmol L^{-1} of each primer and 0.4 U of Taq polymerase (Invitrogen, San Diego, Calif.). The thermocycler (PTC-100 MJ Research, Watertown, Mass.) was programmed for an initial denaturalization step of 94°C for 1 min and then 35 cycles of 94°C for 30 sec, 51°C for 1 min and 72°C for 2 min. The PCR products were separated in a 1.5% agarose gel and stained with ethidium bromide. The stained gel was photographed with the Molecular Analyst Gel Photo documentation System (Bio-Rad, Calif.) and analyzed with the Image J software. The quantity of each PCR product was expressed as the ratio of the sample densitometry measurement to the corresponding internal standard (action). To evaluate the changes in relative mRNA levels, densitometry values for the target genes were normalized against those of the housekeeping action mRNA. Data were statistically analyzed with the Info Stat software (Córdoba, Argentina).

RESULTS

Phenotypic characterization of a *ufm1* mutant: To analyze *ufm1* function, a T-DNA insertion homozygous mutant of *Arabidopsis thaliana* (SALK 040955C) and wild-type plants were used. Analyzed growth stages included the development of the plant from seed through the completion of flowering and seed maturation. Figure 1 and 2 shows the vegetative growth of mutant and wild-type plants was not statistically different. Flower morphology of the mutant was normal. However, silique of the homozygous mutant plants contained approximately 75% of aborted seeds (Figure 3).

Based on these findings, embryo development at different stages was investigated. In the mutants, the aborted seeds had embryos arrested at globular or torpedo stages. The rest of the mutant seeds presented embryos with delayed development respect to the wild type. At 6 Days After Pollination (DAP), mutant embryos were at globular, heart or torpedo stage, while wild type embryos were at torpedo stage. Between 8-12 DAP, the growth of the mutant embryos was retarded respect to the wild type. Mutant embryos at 8 DAP were arrested at heart or torpedo stage while the wild type was at cotyledon stage. At 12 DAP viable mutant seeds presented embryos at cotyledonal stage with several abnormalities, like enlarged cells and nuclei, disorganized cells in the tissue and flatten and smaller cotyledons. The mature embryos were significantly smaller than the wild type (Figure 4a, b).

Germination: Tetrazolium staining showed that the mutant embryos had reduced germination potential, showing a reduce coloration and only morphologically mature embryos exhibited full staining by tetrazolium in wild type plants.

Consistent with the viability of the embryos using the tetrazolium staining method the seed **GERMINATION PERCENTAGE** was significantly lower in the mutant compared to the wild type (Figure 5).

Embryo DNA content: The DNA content in nuclei of wild type and mutant embryos was estimated by considering intensity of the DAPI staining.

Figure 1(a-l): Representative mutant growth stages, (a) Imbibition, (b) Radicle emergence, (c) Hypocotyl and cotyledons emerged from seed coat, (d, e) Fully opened cotyledons, (f) Two rosette leaves of 1 mm in length, (g) Four rosette leaves of 1 mm in length, (h) Ten rosette leaves of 1 mm in length, (i) First visible flower buds (indicated by arrow), (j) First open flowers, (k) Flowering complete and (l) Mature silique

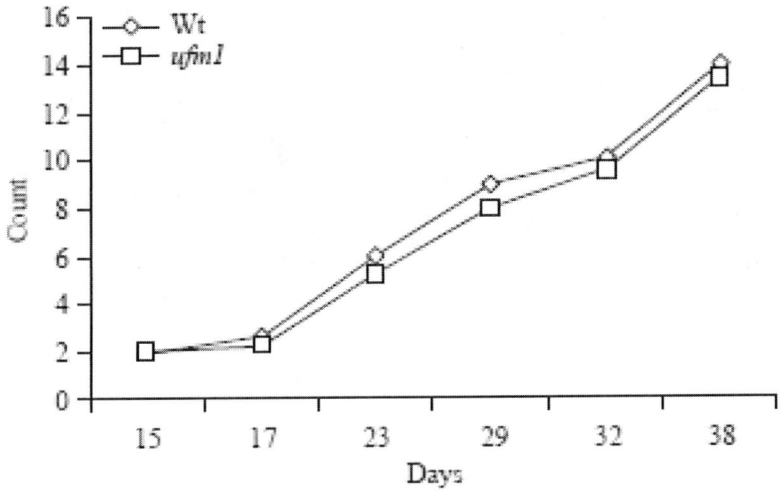

Figure 2: Representative pdata of vegetative parts of wild-type and mutant plants, maximum rosette radius, length of the longest leaf over time and number of rosette leaves >1 mm in length over time. Data are given as average±SE for 24 individual plants. Days including a 3 day stratification at 4°C to synchronize seed germination. No significant differences were observed ($p<0.01$) between wild type and mutant plants

The signal was 1.7 times stronger in the mutant embryos (45929 U per nucleus±1312,8, n = 60) compared with wild type embryos (25883 U per nucleus±1468,7, n = 60) (Figure 6a, b).

Gene expression analysis: To verify the expression patterns, specific primers were designed to amplify the gene *ufm1*. The RT-PCR expression analysis was carried out on mature seeds of *ufm1* mutants and wild type plants. Figure 7 shows the results indicate that *ufm1* was expressed in seed of the wild type plants, but its expression was null in seeds of the mutant plants, which were characterized by the embryo retarded growth and seed abortion.

DISCUSSION

Post-translational modification by ubiquitin-like proteins (UBLs) plays a critical role in controlling stress responses, signaling, plant growth, cell cycle, transcription, abnormalities and show reduced viability.

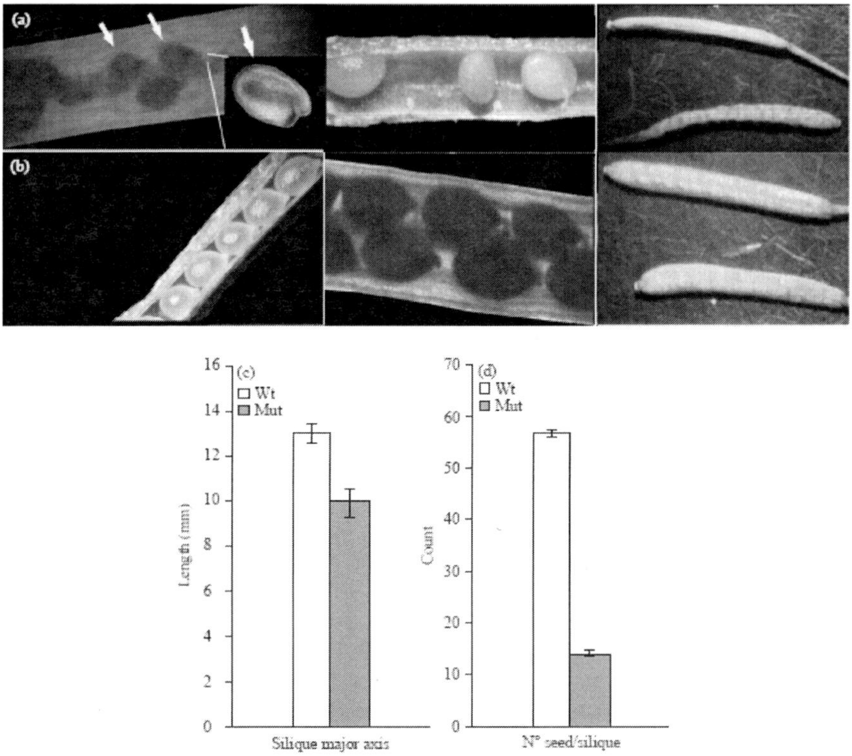

Figure 3(a-d): Comparative analysis of siliques, (a) Siliques of homozygous ufm1 mutant plants with early aborted seeds (arrows), (b) Siliques of wild type plants, (c) Comparison of major axis of siliques between wild type and mutant plants. Data were average from three siliques per plant and (d) Comparison of seeds per silique between wild type and mutant. The number of seeds was observed in mature siliques. Data were average from three siliques per plant. Significant differences were observed ($p<0.01$). Error bars indicate ±SE

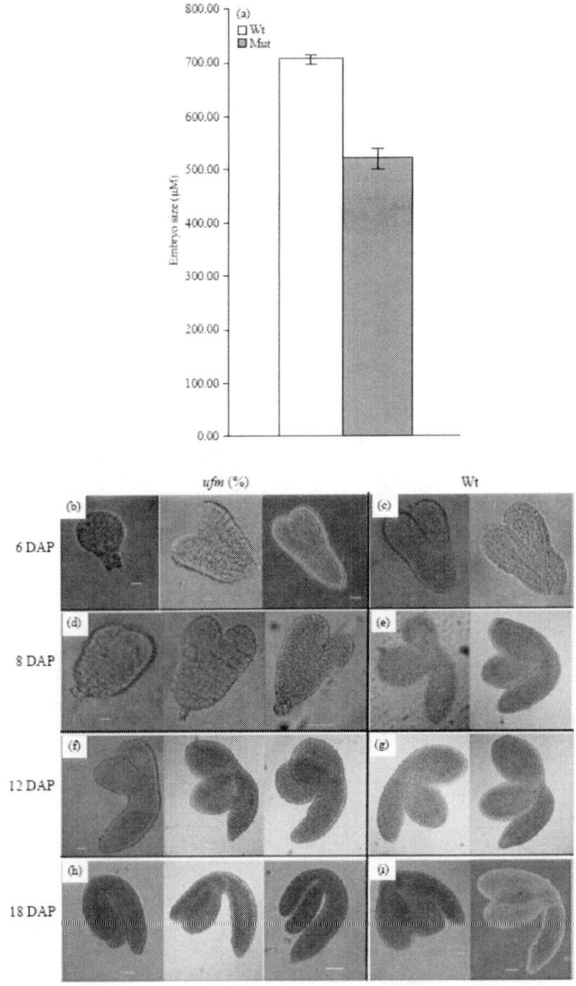

Figure 4. Comparative analysis of embryos, (a) Comparative wild type and mutant embryo size. Significant differences were observed ($p<0.01$). Error bars indicate ±SE, (b-i) Image of cleared embryos of wild type and mutant seeds, (b) Embryos of the mutant to 6 DAP arrested at globular, heart and torpedo stage, (c) Wild type embryos to 6 DAP are uniformly developed at torpedo stage, (d) Mutant embryos to 8 DAP, (e) Wild type embryos to 8 DAP, (f) Mutant embryos to 12 DAP, (g) Wild type embryos to 12 DAP are mature, (h) Mutant mature embryos to 18 DAP present abnormalities and (i) Wild type embryos to 18 DAP are normal and mature

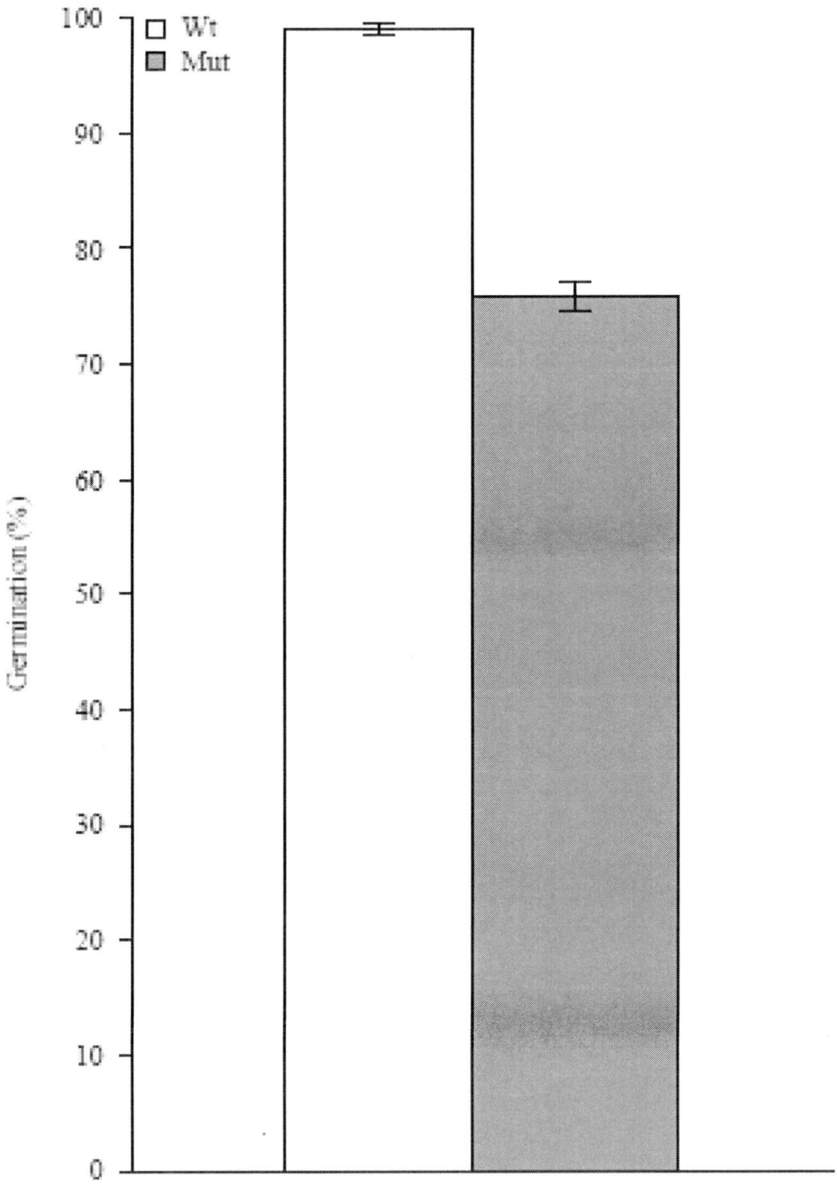

Figure 5. Germination assay of mutant and wild type seeds. Germination was scored 5 days after imbibition. Significant differences were observed ($p<0.01$). Error bars indicate ±SE

A possible embryogenesis, circadian rhythms and other pathways[3,14,26-29]. Recently, the ubiquitin fold modifier 1 (*ufm1*) pathway described in mammalian cells, involves E1 activating (*Uba5*), E2 conjugating (*ufc1*) and E3 ligating (*ufl1*) enzymes, which will conjugate *ufm1* to targets both ***IN VITRO*** and *in vivo*[18,19]. The abnormal *ufm1* cascade is associated with many human diseases including cancer, ischemic heart diseases, diabetes, atherosclerosis, hip dysplasia and schizophrenia[30]. Studies on mutant mice of the enzyme *uba5* resulted lethal for the embryos. It was demonstrated that *Uba5* was indispensable for erythroid differentiation[23]. Also, cultured human cells silenced for *ufm1* and *ufl1* showed accelerated apoptosis[22].

Several genes have an essential role in plants life cycle, the absence of a functional gene product becomes critical or lethal. The distinction between embryo and gametophyte mutants is subjective. Some mutants with defects in embryos may also have defects in the gametophyte development[31]. Plant embryogenesis involves a strict synchronization of cell division, elongation and differentiation[32].

An important objective in *Arabidopsis* research is to identify minimal number of genes necessary to have a viable plant under normal growth conditions[33].

To understand the molecular bases of the embryonic development in plants, in the present study, an evidence was provided on the *ufm1* gene function and seed phenotypes. The RT-PCR analysis showed that *ufm1* is expressed in the seeds of the wild type plants and the expression is suppressed in the mutants. The use of null mutants revealed that *ufm1* is involved in the normal development of the *Arabidopsis* embryo.

To understand the role of *ufm1* in plants, reverse genetics to search for changes in phenotype of plants at different growth stages, morphology and DNA content in the nucleus of embryos in wild type (ecotype Col-0) and homozygous mutant (SALK 040955C) lines was used. Interestingly, phenotypic analysis reveal no significant differences in the vegetative parts but significant differences were observed in the number of viable seeds and size of the embryos, the mutant embryos are smaller and the number of seeds per silique is lower than the wild type.

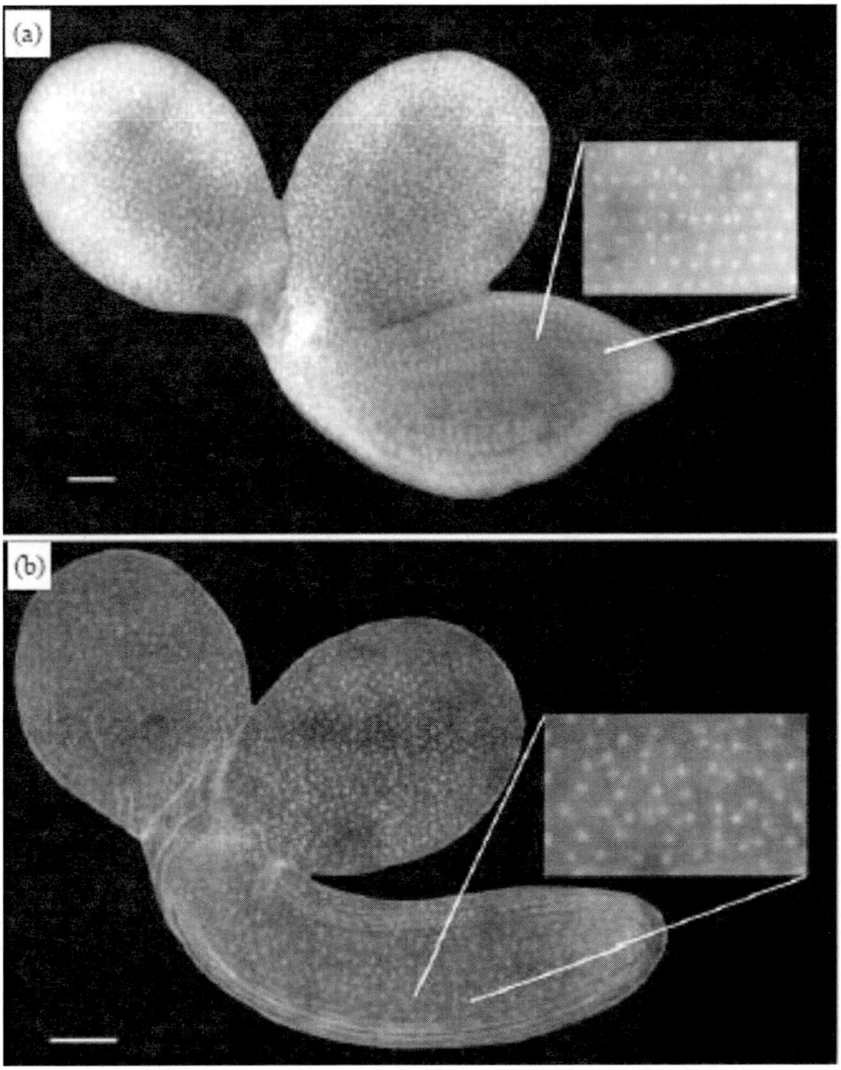

Figure 6. Analysis of nuclear DNA content in wild type and mutant embryos. The DAPI stained nuclei in embryos were recorded by fluorescence microscopy and their nuclear DNA, defined as total pixel intensities of DAPI signals per individual nucleus, was quantified using Image J (a) Wild type embryo and (b) Mutant embryo *ufm1*. Bar = 50 μm. The horizontal lengths of the rectangle zoom section correspond to 50 μm

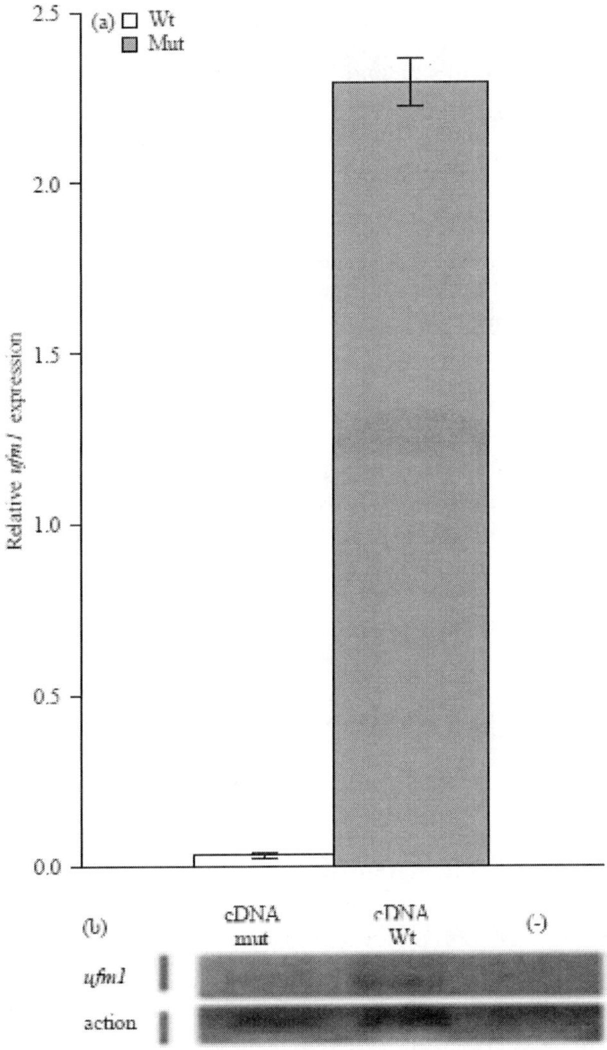

Figure 7. RT-PCR analysis of mutant and wild type seeds of *Arabidopsis thaliana*, (a) Relative expression level of *ufm1* gene and (b) The wild type seeds amplified the expected product of 506 bp. The actin gene was used as a positive control

Also was observed that mutant embryos have delayed development, present abnormalities and show reduced viability. A possible explanation for how some embryo defective mutant survive and develop normally, to complete their life cycle, involve the presence of genes with redundant function.

The results showed an increment in size and DNA content of embryo nuclei in *ufm1* mutants, denoting an alteration in the cell cycle and cellularization possibly by alteration of the ubiquitin pathway. Similar results, although more severe, were obtained in TTN6 seed mutants in *Arabidopsis*[33,34] and in seeds of incompatible crosses in potato[35-38]. This observation establishes an essential role for *ufm1* in plant development, demonstrating that *ufm1* is primordial for plant embryo genesis as observed in mammals[23,39,40]. Further studies of *ufm1* interactions and pathway should clarify the molecular basis for the phenotypes resulting in mutant seeds.

CONCLUSION

It is observed that mutant embryos, which do not express *ufm1* have delayed development, present abnormalities and show reduced viability, relative to control. This is the 1st report to establish the role of *ufm1* in plant embryogenesis. These results are important because they contribute to the understanding of the mechanisms involved in seed development.

SIGNIFICANCE STATEMENT

Little is known about the role of the *ufm1* gene in plants. This study sets the role played by the *ufm1* gene in plant development, remains essential in embryogenesis of plants as well as in mammalian.

ACKNOWLEDGMENTS

This study was funded by Agencia Nacional de Promoción Científica y Tecnológica, Argentina (ANPCyT) [BID 1201-OC-AR PICT-2007-01294]. We are greatful with Carina González member of the laboratory of deceased H. Boccalandro for helpful device and for plant material.

REFERENCES

1. Alonso, J.M., A.N. Stepanova, T.J. Leisse, C.J. Kim and H. Chen et al., 2003. Genome-wide insertional mutagenesis of Arabidopsis thaliana. Science, 301: 653-657.
2. Burroughs, A.M., L.M. Iyer and L. Aravind, 2012. The natural history of ubiquitin and ubiquitin-related domains. Front. Biosci., 17: 1433-1460.
3. Cai, Y., W. Pi, S. Sivaprakasam, X. Zhu and M. Zhang et al., 2015. UFBP1, a key component of the Ufm1 conjugation system, is essential for ufmylation-mediated regulation of erythroid development. PLoS Genet. 10.1371/journal.pgen.1005643
4. Callis, J., T. Carpenter, C.W. Sun and R.D. Vierstra, 1995. Structure and evolution of genes encoding polyubiquitin and ubiquitin-like proteins in Arabidopsis thaliana ecotype Columbia. Genetics, 139: 921-939.
5. Cornejo, P., E.L. Camadro and R.W. Masuelli, 2006. The importance of Arabidopsis seed mutants in the elucidation of the molecular basis of Endosperm Balance Number in tuber-bearing Solanum species. Biocell, 30: 413-421.
6. Cornejo, P., E.L. Camadro and R.W. Masuelli, 2012. Molecular bases of the postzygotic barriers in interspecific crosses between the wild potato species Solanum acaule and Solanum commersonii. Genome, 55: 605-614.
7. Daniel, J. and E. Liebau, 2014. The ufm1 cascade. Cells, 3: 627-638.
8. Downes, B. and R.D. Vierstra, 2005. Post-translational regulation in plants employing a diverse set of polypeptide tags. Biochem. Soc. Trans., 33: 393-399.
9. Downes, B.P., S.A. Saracco, S.S. Lee, D.N. Crowell and R.D. Vierstra, 2006. MUBs, a family of ubiquitin-fold proteins that are plasma membrane-anchored by prenylation. J. Biol. Chem., 281: 27145-27157.
10. Herhaus L. and I. Dikic, 2015. Expanding the ubiquitin code through post-translational modification. EMBO Rep., 16: 1071-1083.
11. Hershko, A. and A. Ciechanover, 1998. The ubiquitin system. Annu. Rev. Biochem., 67: 425-479.
12. Hochstrasser, M., 2009. Origin and function of ubiquitin-like proteins. Nature, 458: 422-429.
13. Iwai, K. and F. Tokunaga, 2009. Linear polyubiquitination: A new regulator of NF-κB activation. EMBO Rep., 10: 706-713.
14. Kang, S.H., G.R. Kim, M. Seong, S.H. Baek and J.H. Seol et al., 2007. Two novel ubiquitin-fold modifier 1 (Ufm1)-specific proteases, UfSP1 and UfSP2. J. Biol. Chem., 282: 5256-5262.
15. Kerscher, O., R. Felberbaum and M. Hochstrasser, 2006. Modification of proteins by ubiquitin and ubiquitin-like proteins. Annu. Rev. Cell. Dev. Biol., 22: 159-180.
16. Komatsu, M., T. Chiba, K. Tatsumi, S.I. Iemura and I. Tanida et al., 2004. A novel protein-conjugating system for Ufm1, a ubiquitin-fold modifier. EMBO J., 23: 1977-1986.
17. Lemaire, K., R.F. Moura, M. Granvik, M. Igoillo-Esteve and H.E. Hohmeier et al., 2011. Ubiquitin fold modifier 1 (UFM1) and its target UFBP1 protect pancreatic beta cells from ER stress-induced apoptosis. PLoS ONE, Vol. 6. 10.1371/journal.pone.0018517
18. Li, F. and R.D. Vierstra, 2012. Autophagy: A multifaceted intracellular system for bulk and selective recycling. Trends Plant Sci., 17: 526-537.
19. Masuelli, R.W. and E.L. Camadro, 1997. Crossability relationships among wild potato species with different ploidies and Endosperm Balance Numbers (EBN). Euphytica, 94: 227-235.
20. Masuelli, R.W., 2001. Mitotic-cycle time and the development of embryo and endosperm in compatible and incompatible crosses in tuber-bearing Solanum species. Genome, 44: 426-431.
21. Matunis, M.J., E. Coutava and G. Blobel, 1996. A novel ubiquitin-like modification modulates the partitioning of the Ran-GTPase-activating protein RanGAP1 between the cytosol and the nuclear pore complex. J. Cell Biol., 135: 1457-1470.

22. Meinke, D.W., 1994. Seed Development in Arabidopsis Thaliana. In: Arabidopsis, Meyerowitz, E.M. and C.R. Somerville (Eds.). Cold Spring Harbor Laboratory Press, Cold Spring Harbor, NY., pp: 253-295.
23. Meng, L. and L. Feldman, 2010. A rapid TRIzol-based two-step method for DNA-free RNA extraction from Arabidopsis siliques and dry seeds. Biotechnol. J., 5: 183-186.
24. Miura, K., J.B. Jin and P.M. Hasegawa, 2007. Sumoylation, a post-translational regulatory process in plants. Curr. Opin. Plant Biol., 10: 495-502.
25. Muralla, R., J. Lloyd and D. Meinke, 2011. Molecular foundations of reproductive lethality in Arabidopsis thaliana. PLoS One, Vol. 6. 10.1371/journal.pone.0028398
26. Ohsumi, Y., 2001. Molecular dissection of autophagy: Two ubiquitin-like systems. Nat. Rev. Mol. Cell Biol., 2: 211-216.
27. Okumoto, K., S. Misono, N. Miyata, Y. Matsumoto, S. Mukai and Y. Fujiki, 2011. Cysteine ubiquitination of PTS1 receptor Pex5p regulates Pex5p recycling. Traffic, 12: 1067-1083.
28. Sahtoe, D.D. and T.K. Sixma, 2015. Layers of DUB regulation. Trends Biochem. Sci., 40: 456-467.
29. Sasakawa, H., E. Sakata, Y. Yamaguchi, M. Komatsu and K. Tatsumi et al., 2006. Solution structure and dynamics of Ufm1, a ubiquitin-fold modifier 1. Biochem. Biophys. Res. Commun., 343: 21-26.
30. Shimizu, Y., Y. Okuda-Shimizu and L.M. Hendershot, 2010. Ubiquitylation of an ERAD substrate occurs on multiple types of amino acids. Mol. Cell, 40: 917-926.
31. Smalle, J. and R.D. Vierstra, 2004. The ubiquitin 26S proteasome proteolytic pathway. Annu. Rev. Plant Biol., 55: 555-590.
32. Tatsumi, K., H. Yamamoto-Mukai, R. Shimizu, S. Waguri and Y.S. Sou et al., 2011. The Ufm1-activating enzyme Uba5 is indispensable for erythroid differentiation in mice. Nat. Commun., Vol. 2. 10.1038/ncomms1182
33. Tatsumi, K., Y.S. Sou, N. Tada, E. Nakamura and S.I. Iemura et al., 2010. A novel type of E3 ligase for the Ufm1 conjugation system. J. Biol. Chem., 285: 5417-5427.
34. Tompa, P., N.E. Davey, T.J. Gibson and M.M. Babu, 2014. A million peptide motifs for the molecular biologist. Mol. Cell, 55: 161-169.
35. Tzafrir, I., J.A. McElver, C.M. Liu, L.J. Yang and J.Q. Wu et al., 2002. Diversity of TITAN functions in Arabidopsis seed development. Plant Physiol., 128: 38-51.
36. Tzafrir, I., R. Pena-Muralla, A. Dickerman, M. Berg and R. Rogers et al., 2004. Identification of genes required for embryo development in Arabidopsis. Plant Physiol., 135: 1206-1220.
37. Vierstra, R.D., 2009. The ubiquitin-26S proteasome system at the nexus of plant biology. Nat. Rev. Mol. Cell. Biol., 10: 385-397.
38. Vierstra, R.D., 2012. The expanding universe of ubiquitin and ubiquitin-like modifiers. Plant Physiol., 160: 2-14.
39. Yoo, H.M., S.H. Kang, J.Y. Kim, J.E. Lee and M.W. Seong et al., 2014. Modification of ASC1 by UFM1 is crucial for ERα transactivation and breast cancer development. Mol. Cell, 56: 261-274.
40. Zhang, M., X. Zhu, Y. Zhang, Y. Cai and J. Chen et al., 2015. RCAD/Ufl1, a Ufm1 E3 ligase, is essential for hematopoietic stem cell function and murine hematopoiesis. Cell Death Differ., 22: 1922-1934.

CHAPTER 19

Physiological and Pathological Impacts of Potassium Silicate on Storability of Anna Apple Fruits

M.E. Tarabih1, E.E. EL-Eryan1 AND M.A. EL-Metwally2

[1]Department of Fruit Handling, Horticulture Research Institute,
[2]Department of Mycological Research, Plant Pathology Research Institute, Agricultural Research Center, Giza, Egypt

ABSTRACT

Exogenous application of silicon (Si) in the form of potassium silicate at different concentrations (0.1, 0.2 and 0.3%) were investigated for maintaining quality and control disease development caused by Penicillium expansum on Anna apple fruits stored 60 days at 0°C±1 with 90-95% R.H (Experiment 1) and held for 6 days at room temperature conditions at 28°C±2 with 65-70% R.H (Experiment 2). It was noticed that the reduction in linear growth and dry weight were positively correlated to the increase in potassium silicate concentrations. So, potassium silicate at 0.3% treatment indicated complete inhibition of the linear growth and dry weight of P. expansum. The disease infection decreased as storage period advanced at cold storage and under marketing conditions. The lowest significant values of disease infection percentage of P. expansum were recorded by dipping fruit at potassium silicate at 0.3% after 60 days of cold storage and 6 days at marketing in the two seasons. Generally, significant changes were observed in potassium silicate at 0.3% which reduced the loss of weight, decay, total loss, respiration rate and PPO activity with respect to the other treatments or the control. Meanwhile, potassium silicate at 0.2% gave a higher fruit firmness. In addition, all silicon treatments reduced SSC, acidity, SSC/acid ratio and total sugar compare to the untreated fruits. The values of hue angle increased with the progress of potassium silicate concentration. These results show that potassium silicate can be used to delay ripening, keep quality and control disease development caused by P. expansum on apple fruits.

INTRODUCTION

Anna apple variety is considered as a hybrid between Red Hadassiya and Golden Delicious. Anna is one of the leading apple cultivars in Egypt, considered the standard in low-chilling apples. The limited shelf-life of the fruits is related to rapid softening and also to the tendency to bruise and brown easily. Postharvest pathogens cause major losses in apple production. More than 90 fungal species have been described as causative agents of apple decay during storage (Pianzzola et al., 2004).

Blue mould caused by P. expansum is destructive postharvest disease of apples. This disease causes shortage of shelf-life and economic loss of apple. The use of synthetic fungicides is a primary method of control for this disease. However, concerns about public health and the development of resistant pathogens have increased the need to search for alternative methods (Gholamnejad et al., 2009).

Silicon (Si) is the second richest element found on the surface of the earth's crust and its content in plants is 0.1-10% of their dry weight (Liang et al., 2008). Silicon applications might become an alternative to currently used fungicides. Silicon has been used to minimize the adverse effects of biotic and abiotic stresses on various fruit crops by stimulating defense reaction mechanisms (Brecht et al., 2003). Also, excess of Si has no toxic reports so far (Ma and Yamaji, 2006).

The concentrations of silicon especially combination with hot water, affects apple's responses towards P. expansum and it could be an important method for control of apple blue mould (Etebarian et al., 2013).

There is much experimental evidences suggesting that Si affects the activity of major antioxidant enzymes, involved in plant stress defense systems (Epstein, 2009). Depositions of Si into epidermal cells may form an effective mechanical barrier against fungal penetration. Plants harden physically as a result of Si accumulation resulting in additional protection, preventing fungi from entering plant cells (Bosse et al., 2011). Farahani et al. (2012), used three concentrations of silicon (Si) at 0.1, 0.3 and 0.5% against apple blue mould caused by P. expansum and proposed two mechanisms for Si-enhanced resistance to diseases, inwhich Si can act as a physical barrier as it is deposited beneath the cuticle to form a cuticle-Si double layer, while it also acts as a modulator of host resistance to pathogens.

Potassium silicate the most commonly used form of Si was therefore applied in order to determine whether such practice could increase the concentration of antifungal compounds and/or the enzyme PAL to be able to increase the concentration of phenolic compounds present at later ripening stages in order to decrease disease incidence.

Potassium silicate is a source of highly soluble potassium and silicon. It is used in agricultural production systems primarily as a silica amendment and has added the benefit of supplying small amounts of potassium.

Kaluwa et al. (2010) used different sources of silicon (potassium silicate, Nontox-silica, calcium silicate, sodium metasilicate and Biosilicate) as post-harvest dips at 2940 ppm on avocado fruits and found that potassium silicate seem to be most beneficial to maintain avocado fruit quality, probably due to a suppression of respiration and a reduction in ethylene evolution. Moreover, recent post-harvest studies on avocado have proved, Si to be a safe and effective antioxidant source (Tesfay et al., 2011). Mditshwa et al. (2013) investigated the ability of K2SiO3 dips at 50, 150 and 250 mg L-1 solutions for 30 min to reduce fruit weight loss and enhance the phenolic content in order to reduce the incidence of chilling injury in lemon fruit.

The objective of this study was to evaluate the use of potassium silicate to extend storability, delay ripening, maintain fruit quality and enhance systemic resistance of Anna apple fruits during cold storage and marketing at room temperature. Also, the effects on the sensory, physical, chemical, pathological and physiological characteristics were examined.

MATERIALS AND METHODS

Isolation and identification of the pathogen: Penicillium expansum was isolated from naturally infected Anna apple fruits. This isolate was the most aggressive one in our collection and produced the largest lesions on inoculated fruits. These fungi were purified and maintained on Potato Dextrose Agar (PDA) and stored at 4°C, with periodic transfers through citrus fruits to maintain its aggressiveness. Seedless grape were ready for examination under a stereoscopic binocular microscope (6-50 X) for the presence of fungi and to study their habit characters. When necessary the compound microscope was used for confirming the identification after examining the morphology of conidia and conidiophores. Fungi were identified by means of comparison with the description sheets of Commonwealth Mycological Institute, Kew, Surrey, England (CMI), Danish Government Institute of Seed Pathology (DGISP) publications as well as publications of Ellis (1971) and Singh et al. (1991).

Effect of potassium silicate on growth of fungi isolated from Anna apple fruits and disease infection percentage

Linear growth: Potassium silicate was tested in vitro on the linear growth of the pathogenic fungi. Different concentrations were added to 10 mL of sterilized PDA before solidification and then poured in sterile petri-dishes. After solidification, the plates were inoculated with fungal disc (5 mm) in the center of the plate and incubated at 27±1°C. Three plates for each particular treatment for each fungus were used as replicates; three plates were prepared to serve as control for each fungus. Linear growth was observed daily and diameter of fungal colonies were recorded when plates of any treatment were filled with the fungal growth.

Dry weight: One hundred milliliter of liquid PD medium in 250 mL Erlenmeyer flasks were amended with different concentrations of the tested compounds after autoclaving. Each flask was inoculated using two discs of 0.6 mm in diameter of fungal culture and then incubated at 20±2°C for 7 days. Control flasks without any concentration of these compounds. Three replicates were used for each concentration. At the end of incubation period, the mycelium was filtered off and washed several times with distilled water, then dried in an oven at 80°C for 48 h till constant weight (El-Morsy, 1993).

Field study: This study was carried out during the two successive seasons of 2012 and 2013 on Anna apple fruits to evaluate the effect of potassium silicate as a post-harvest treatment for keeping quality of Anna apple fruits under cold storage condition and marketing at room temperature. Fruits were picked from trees about five years old, grown in sandy soil in a private orchard at EL-Khatatba city, Monifia, Governorate. These trees were healthy and similar in their vigor as possible and were treated with common agriculture practices in both seasons. Mature apple fruits were picked when the red color reached over 50%, fruit firmness was about 11-12% according to Drake and Kupferman (2000). Selected fruits were directly transported to the Laboratory and defective fruits were almost equal in size and apparently insect and pathogen injury free. At the beginning of the experiments, samples of 15 fruits were taken to determine the initial fruit properties and then received the following treatments:

- Dipping fruit with potassium silicate (K2SiO4) 0.1% for 5 min

- Dipping fruit with potassium silicate ($K2SiO4$) 0.2% for 5 min
- Dipping fruit with potassium silicate ($K2SiO4$) 0.3% for 5 min
- Control (Dipping fruits in tap water) for 5 min

Fruits of all treatments were sorted to remove any infected and damaged and then stored in perforated plastic bags (each contain 5 fruits). All bags with fruits were weighted and every three bags were put in ventilated carton box.

Experiment 1: For storage study, fruits of each treatment were stored 60 days at 0±1°C with 90-95% R.H and then the fruits were taken 20 days intervals to determine fruits characteristics.

Experiment 2: For marketing study, fruits for each treatment were held 6 days at room temperature conditions at 28±2°C with 65-70% R.H and then fruits were taken 2, 4 and 6 days intervals to determine the following parameters.

Disease infection: It was determined according to the following equation:

$$\text{Disease infection } (\%) = \frac{\text{No. of naturally infected fruits}}{\text{No. of total fruits}} \times 100$$

Effect of potassium silicate on Anna apple fruits quality: It was determined by the follwing factors.

Loss in fruit weight: It was determined according to the following equation:

$$\text{Loss in fruit weight } (\%) = \frac{\text{Initial weight} - \text{weight at sampling date}}{\text{Initial fruit weight}} \times 100$$

Decay: It was determined according to the following equation:

$$\text{Decay } (\%) = \frac{\text{Weight of decayed fruits}}{\text{Initial fruit weight}} \times 100$$

Total loss in fruit (%): Following equation is used to determine the percentage of total loss in fruits:

Total loss in fruit (%) = Loss weight (%)+Decayed fruits weight (%)

Respiration rate (mg CO2 kg-1 h-1): Carbon dioxide produced by apple fruits was determined as mg CO2 kg-1 h-1 according to AOAC (1970).

Skin hue color: Skin color was measured using a hand-held colorimeter (CR-10; Minolta Co., Ltd., Osaka, Japan). Color changes from green to yellow were indicated by calculating the hue angle (h°), from (a*, b*) using the methods described by McGuire (1992).

Fruit firmness (Ib inch-2): It was determined on the two opposite sides of fruit using a hand Effegi-Penetrometers and the average was estimated as Ib inch-2.

Soluble solids content (SSC%): Soluble solids content in fruit juice was measured using a Carl-Zeiss hand refractometer according to AOAC (2005).

Titratable acidity (TA%): It was determined in10 mL of fruit juice as a percentage of malic acid according to AOAC (2005).

SSC/acid ratio (%): It was calculated by dividing the value of SSC over the value of titratable acidity of each sample.

Polyphenol oxidase (PPO): The specific activity was expressed as unit per milligram according to Meng et al. (2008).

Total sugar (%): The extract was prepared by taking 0.5 g of fresh pulp and extracting with 80% ethanol according to Ranganna (1979).

Total starch: Starch content was calculated with starch-iodine test according to Generic chart scores (1-8), where 1 represents the least and 8 the highest starch scores.

Statistical analysis: Data of both seasons of the study were analyzed using Analysis of Variance (ANOVA) technique. Differences among treatment means were statistically compared using Duncan's multiple range tests at a level of 0.05 using the CoStat v6.4 program.

RESULTS

Effect of potassium silicate on linear growth (cm), dry weight (g) of fungi isolated from Anna apple fruits and percentage disease infection as post-harvest treatments: Data in Table 1 indicated that the effect of treatments by dipping with potassium silicate on linear growth and dry weight of P. expansum isolated from Anna apple fruits. It was also noticed that the reduction in linear growth and dry weight were positively correlated to the increase in potassium silicate concentrations. Potassium silicate at 0.3% treatment showed complete inhibition of the linear growth and dry weight of P. expansum.

The disease infection at cold storage and 6 days during marketing are presented in Table 2. The data showed that, the disease infection decreased as storage period advanced under cold storage and during marketing at room temperature.

Table 1: Effect of potassium silicate on linear growth (cm) and dry weight (g) of fungi isolated from Anna apple fruits

	Penicillium expansum	
Treatments	Linear growth (cm)	Dry weight (g)
Potassium silicate 0.1%	5.63b	0.71b
Potassium silicate 0.2%	2.96c	0.39c
Potassium silicate 0.3%	0.00d	0.00d
Control (water)	9.00a	1.86a

Means followed by the same letters are not significantly different by Duncan's multiple range test at 0.05 level

Table 2: Effect of potassium silicate on disease infection during cold storage and under market conditions during 2012 and 2013 seasons

	Cold storage (days)				Marketing 6 days at room temperature			
Treatments	0	20	40	60	0	2	4	6
Season 2012								
Potassium silicate 0.1%	0.00n	2.96l	7.69h	11.61f	0.00n	4.02k	9.17g	13.77d
Potassium silicate 0.2%	0.00n	2.18m	4.60j	6.53i	0.00n	2.35lm	4.98j	7.93h
Potassium silicate 0.3%	0.00n	0.00n	0.00n	0.00n	0.00n	0.00n	0.00n	0.00n
Control	0.00n	4.52jk	12.94e	26.57b	0.00n	6.14i	15.28c	29.67a
Season 2013								
Potassium silicate 0.1%	0.00m	2.58l	5.71i	9.13f	0.00m	3.47k	7.91g	11.87d
Potassium silicate 0.2%	0.00m	2.08l	4.34j	5.70i	0.00m	2.03l	4.29j	6.84h
Potassium silicate 0.3%	0.00m	0.00m	0.00m	0.00m	0.00m	0.00m	0.00m	0.00m
Control	0.00m	3.96jk	10.29e	23.87b	0.00m	5.29i	13.18c	25.58a

Means followed by the same letters are not significantly different by Duncan's multiple range test at 0.05 level

Thus, all the applied treatments reduced the disease infection than the control. Since, the percentage of disease infection of the untreated fruits were 26.57 and 23.87% after 60 days of cold storage and reached 29.67 and 25.58% as marketing in the both seasons, respectively. The lowest significant values of disease infection percentage of P. expansum were recorded by dipping fruit at potassium silicate at 0.3% ranged 0.00% in both seasons after 60 days of cold storage and 6 days of marketing in the two seasons.

Effect of potassium silicate on characteristics of Anna apple fruits under cold storage and through marketing at room temperature

Loss in fruit weight percentage: The loss in weight of Anna apple fruits at cold storage and 6 days during marketing are presented in Table 3. The data reveal that, the loss in fruit weight increased as storage period advanced under cold storage and in the second experiment, at room temperature. Thus, all the applied treatments reduced the loss in fruit weight than the control. Since, the percent of loss in fruit weight of the untreated fruits were 10.22 and 11.04% after 60 days of cold storage and it were 6.73 and 6.79% in the second experiment, after six days as marketing in both seasons, respectively. The lowest significant values of weight loss percentage were recorded by dipping fruits in potassium silicate at 0.3% ranged 3.20 and 3.90% after 60 days of cold storage and it were 2.91 and 3.10% after 6 days of marketing in the two seasons, respectively.

Decay percentage: It is clear from Table 1 that all dipping treatments did not present any decayed fruits till 20 days of cold storage through both seasons. Since, all treatments significantly reduced the percentage of decayed fruits than the untreated fruits either after 60 days of cold storage or 6 days during marketing in the second experiment at room temperature in both seasons. Thus, the percentage of decayed fruits for the control were 27.23 and 28.41% after 60 days of cold storage but it reached about 18.94 and 19.10% through marketing in the second experiment in both seasons. Yet, dipping Anna apple fruits in potassium silicate at 0.3% significantly reduced decay percentage than all treatments applied after 60 days of cold storage (5.10 and 5.54%) however it were 6.00 and 6.20% during marketing in both seasons, respectively.

Total loss (%): Total loss in fruit weight is mainly due to loss in fruit weight and decay percentages are presented in Table 1. It is clear from this table that dipping Anna apple fruits in potassium silicate at 0.3% significantly reduced the percentage of total loss in fruit weight in both seasons than the other treatments or the control. Since, potassium silicate at 0.3% presented about 8.30% and 9.44%, respectively, after 60 days of cold storage whereas, the loss percentage reached 8.91 and 9.30% after 6 days at room temperature in the second experiment as mean of two seasons, respectively.

Moreover, the percentage of total loss in fruit weight was gradually increased during cold storage or at marketing as storage period advanced. Since, the percentage of total loss of the untreated fruits were about 37.45 and 39.45% after 60 days of cold storage but reached 25.67 and 25.89% when held 6 days at room temperature in the second experiment as mean of both seasons.

Respiration rate (mg CO2 kg-1 h-1): Results of Table 4 show that, all evaluated treatments succeeded in reducing respiration rate of Anna apple fruits during durations in comparison with the control treatment.

Table 3: Effect of potassium silicate on weight loss, decay and total loss percentage of Anna apple fruits during cold storage and under market conditions during 2012 and 2013 seasons

Treatments at different storage periods (days)	Cold storage			Marketing 6 days at room temperature		
	Loss weight (%)	Decay (%)	Total loss (%)	Loss weight (%)	Decay (%)	Total loss (%)
Season 2012						
Potassium silicate 0.1%						
0	0.00l	0.00g	0.00l	0.00j	0.00l	0.00l
20	2.11j	0.00g	2.11j	2.00h	5.21i	7.21i
40	3.93f	5.13f	9.06f	3.70e	10.13f	13.83f
60	6.13c	13.00d	19.13d	4.41c	12.19d	16.60d
Potassium silicate 0.2%						
0	0.00l	0.00g	0.00l	0.00j	0.00l	0.00l
20	2.29i	0.00g	2.29i	2.10g	5.10j	7.20i
40	4.17e	6.27e	10.44e	3.95d	10.24e	14.19e
60	7.00b	15.12b	22.12b	4.60b	12.90c	17.50c
Potassium silicate 0.3%						
0	0.00l	0.00g	0.00l	0.00j	0.00l	0.00l
20	1.96k	0.00g	1.96k	1.72i	0.00l	1.72k
40	2.34i	0.00g	2.34i	2.09gh	4.49k	6.58j
60	3.20h	5.10f	8.30g	2.91f	6.00h	8.91h
Control						
0	0.00l	0.00g	0.00l	0.00j	0.00l	0.00l
20	3.58g	0.00g	3.58h	3.00f	9.18g	12.18g
40	6.02d	14.27c	20.29c	4.33c	10.04b	18.37b
60	10.22a	27.23a	37.45a	6.73a	18.94a	25.67a
Season 2013						
Potassium silicate 0.1%						
0	0.00m	0.00h	0.00k	0.00k	0.00l	0.00l
20	2.24k	0.00h	2.17i	2.00i	5.44i	7.44i
40	4.40f	8.90e	13.30e	4.48e	9.54f	14.02f
60	6.59c	13.39d	19.98d	4.63d	13.10d	17.73d
Potassium silicate 0.2%						
0	0.00m	0.00h	0.00k	0.00k	0.00l	0.00l
20	2.49j	0.00h	2.49h	2.16h	5.10j	7.26j
40	4.48e	8.73f	13.21e	4.74c	10.10e	14.84e
60	7.30b	15.43b	22.73b	4.79c	13.45c	18.24c
Potassium silicate 0.3%						
0	0.00m	0.00h	0.00k	0.00k	0.00l	0.00l
20	2.00l	0.00h	2.00j	1.83j	0.00l	1.83k
40	2.60i	0.00h	2.60h	2.97g	4.23k	7.20j
60	3.90h	5.54g	9.44f	3.10f	6.20h	9.30h
Control						
0	0.00m	0.00h	0.00k	0.00k	0.00l	0.00l
20	3.99g	0.00h	3.99g	3.10f	8.48g	11.58g
40	6.45d	14.91c	21.36c	6.70b	14.22b	20.92b
60	11.04a	28.41a	39.45a	6.79a	19.10a	25.89a

Means followed by the same letters are not significantly different by Duncan's multiple range test at 0.05 level

Whereas, fruits dipping in potassium silicate at 0.3% proved to be the most efficient treatment (13.91 and 13.32 mg CO2 kg-1 h-1) till 60 days under cold storage and (12.93 and 12.70 mg CO2 kg-1 h-1) 6 days during marketing in the second experiment through both seasons, respectively. On the other hand, the highest respiration rate were obtained by treatment of control (15.11 and 14.95 mg CO2 kg-1 h-1) till 60 days under cold storage and (14.45 and 14.40 mg CO_2 kg-1 h-1) 6 days during marketing in both seasons, respectively.

Firmness (lb inch¯2): Data from Table 4 show clearly that, fruit firmness was reduced as storage period advanced under cold storage or through marketing at room temperature. The data also confirm that all treatments used significantly reduced the loss in fruit firmness than the control at cold storage or marketing in the second experiment at room temperature through the two seasons. Thus, fruit firmness for the control treatment was 6.55 and 6.79 lb inch-2 after 60 days of cold storage but it reached about 6.00 and 6.10 lb inch-2through marketing in both seasons. Furthermore, treated fruits with potassium silicate at 0.3% showed a higher fruit firmness (8.10 and 8.29 lb inch-2) after 60 days of cold storage while, reached 7.90 and 8.33 lb inch-2 from fruits held 6 days at room temperature in the second experiment of the two seasons, respectively.

Polyphenol oxidase (PPO unit mg-1): The great relationship among the rate of respiration, firmness and PPO activity of Anna apple fruits during cold storage and through marketing at room temperature, as soon as increasing the fruit respiration rate resulting decrease in fruits firmness and PPO activity was observed.

The inactivation and reactivation of PPO residual activity in the flesh of Anna apple juice subjected to potassium silicate treatments for 60 days was showed in Table 4. The activity of PPO in Anna apple juice of fruits treated by potassium silicate at 0.3% showed a statistically significant decrease with the advanced storage period, the residual activity of PPO was 0.390 and 0.393 U mg-1 after 60 days under cold storage and 6 days during marketing (0.361 and 0.377 U mg-1) in the second experiment of two seasons, respectively. The higher residual activity of PPO was 0.430 and 0.433 U mg-1 after 60 days under cold storage and 6 days at marketing in the second experiment (0.423 and 0.429 U mg-1) at the untreated ones during the two seasons, respectively.

Soluble solids content (SSC%): Concerning to the effect on SSC, data from Table 5 showed that soluble solids content in fruit juice of Anna apple was gradually increased as storage period prolonged either after cold storage or during shelf life at room temperature. It would be expected that increased weight loss in control fruits would lead to the increase of SSC due to disappearance of water from the fruits and concentration of soluble solids. Since, all treatments gave significantly lower values of SSC in fruit juice than the control fruit which ranged 14.15 and 13.90% after 60 days of cold storage and it were 14.90 and 14.69% after six days, marketing in both seasons under the study.

Titratable acidity (TA%): From Table 5, its clear that the content of total acidity in fruit juice was decreased with the progress in storage period from harvest till 60 day at cold storage or during marketing at room temperature. Since, all treatments gave significantly higher values of SSC in fruit juice than the control fruit. The highest value was obtained with potassium silicate at 0.3% (0.654 and 0.678%) after 60 days of cold storage and it was 14.90, 0.520 and 0.540% after six days, marketing in both seasons under the study. Moreover, control treatment produced lower significant acidity after 60 days of cold storage (0.630 and 0.654%), while after 6 days as marketing in the two seasons ranged 0.490 and 0.505% in the two seasons, respectively.

Table 4: Effect of potassium silicate on respiration rate (mg CO2 kg-1 h-1), firmness (lb inch-2) and polyphenol oxidase (PPO) U mg-1 of Anna apple fruits during cold storage and under market conditions during 2012 and 2013 seasons

Treatments at different stroage periods (days)	Cold storage			Marketing 6 days at room temperature		
	Respiration rate	Firmness	Polyphenol oxidase	Respiration rate	Firmness	Polyphenol oxidase
Season 2012						
Potassium silicate 0.1%						
0	12.10g	10.70a	0.631a	12.20e	10.70a	0.633a
20	12.00gh	9.56b	0.629a	12.49d	9.90c	0.551c
40	13.20e	8.20d	0.542c	12.85c	8.00f	0.443h
60	14.33b	6.97f	0.421fg	13.36b	7.12i	0.405jk
Potassium silicate 0.2%						
0	12.10g	10.70a	0.631a	12.20e	10.70a	0.633a
20	11.91hi	9.70b	0.629a	12.55d	9.90c	0.539d
40	13.13e	8.83c	0.530d	12.96c	8.00f	0.456g
60	14.20c	7.11f	0.418g	13.25b	7.23h	0.396k
Potassium silicate 0.3%						
0	12.10g	10.70a	0.631a	12.20e	10.70a	0.633a
20	11.95hi	9.74b	0.625a	12.21e	10.00b	0.512e
40	12.92f	9.00c	0.510e	12.43d	8.95e	0.411j
60	13.91d	8.10d	0.390h	12.93c	7.90g	0.361l
Control						
0	12.10g	10.70a	0.631a	12.20e	10.70a	0.633a
20	11.83i	9.00c	0.630a	12.89c	9.10d	0.579b
40	13.83d	7.65e	0.587b	13.24b	7.88g	0.469f
60	15.11a	6.55g	0.430f	14.45a	6.00j	0.423i
Season 2013						
Potassium silicate 0.1%						
0	12.00g	10.93a	0.639a	12.00h	10.93a	0.639a
20	11.90hi	9.61c	0.631ab	12.20fg	10.00c	0.567c
40	13.00e	8.27f	0.554d	12.33ef	8.55f	0.458h
60	13.82b	7.03i	0.427fg	13.20b	7.73i	0.418j
Potassium silicate 0.2%						
0	12.00g	10.93a	0.639a	12.00h	10.93a	0.639a
20	11.82ij	9.70c	0.626b	12.29ef	9.95c	0.544d
40	12.92e	8.98e	0.540e	12.40e	8.60f	0.469g
60	13.57c	7.23h	0.421g	12.98c	7.90h	0.403k
Potassium silicate 0.3%						
0	12.00g	10.93a	0.639a	12.00h	10.93a	0.639a
20	11.93gh	10.11b	0.623b	12.13gh	10.15b	0.524e
40	12.79f	9.43d	0.560d	12.22fg	9.10e	0.428i
60	13.32d	8.29f	0.393h	12.70d	8.33g	0.377l
Control						
0	12.00g	10.93a	0.639a	12.00h	10.93a	0.639a
20	11.80j	9.10e	0.631ab	12.60d	9.70d	0.588b
40	13.61c	8.00g	0.593c	13.10bc	7.90h	0.483f
60	14.95a	6.79j	0.433f	14.40a	6.10j	0.429i

Means followed by the same letters are not significantly different by Duncan's multiple range test at 0.05 level

SSC/acid ratio (%): Considering the effect on SSC/acid ratio, data in Table 5 reveal that the values of SSC/acid ratio were progressively increased by the storage period advanced from harvest till 60 days either at cold storage or during marketing at room temperature.

Table 5: Effect of potassium silicate on SSC (%), acidity (%) and SSC/acid ratio (%) of Anna apple fruits during cold storage and under market conditions during 2012 and 2013 seasons Table

Treatments at different storage periods (days)	Cold storage			Marketing 6 days at room temperature		
	SSC (%)	Acidity (%)	SSC/acid ratio (%)	SSC (%)	Acidity (%)	SSC/acid ratio (%)
Season 2012						
Potassium silicate 0.1%						
0	12.90h	0.750a	17.19j	12.90h	0.750a	17.19j
20	13.00gh	0.725c	17.92gh	13.39f	0.712c	18.80h
40	13.39d	0.680e	19.68e	13.83de	0.612fg	22.59f
60	13.85b	0.639g	21.67b	14.24b	0.500i	28.48b
Potassium silicate 0.2%						
0	12.90h	0.750a	17.19j	12.90h	0.750a	17.19j
20	12.97h	0.732bc	17.72hi	13.10g	0.719bc	18.21i
40	13.26e	0.682de	19.44e	13.99cd	0.620f	22.56f
60	13.61c	0.641g	21.23c	14.00c	0.509i	27.50c
Potassium silicate 0.3%						
0	12.90h	0.750a	17.19j	12.90h	0.750a	17.19j
20	12.93h	0.741ab	17.44ij	12.92h	0.724b	17.84i
40	13.10fg	0.694d	18.87f	13.69e	0.632e	21.66g
60	13.24e	0.654f	20.24d	13.79e	0.520h	26.52d
Control						
0	12.90h	0.750a	17.19j	12.90h	0.750a	17.19j
20	13.15ef	0.720c	18.26g	13.44f	0.700d	19.20h
40	13.69c	0.670e	20.43d	14.17b	0.603g	23.50e
60	14.15a	0.630g	22.45a	14.90a	0.490j	30.41a
Season 2013						
Potassium silicate 0.1%						
0	12.63f	0.776a	16.27g	12.63i	0.776a	16.27m
20	12.75ef	0.765ab	16.66fg	13.00g	0.713d	18.23j
40	13.00c	0.710cd	18.31d	13.65d	0.629f	21.83f
60	13.43b	0.663fg	20.25b	14.10b	0.519i	27.16b
Potassium silicate 0.2%						
0	12.63f	0.776a	16.27g	12.63i	0.776a	16.27m
20	12.70ef	0.765ab	16.59g	12.90gh	0.729c	17.69k
40	12.97cd	0.719c	18.03de	13.40e	0.631f	21.23g
60	13.32b	0.669ef	19.91b	13.92c	0.526i	26.46c
Potassium silicate 0.3%						
0	12.63f	0.776a	16.27g	12.63i	0.776a	16.27m
20	12.68f	0.761b	16.65fg	12.81h	0.741b	17.28l
40	12.83de	0.720c	17.82e	13.19f	0.649e	20.32h
60	13.00c	0.678e	19.16c	13.63d	0.540h	25.23d
Control						
0	12.63f	0.776a	16.27g	12.63i	0.776a	16.27m
20	12.96cd	0.760b	17.05f	13.26f	0.709d	18.70i
40	13.33b	0.705d	18.90c	14.00bc	0.616g	22.72e
60	13.90a	0.654g	21.24a	14.69a	0.505j	29.08a

With regard to the effect of these treatments on SSC/acid ratio the data reveal that, control treatment produced a higher value of SSC/acid ratio at cold storage, since the values averaged about 22.45 and 21.24% while, in the second experiment after 6 days in marketing ranged 30.41 and 29.08% during the two seasons, respectively.

Moreover, after 60 days during cold storage potassium silicate at 0.3% treatment gave a lower values of SSC/acid ratio (20.24 and 19.16%) while, after 6 days as marketing ranged 26.52 and 25.23% during the two seasons, respectively.

Skin hue color (h°): The decline in hue angle (change in color from green to red) occurred rapidly with storage period advanced during cold storage or under room temperature. From Table 6 data presented that all applied treatments had a delay in the development of fruits skin color when compared with the untreated fruits. The loss of the green color in apple skin was expressed as lower hue angle (h°). The values of hue angle increased with the progress of potassium silicate concentrations. In control fruit, hue angle decreased rapidly during storage, indicated a losing green color, either at cold storage (72.33 and 68.40 h°) or during marketing in the second experiment (64.30 and 61.54 h°). Furthermore, the values of hue angle (h°) during marketing were almost lower than those obtained at cold storage during the both seasons of study.

Moreover, potassium silicate at 0.3% produced a higher hue angle (h°) than all treatments or the control after 60 days of cold storage and 6 days during marketing in both season. The increment due using these treatments reached about 79.40 and 76.85 h° after 60 days of cold storage, respectively during both seasons. While, after 6 days during marketing in the second experiment the values averaged 69.90 and 66.50 h° in both seasons, respectively.

Total sugar (%): Concerning to the effect on total sugar, Data from Table 6 showed that total sugar content in fruit juice of Anna apple fruits was gradually increased as storage period prolonged either after cold storage or during marketing at room temperature.

Since, the untreated fruits produced higher significant values of total sugar in apple fruits than all treatments which ranged 13.76 and 13.55% after 60 days of cold storage and it were 14.30 and 14.05% after six days as marketing in the second experiment during both seasons under the study. The data also disclose that, application of potassium silicate at 0.3% presented lower values of total sugars compare to the applied treatments averaged 13.32 and 13.27% after 60 days of cold storage and it were 13.71 and 13.63% after six days as marketing in both seasons, respectively.

Total starch: Starch iodine index indicates the gradual change of starch to sugars in the fruit. Considering to the effect on total starch, data from Table 6 showed that total starch content in fruit pulp of Anna apple fruits were gradually decreased as storage period prolonged either after cold storage or during marketing at room temperature. The maximum starch score (6.31 and 6.49) recorded in fruit dipping at potassium silicate at 0.3% after 60 days of cold storage and it were 4.79 and 4.97% after six days as marketing in the second experiment of both seasons, respectively. While, the minimum starch score (5.86 and 6.21) recorded in control fruit after 60 days of cold storage and it were 4.34 and 4.65% after six days as marketing in the second experiment of both seasons, respectively.

DISCUSSION

The reduction in linear growth and dry weight were positive correlated to the increase in potassium silicate concentrations. Potassium silicate at 0.3% treatment showed complete inhibition of the linear growth and dry weight of P. expansum.

Table 6: Effect of potassium silicate on hue angle (h°), sugar (%) and starch (iodine index) of Anna apple fruits during cold storage and under market conditions during 2012 and 2013 seasons

Treatments at different storage periods (days)	Cold storage			Marketing 6 days at room temperature		
	Hue angle	Total sugar	Total starch	Hue angle	Total sugar	Total starch
Season 2012						
Potassium silicate 0.1%						
0	95.15a	13.16g	8.06a	95.13a	13.16g	8.00a
20	91.50bc	13.03i	7.20d	84.95d	13.30f	6.17c
40	84.60ef	13.25e	6.94e	76.30h	13.68d	5.30f
60	74.93i	13.61b	6.00g	65.00l	14.09b	4.40j
Potassium silicate 0.2%						
0	95.15a	13.16g	8.06a	95.13a	13.16g	8.00a
20	92.20b	13.10h	7.23d	85.90c	13.27fg	6.23c
40	86.00e	13.23ef	7.16d	79.45g	13.50e	5.36f
60	76.53h	13.54c	6.19f	67.44k	13.86c	4.59i
Potassium silicate 0.3%						
0	95.15a	13.16g	8.06a	95.13a	13.16g	8.00a
20	93.73a	13.03i	7.41c	89.44b	13.23fg	6.44b
40	89.64d	13.19fg	7.59b	82.63f	13.33f	5.77e
60	79.40g	13.32d	6.31f	69.90j	13.71d	4.79h
Control						
0	95.15a	13.16g	8.06a	95.13a	13.16g	8.00a
20	90.40cd	13.10h	7.13d	83.60e	13.33f	6.00d
40	84.30f	13.30d	6.80e	74.73i	13.79cd	5.16g
60	72.33j	13.76a	5.86g	64.30m	14.30a	4.34j
Season 2013						

Season 2013						
Potassium silicate 0.1%						
0	92.20a	13.02a	8.20a	92.20a	13.06l	8.22a
20	89.81c	13.00a	7.53d	82.33d	13.23gh	6.37d
40	80.31g	13.14a	7.19g	72.00h	13.46de	5.50h
60	72.37k	13.49a	6.30k	63.10l	13.90b	4.75l
Potassium silicate 0.2%						
0	92.20a	13.02a	8.20a	92.20a	13.06l	8.22a
20	89.90c	13.01a	7.68c	84.50c	13.21gh	6.49c
40	83.10f	9.79b	7.29f	73.40g	13.40ef	5.59g
60	74.10j	13.41a	6.39j	64.90k	13.82b	4.83k
Potassium silicate 0.3%						
0	92.20a	13.02a	8.20a	92.20a	13.05l	8.22a
20	90.99b	12.96a	7.80b	90.00b	13.15hi	6.70b
40	86.73e	13.09a	7.43e	78.13f	13.29fg	5.86f
60	76.85i	13.27a	6.49i	66.50j	13.63c	4.97j
Control						
0	92.20a	13.02a	8.20a	92.20a	13.05l	8.22a
20	87.40d	13.00a	7.47de	80.75e	13.29fg	6.28e
40	78.50h	13.18a	7.08h	71.25i	13.55cd	5.36i
60	68.40l	13.55a	6.21l	61.54m	14.05a	4.65m

Means followed by the same letters are not significantly different by Duncan's multiple range test at 0.05 level

Means followed by the same letters are not significantly different by Duncan's multiple range test at 0.05 level

This result is in agreement with the findings of Tian et al. (2005) and Etebarian et al. (2013) on blue mould caused by P. expansum in apple, the results of in vitro studies showed that silicon in concentrations of higher than 0.6% inhibited the growth of pathogen completely. Also, Si at concentration above 0.6% w/v inhibited the growth of the both isolates P1 and P2 completely, while Si at 0.2% had the least effect on the both isolates. There were no significant differences between 0.6, 0.8 and 1% w/v (Farahani et al., 2012). Furthermore, Li et al. (2009) found that Si at 200 mL mol-1 inhibited the germination and in vitro mycelial growth of P. expansum.

Etebarian et al. (2013) showed in vivo experiments that, the lesion diameters in apples which were immersed in hot water at 40 or 50°C after inoculation with pathogen and then treated with silicon at concentrations of 0.2, 0.4, 0.6 and 1% in wounds were less than pathogen control after 15 and 45 days incubation at 20 and 4°C, respectively. In another experiment, the lesion diameters in apples which were immersed in hot water at 40 or 50°C after inoculation with pathogen and then treated with silicon were from 0-4.99 mm compared to 12.5 mm in pathogen control.

The results of initial Si experiments indicate that silicon affects plant growth and crop quality, stimulates photosynthesis, reduces transpiration rate and enhances plant resistance to a series of both abiotic and biotic stresses such as water and chemical stresses, nutrient imbalances, metal toxicities, diseases and pest's problems (Zhou et al., 2002). Liang et al.

(2008) reviewed that silicon can stimulate the antioxidant system in plants and affect the structure of the plasma membrane so as to alleviate abiotic stresses. Recently, there has been an increased interest for sustainable crop production and Si can contribute to that direction with its prophylactic properties and promotion of plant health. Silica accumulation in cell walls was believed to be physically responsible for plant disease resistance it has a potential signal that activates defense mechanisms. Bekker et al. (2007) suggested that the positive effect of Si on plant growth and performance is only evident when plants are under some stress. So, Si has been found to offer protection against fungal infections by strengthening cell walls, thus making it more difficult for the fungi to penetrate and colonize the plant.

Si may enhance activity of chitinases, peroxidases and polyphenyloxidases and increased formation deposition of callose and hydrogen peroxide (Shetty et al., 2012). Wei et al. (2004) reported that soluble silicon application stimulated the activity of the enzymes PAL, PPO and POD in hydroponically grown Chinese white-flowered gourd and increased silicon content by 1.43 times. Therefore, by increasing PAL activity by the addition of potassium silicate in these experiments, the fruit's resistance to disease can be increased. This may be especially important as fruit softens.

Zhu et al. (2004) indicated that silicon induced stress tolerance in plants may be caused, at least in part, by increased antioxidant enzymes activity, which in turn decrease oxidative damage to membrane and enzyme activity. Mditshwa et al. (2013) reported that chilling injury was exacerbated by 150 and 250 mg L-1 K2SiO3. This was probably caused by the glassy characteristics of Si that can potentially damage the cell hence increasing fruit water loss. The reduced fruit weight loss and electrolyte leakage following treatment with Si also showed the potential of Si to retard stress. The reduction in fruit weigh loss at 50 mg L-1 K2SiO3 was probably due to the modification of cell membranes after Si application that led to reduction of water loss and subsequently reduced fruit weight loss (Epstein, 2009).

It is also a known fact that treating plants with soluble Si enhances fungal resistance, inhibits fungal diseases through modifications of the epidermal layer of the leaves and fruits as well as by increasing presence of low-molecular-weight metabolites (Gillman et al., 2003).

Texture is one of the most important factors affecting apple quality. Sugar and acid ratio is also deemed important as long as the acidity does not decline too far. During ripening, acidity and firmness decline concurrently. Apple skin color results from the blending of chlorophyll, carotenoid and anthocyanin pigments, with anthocyanins being primarily responsible for the red color (Lancaster and Dougall, 1992). The storage durations significantly affected the starch content of apple. Since starch is the major storage carbohydrates in apple fruit, it is converted to sugars at the onset of ripening and during storage to meet the respiratory demand of the fruit (Crouch, 2003). Maksimovic et al. (2007) found increment in flavonoid and phenolic content in plants after silicon application. Covering fruit with a silicon layer seemed to increase gas exchange (respiration) at the two lower Si concentrations, thereby potentially hastening the ripening process (Tesfay et al., 2011). The contents of total solid solutes of the fruit were significantly increased by Si in the form of K2SiO3, also enhanced the fruit firmness and vitamin C in the tomato fruits (Stamatakis et al., 2003). The Si has beneficial effects on plant metabolism as increased contents of malic acid while, sucrose, glucose and fructose were reduced (Wang and Galletta, 1998). Jamali and Rahemi (2011) was confirmed that the inhibitory effect of Si on ethylene production increases the postharvest quality and longevity of carnations.

Kaluwa et al. (2010) concluded that post-harvest applications of 2940 ppm Si in the form of KSi seem to be most beneficial in suppressing respiration, reduction in ethylene evolution and would result in a slower decrease in the plant's carbon reserve compounds. The Si concentration applied had, however, no effect on the deposition of Si in the mesocarp tissue but the amount of Si found in the exocarp was higher in fruit treated with high Si concentrations. It would also be beneficial to investigate if Si increases antioxidant and total phenolics accumulation in the fruit, thereby increasing the stress-relieving ability of the fruit, producing fruit with a higher ability to withstand long-term storage.

CONCLUSION

The above results show that, dipping Anna apple fruits in potassium silicate has a great potential to enhance quality during cold storage (at 0±1°C with 90-95% R.H) and through marketing at room temperature. All silicon treatments has an ability to inhibit the radial mycelial growth of Penicillium expansum development, reduced SSC, acidity, SSC/acid ratio and total sugar compare to the untreated fruits. The present study showed that, potassium silicate at 0.3% gave the best results for reduced loss weight, decay, total loss weight, respiration rate and PPO activity than the other treatments or the control. Therefore, from this study the treatment with potassium silicate as a promising material to maintain the quality of the apples during storage and marketing for the longest possible period was found.

REFERENCES

1. AOAC., 1970. Official Methods of Analysis. 9th Edn., Association of Official Agricultural Chemists, Washington, DC., USA., pp: 789.
2. AOAC., 2005. Official Methods of Analysis. 16th Edn., Association of Official Analytical Chemists, Washington DC., USA.
3. Bekker, T.F., N. Labuschagne, T. Aveling and C. Kaiser, 2007. Efficacy of water soluble potassium silicate against phytophthora root rot of avocado under field conditions. S. Afr. Avocado Growers Assoc., 30: 39-48.
4. Bosse, R.J., J.P. Bower and I. Bertling, 2011. Pre- and post-harvest treatments on Fuerte avocados to control anthracnose (Colletotrichum gloeosporioides) during ripening. S. Afr. Avocado Growers Assoc., 34: 65-69.
5. Brecht, M., L. Datnoff, R. Nagata and T. Kucharek, 2003. The Role of Silicon in Suppressing Tray Leaf Spot Development in St. Augustine Grass. University of Florida, Florida.
6. Crouch, I., 2003. 1-Methylcyclopropene (SmartfreshTM) as an alternative to modified atmosphere and controlled atmosphere storage of apples and pears. Acta Hortic., 600: 433-436.
7. Drake, S.R. and E.M. Kupferman, 2000. Maturity and storage quality of Jonagold apples related to starch index. J. Am. Pomol. Soc., 54: 213-218.
8. El-Morsy, T.H.A., 1993. Ecological and physiological studies on fungi present in water and its relation to pollutants in Dakahlia province. M.Sc. Thesis, Botany Department, Faculty of Science, Mansoura University, Egypt.
9. Ellis, M.B., 1971. Dematiaceous Hyphomycetes. 1st Edn., Commonwealth Mycological Institute, Kew, Surrey, UK., ISBN-13: 978-0851986180, Pages: 608.
10. Epstein, E., 2009. Silicon: Its manifold roles in plants. Ann. Applied Biol., 155: 155-160.
11. Etebarian, H.R., S.F. Kashani and L. Ebrahimi, 2013. Combination of silicon and hot water to control of postharvest blue mould caused by Penicilum expansum in apple. Int. J. Agric. Res. Rev., 3: 72-79.
12. Farahani, L., H.R. Etebarian, N. Sahebani and H. Aminian, 2012. Effect of two strains of antagonistic yeasts in combination with silicon against two isolates of Penicillium expansum on apple fruit. Int. Res. J. Applied Basic Sci., 3: 18-23.
13. Gholamnejad, J., H. Etebarian, A. Roustaee and N. Sahebani, 2009. Biological control of apples blue mold by isolates of Saccharomyces cerevisiae. J. Plant Protect. Res., 49: 270-275.
14. Gillman, J.H., D.C. Zlesak and J.A. Smith, 2003. Applications of Potassium silicate decrease black spot infection in Rosa hybrid Meipelta (Fuschia Meidiland). HortScience, 38: 1144-1147.
15. Jamali, B. and M. Rahemi, 2011. Carnation flowers senescence as influenced by nickel, cobalt and silicon. J. Biol. Environ. Sci., 5: 147-152.
16. Kaluwa, K., I. Bertling, J.P. Bower and S.Z. Tesfay, 2010. Silicon application effects on Hass avocado fruit physiology. S. Afr. Avocado Growers Assoc., 33: 44-47.
17. Lancaster, J.E. and D.K. Dougall, 1992. Regulation of skin color in apples. Crit. Rev. Plant Sci., 10: 487-502.
18. Li, Y.C., Y. Bi, Y.H. Ge, X.J. Sun and Y. Wang, 2009. Antifungal activity of sodium silicate on Fusarium sulphureum and its effect on dry rot of potato tubers. J. Food Sci., 74: M213-M218.
19. Liang, Y., J. Zhu, Z. Li, G. Chu, Y. Ding, J. Zhang and V. Sun, 2008. Role of silicon in enhancing resistance to freezing stress in two contrasting winter wheat cultivars. Environ. Exp. Bot., 64: 286-294.
20. Ma, J.F. and N. Yamaji, 2006. Silicon uptake and accumulation in higher plants. Trends Plant Sci., 11: 392-397.

CHAPTER 20

Crop Management as a Driving Force of Plant Growth Promoting Rhizobacteria Physiology

Juliana Melo[1,2], Manuela Carolino[2] , Luís Carvalho[2] , Patrícia Correia[2] , Rogério Tenreiro[3] , Sandra Chaves[3] , Ana I. Meleiro[2] , Sávio B. de Souza[4] , Teresa Dias[2] , Cristina Cruz[2] and Alessandro C. Ramos[4]*

[1] *Ecosystems Ecology Unit, Universidade Vila Velha (UVV), Vila Velha, ES 29102-920, Brazil.*

[2] *Center for Ecology, Evolution and Environmental Changes, Faculdade de Ciências, Universidade de Lisboa, 1649-004 Lisbon, Portugal.*

[3] *Center for Biodiversity, Functional and Integrative Genomics, Faculdade de Ciências, Universidade de Lisboa, 1649-004 Lisbon, Portugal.*

[4] *Physiology and Biochemistry of Microorganisms Lab., Center of Biosciences and Biotechnology, Universidade Estadual do Norte Fluminense Darcy Ribeiro (UENF), Rio de Janeiro 28013-620, Brazil.*

ABSTRACT

Crop management systems influence plant productivity and nutrient use efficiency, as well as plant growth-promoting rhizobacteria (PGPR), which are known to influence the growth of plants via phytohormone production, phosphate solubilization, nitrogen (N) fixation and antimicrobial activity. The objective of this study was to compare the influence of two crop management system on microbial PGPR features. PGPR isolated from the rhizospheres of *Carica papaya* L. grown under two distinct management systems (conventional and organic) were identified and characterized. The 12 strains most efficient in solubilizing inorganic phosphate belonged to the genera *Burkholderia, Klebsiella,* and *Leclercia.* N fixation was observed in the strains *B. vietnamiensis* from the conventional farming system and *B. vietnamiensis, B. cepacia* and *Leclercia* sp. from the organic farming system. The *B. vietnamiensis, B. cepacia, Klebsiella* sp. and *Klebsiella* sp. isolates showed antifungal activity, while *Leclercia* sp. did not. The strains *B. vietnamiensis* and *Enterobcter* sp. (isolated from the conventional farming system) and *Klebsiella* sp. (isolated from the organic farming system) were efficient at solubilizing phosphate, producing phytohormones and siderophores, and inhibiting the mycelial growth of various phytopathogenic fungi (*Botrytis cinerea, Pestalotia* sp., *Alternaria* sp., *Phoma* sp., *Fusarium culmorum, Geotrichum candidum*). Physiological differences between the isolates from the two crop management regimes were distinguishable after 10 years of distinct management.

Keywords: Biocontrol Organic farming Conventional farming Fungi Nitrogen fixation Rhizosphere

BACKGROUND

Agriculture is struggling to meet the enormous challenge of producing enough food for an ever-expanding world population. To maintain this high productivity, great quantities of synthetic fertilizers are required (Avis et al. 2008; Godfray et al. 2010), which can damage ecosystem structures and functions, including the soil microbial community which plays an important role in agriculture sustainability (Ahemad and Khan 2010; Avis et al. 2008; Srinivas et al. 2008).

The structure and function of the soil microbial community are directly and indirectly influenced by the management regime (Barea et al. 2005; Araújo et al. 2009). On one hand, intensive conventional farming systems are dependent on large fertilizer inputs and characterized by low nutrient use efficiencies, which result in environmental threats (Marinari et al. 2010; Radíc et al. 2014) including environmental pollution, soil erosion (Marja et al. 2014) and loss of biodiversity (Dai et al. 2014). On the other hand, organic farming is defined as sustainable since it relies on diversified inputs, most of them organic and not immediately available to the crop, which stimulate the networking and complementarity of the soil microbial community, and reduce environmental threats (Beltrán-Esteve and Reig-Martínez 2014; Marja et al. 2014; Perez et al. 2014). As nitrogen (N) and phosphorus (P) are the main fertilizers used worldwide, and have biogeochemical cycles involving many bio-transformations, the efficiency of their use by plants is intrinsically related with the crop management regime (van Diepeningen et al. 2006; Moeskops et al. 2010; Barea et al. 2014). Many of the plant growth promoting rhizobacteria used as biofertilizers are involved in the soil P and/or N cascades of transformation (Chiarini et al. 1998).

PGPR constitute a heterogeneous group that include organisms of the genera *Pseudomonas, Azospirillum, Burkholderia, Bacillus, Enterobacter, Kle bsiella, Rhizobium, Erwinia, Serratia, Alcaligenes, Arthrobacter, Acinetobacte r*, and *Flavobacterium* (Rodríguez and Fraga 1999; Khan et al. 2009; Sharma et al. 2013; Ahemad and Kibret 2014; Reed et al. 2015).

PGPR exert beneficial effects on plant growth via direct and indirect mechanisms. Direct mechanisms involve the synthesis of substances or modulation at an enzymatic level, which facilitates the absorption of certain nutrients, solubilization of mineral phosphates (Ahemad et al. 2008; Malboobi et al. 2009), biological fixation of nitrogen (Peix et al. 2003; Caballero-Mellado et al. 2007; Jackson et al. 2008) and synthesis of plant hormones such as gibberellic acid, cytokinins, ethylene, and indolic acetic acid (IAA). Indirect mechanisms include PGPR decreasing or preventing the destructive effects of one or more phytopathogens, by the production of antibiotics (Richardson et al. 2009; Bevivino et al. 1994; Rodríguez and Fraga 1999) or siderophores (Reed et al. 2015). Siderophores produced by PGPR have a high affinity with iron III from the rhizosphere and, consequently, retain a most of the iron available, inhibiting the proliferation of phytopathogenic fungi (Bevivino et al. 1998; Laslo et al. 2012).

However, sometimes the effects of PGPR are unstable, and greatly influenced by biotic and abiotic soil conditions, conferring on PGPR a certain degree of specificity in relation to crop improvement. It is consequently recommended that, in order to obtain the best results, PGPR inoculants should be isolated from native PGPR populations (Reed et al. 2015; Santos-Villalobos et al. 2012; Bashan et al. 2014). This poses the question of what is the bigger driver for the rhizobacteria's functional ecology: the crop or management regime? This work was intended to address this question by studying potential PGPR isolated from the rhizosphere of papaya (*Carica papaya* L.) plants grown in farms with similar soil and edaphic conditions, under distinct crop management regimes.

Papaya is one of the most commonly cultivated fruits in almost all tropical American countries, and it is also one of the most consumed fruits in the tropical and sub-tropical regions of the world (FAO 2012). Due to its continuous growth, uninterrupted and simultaneous flowering and fruiting, it needs water and nutrients throughout its growing cycle (Mendonça et al. 2006; Trindade et al. 2006) and therefore is a high nutrient-demanding crop. This work is based on the hypothesis that the different crop management systems represent distinct soil perturbations and influence differently the microbial structure and the mechanisms involved in the promotion of plant growth. The objective of this study was to compare the influence of two crop management system (conventional and organic) on microbial PGPR features.

RESULTS

Isolation and identification of phosphate-solubilizing bacteria using 16S rRNA sequencing

The shortage of P reserves and their rapid loss in the soil even after fertilization promotes the search for PGPR with phosphorus solubilizing capacity which may eventually be used as soil inoculants. Therefore our initial aim was to conduct a screening for selecting bacteria isolates with high ability to solubilize phosphate.

The counts of total cultivable bacteria isolated from the rhizosphere of *C. papaya* were not influenced by the management regime (Figure 1a). However, the number of rock phosphate-solubilizing bacteria (based on the CFU formed in the selective medium) was higher in soils from the organic system (P = 0.02) (Figure 1b).

The 12 CFUs presenting the greater rock phosphate-solubilizing capacity (higher solubilization index), in the NBRIP medium, 5 from the conventional and 7 from the organic farming systems, were selected for molecular identification. All the isolates preserved in vitro conditions, maintained their viability and functionality over time in laboratory assays). The 12 isolates were identified as belonging to two families: Enterobacteriaceae and Burkholderiaceae. Seven isolates were identified as belonging to the genus *Klebsiella*, the predominant genus in both management systems and were very similar to each other. The others were of the genera *Burkolderia* (3, including one *B. cepacia* and two *B. vietnamiensis*), *Enterobacter* (1) and *Leclercia* (Figure 2).

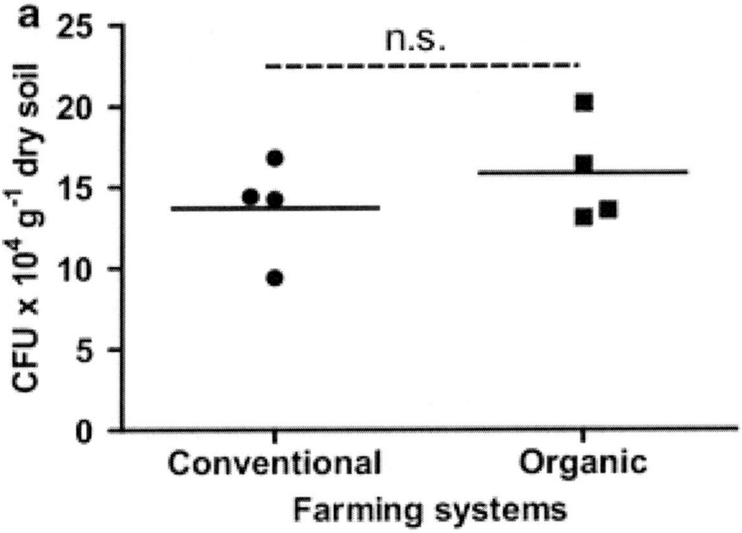

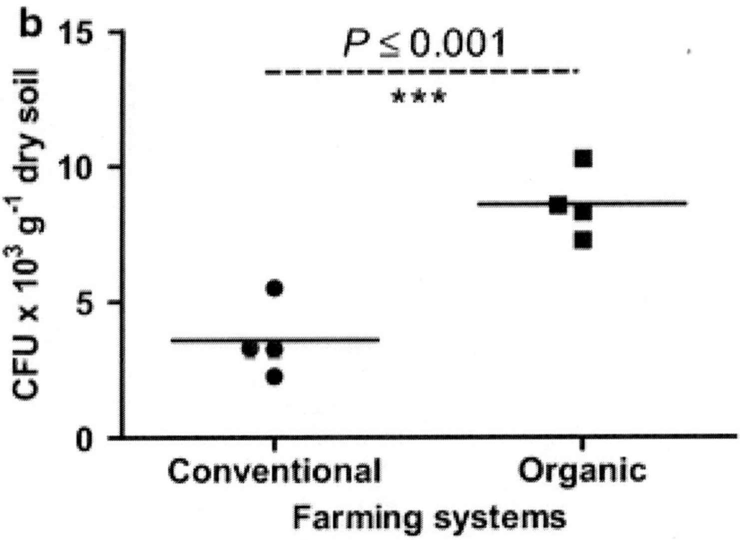

Figure 1. Bacterial total density (**a**) and tricalcium phosphate solubilizing bacterial Density (**b**) in the rhizosphere of *C. papaya* L. plants grown in conventional or organic system (n = 4). n.s. means are not significantly by Student's *t* test at $P \leq 0.05$. *** Means are statistically different by Student's *t* test at $P \leq 0.001$

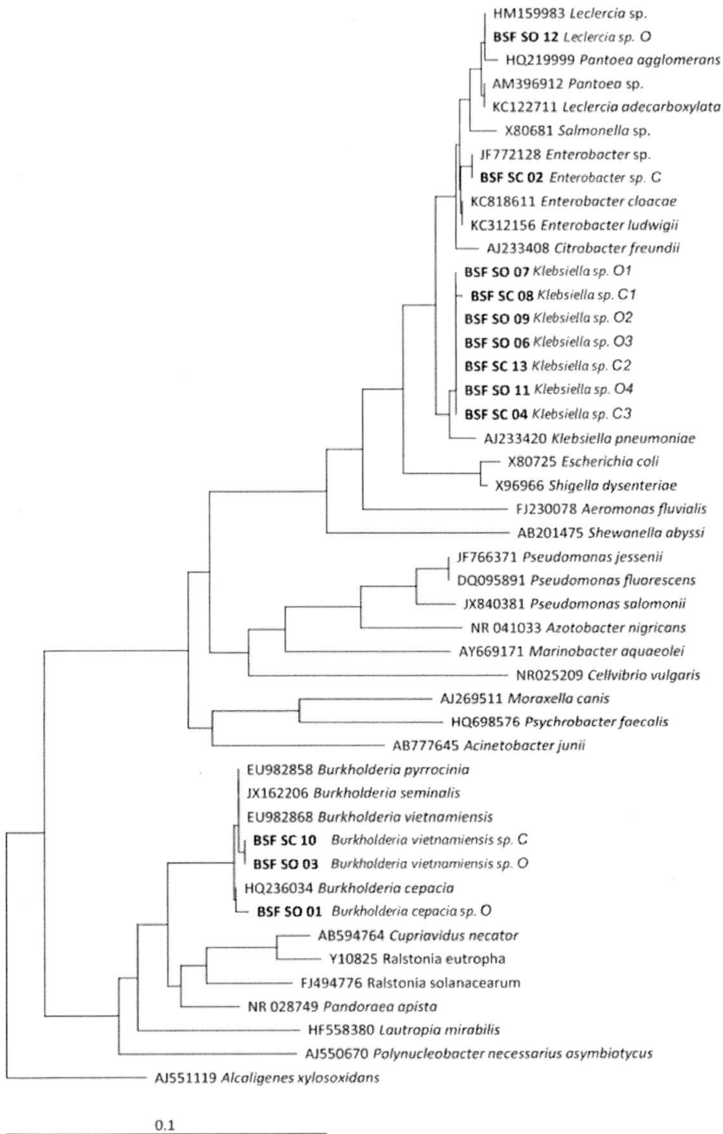

Figure 2. Phylogenetic tree obtained with 16S rDNA partial sequences (518 nucleotide positions), corresponding to the PSB isolates and the most closely related ones retrieved from BLAST search. Phylogeny was inferred using the Neighbor Joining method of aligned 16S rDNA fragments. *Alcaligenes xylosoxidans*, was included to root the tree. Access numbers of GenBank sequences are indicated in the figure and names in *bold face* correspond to sequences determined in this work

The five isolates from the conventional farming system were identified as *Enterobacter* sp., *Klebsiella* sp. and *Burkholderia vietnamiensis*, designated in this work as *Enterobacter* sp. C, *Klebsiella* sp. C1, *Klebsiella* sp. C2, *Klebsiella* sp. C3 and *Burkholderia vietnamiensis* C. The seven isolates from the organic farming system were *Burkholderia cepacia*, *Burkholdeira vietnamiensis*, *Klebsiella* sp. and *Leclercia* sp., here designated *Burkholderia cepacia* O, *Burkholderia vietnamiensis* O, *Klebsiella* sp. O1, *Klebsiella* sp. O2, *Klebsiella* sp. O3, *Klebsiella* sp. O4 and *Leclercia* sp. O (where O stands for the organic management system).

Solubilization index (SI) for different phosphate sources and potential plant growth promoter

Although the isolates from the organic system had the highest capacity to solubilize rock phosphate, the differences between the farming systems were not so clear in relation to their capacity to use phytate (Figure 3), the main organic source of phosphorus in the soil, as a P source. Of the twelve isolates, only four (*B. vietnamiensis* C, *B. vietnamiensis* O, *B. cepacia* O and *Leclercia* sp O) displayed the ability to fix nitrogen. Siderophore production was observed in all the bacterial isolates. With the exception of isolated *Leclercia* sp O, With the exception of *Leclercia* sp O, all the other strains were resistant to ampicillin and methacycline, and were susceptible to chloramphenicol, kanamycin, erythromycin and tetracycline.

Temporal kinetics of IAA production

Regardless of the farming system, the highest production of IAA by all the isolated bacteria (in the culture medium supplemented with 200 ng mL^{-1} of tryptophan) took place at the beginning of the stationary phase of bacterial growth (Figs. 4, 5).

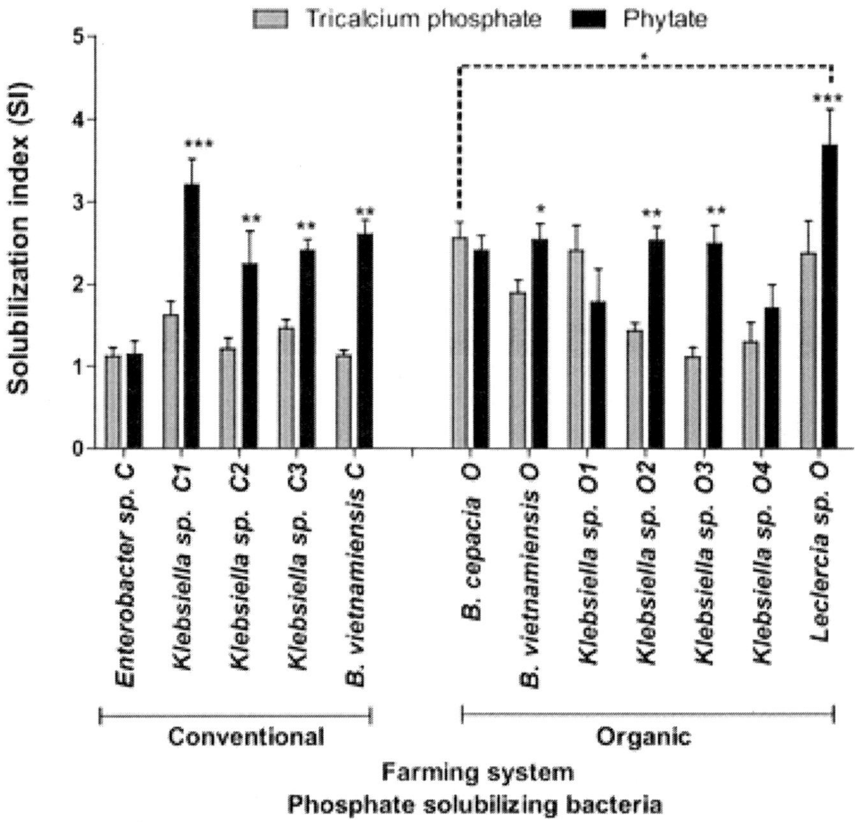

Figure 3. Phosphate solubilization index in solid NBRIP supplemented with tricalcium phosphate or phytate by phosphate solubilizing bacteria isolated from conventional and organic farming systems (n = 4). Means are statistically different by Student's *t* test (*P ≤ 0.05; **P ≤ 0.01; ***P ≤ 0.001)

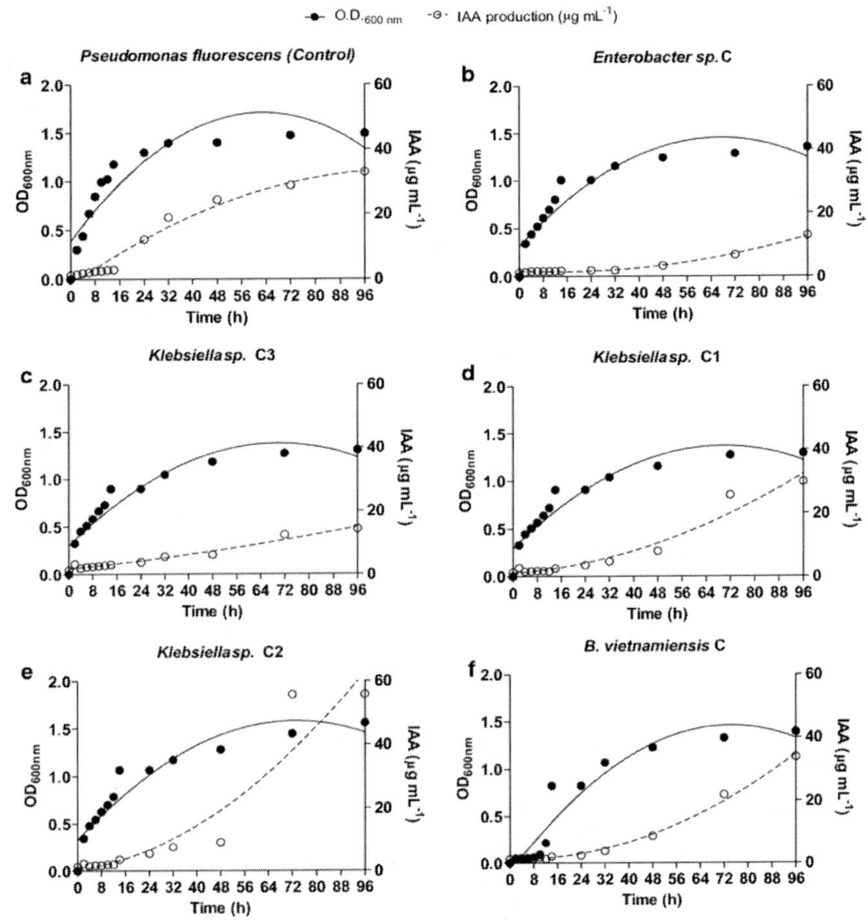

Figure 4. IAA production kinetics and bacterial growth of *Pseudomonas fluorescens* (**a**), *Enterobacter sp.* C (**b**), *Klebsiella sp.* C3 (**c**), *Klebsiella sp.* C1 (**d**), *Klebsiella sp.* C2 (**e**), *B. vietnamiensis* C (**f**) isolated from the rhizosphere of *C. papaya* L. (conventional farming system) (n = 4)

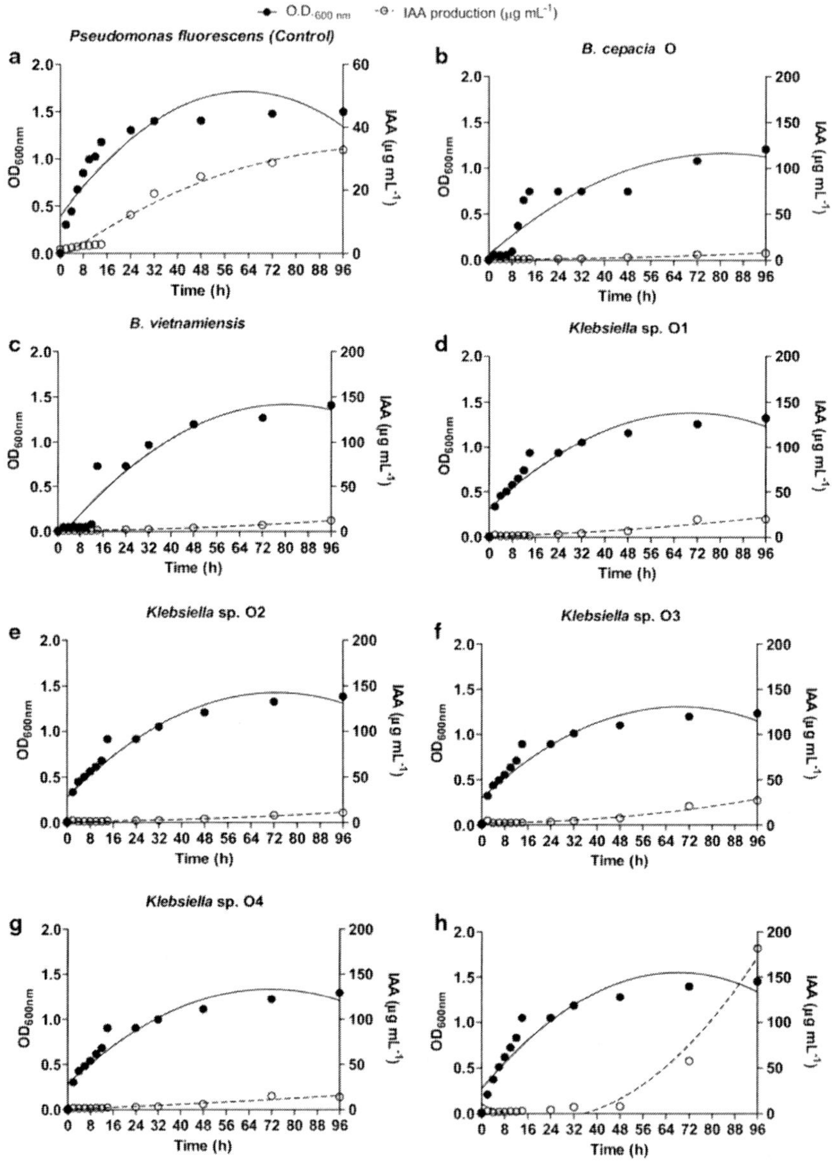

Figure 5. IAA production kinetics and bacterial growth of *Pseudomonas fluorescens* (**a**), *B. cepacia* O (**b**), *B. vietnamiensis* (**c**), *Klebsiella* sp. O1 (**d**), *Klebsiella* sp. O2 (**e**), *Klebsiella* sp. O3 (**f**), *Klebsiella* sp. O4 (**g**) and *Leclercia* sp. O (**h**) isolated from the rhizosphere of *C. papaya* L. (organic farming system) (n = 4)

The colonies assessed in this study presented the largest production of IAA after 96 h of incubation, with an optical density (OD) of 1.3 (Figs. 4, 5). IAA production by the bacterial colonies from the conventional farming system ranged from 15 to 60 µg mL^{-1} (Figure 4). Of these, the lowest IAA production observed was that by *Enterobacter* sp C (15 µg mL^{-1}), with an OD of 1.3 (Figure 4b). The highest production (60 µg/mL) was that by *Klebsiella* sp C2, also with an OD of 1.3 after 96 h (Figure 4d) of incubation. After 96 h, the other isolates from this farming system (*Klebsiella* sp C1, *B. vietnamiensis* C and *Klebsiella* sp C3) produced 30, 35 and 20 µg mL^{-1} of IAA, respectively. The OD was also 1.3 (Figure 4c, e, f).

The production of IAA by isolates from the organic farming system after 96 h ranged from 8 to 180 µg mL^{-1}, the OD being 1.3 (Figure 5). The lowest production observed was that by *B. cepacia* O (8 µg mL^{-1}) (Figure 5b), while the highest was that by *Leclercia* sp. O (180 µg/mL) (Figure 5h). The colonies of the other isolates (*B. vietnamiensis* O, *Klebsiella* sp. O1 *Klebsiella* sp. O2, *Klebsiella* sp. O3 and *Klebsiella* sp. O4) presented, after 96 h, IAA productions of 14, 20, 12, 27 and 14 µg/mL respectively, with an OD of 1.3 (Figure 5b–g).

All the strains isolated from the rhizosphere of *C. papaya* except *Klebsiella* sp. C1, *B. vietnamiensis* C and *Klebsiella* sp. O2, *B. vietnamiensis* O and *B. cepacia* O, increased production of IAA in the presence of 200 µg mL^{-1} of the precursor tryptophan, in comparison to the control with 2 µg mL^{-1} of tryptophan (Figure 6).

Antifungal activity

The isolates displayed antifungal properties against the phytopathogenic fungi *Fusarium culmorum*, *Geotrichum candidum*, *Pestalotia sp.*, *Alternaria sp.*, *Phoma sp.* and *Botrytis cinérea* (Figs. 7). Which are associated with papaya culture, for instances *Fusarium culmorum*, *Phoma* sp. and *Alternaria* sp induce peduncular rot in Papaya (Suzuki et al. 2007; Nery-Silva et al. 2007) seriously affecting papaya export and farmers income. The most effective isolates tested were *Klebsiella* sp. C1 and *B. vietnamiensis* C, which inhibited mycelial growth by 78 and 76 % respectively. *B. vietnamiensis* O inhibited mycelial growth of *Geotrichum candidum* by 81 % (Figure 7a). Inhibition of mycelial growth of *Fusarium culmorum* by the isolates was over 30 %, the greatest inhibition being by *B. vietnamiensis* C (67 %), *Klebsiella* sp. O1, (68 %), *B. vietnamiensis* O (66 %) and *Klebsiella* sp. O3 (63 %) (Figure 7b).

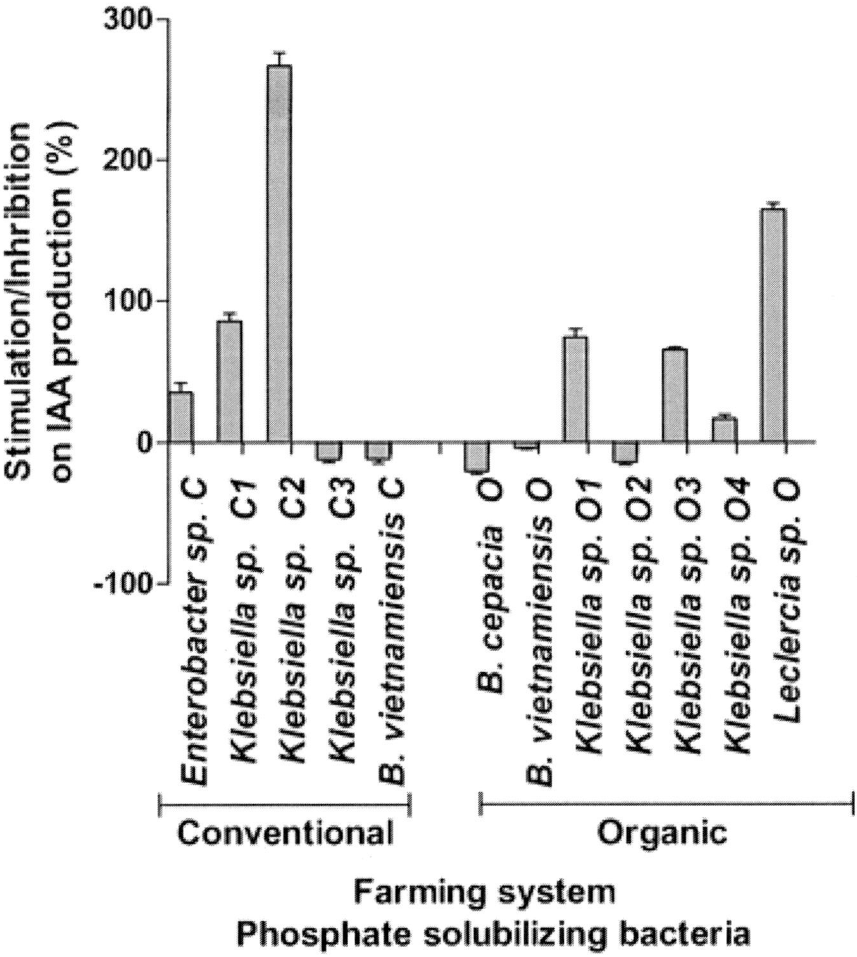

Figure 6. Inhibition and stimulation in the production of IAA in medium supplemented with 2 (control) and 200 mg L^{-1} of tryptophan by bacteria isolated from the rhizosphere of *C. papaya* L. in conventional and organic farming systems (n = 4)

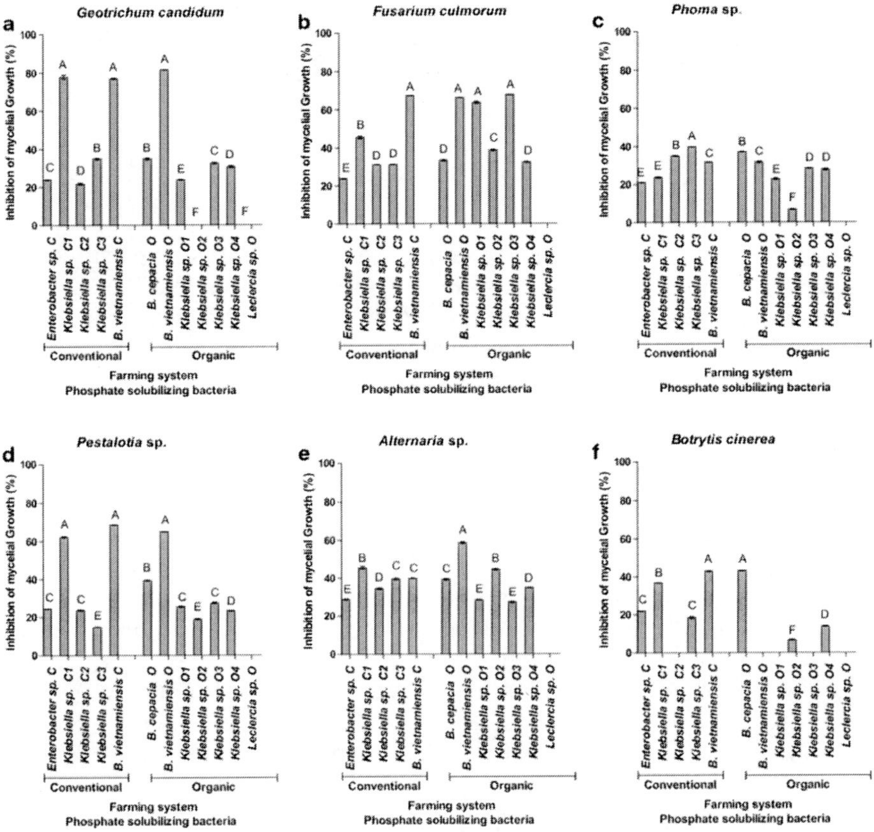

Figure 7. Inhibition of mycelium growth rate of *Geotrichum candidum* (**a**), *Fusarium culmorum* (**b**), *Pestalotia* sp. (**c**), *Alternatia* sp (**d**), *Phoma* sp. (**e**) and *Botrytis cinerea* (**f**) by bacteria isolated from the rhizosphere of *C. papaya* L. (organic and conventional farming system (n = 4). Mean values followed by the same capital letters do not differ significantly by the Tukey test (P ≤ 0.05)

The greatest inhibition of mycelial growth of the fungus *Pestalotia* sp. was by the isolates *B. vietnamiensis* C (68 %) and *Klebsiella* sp. C1 (62 %) (Figure 7c). The phytopathogenic fungus *Alternaria* sp. was inhibited by more than 30 %. The greatest inhibition was by *Klebsiella* sp.C1 (45 %), *B. vietnamiensis* O (58 %) and *Klebsiella* sp. O2 (44 %) (Figure 7a). *Phoma* sp., was inhibited by up to 40 %, the greatest inhibition being by *Klebsiella* sp. C3 (40 %) and *B. cepacia* O (37 %) (Figure 7b).

All the bacterial isolates inhibited the phytopathogenic fungus *Botrytis cinerea* by less than 44 % and the greatest inhibition was that by *B. vietnamiensis* C, *B. cepacia* O and *B. vietnamiensis* O (Figure 7c).

DISCUSSION

The crop management regime did not affect the size of the culturable bacteria community in the rhizosphere. However, it did affect the size and structure of the bacterial community able solubilize rock phosphate (Figs. 1, 2). The effect of the management system on bacterial numbers is controversial. Some studies did not observe differences (Shannon et al. 2002), while others found higher numbers of bacteria in the rhizospheres of organically grown crops (Grantina et al. 2011). It is possible that part of the variation can be explained by soil pH and chemical composition, as well as by the quantities and nature of the fertilizers applied (Naher et al. 2013). The effect of the management system on certain bacterial functional groups (such as the phosphate-solubilizing bacteria) may be direct or indirect. In our study the higher number of rock phosphate-solubilizing bacteria observed in the soils of the organic farming system may be due to the soil pH (6.5), which is more suitable for the functionality of phosphate-solubilizing bacteria than the 5.5 observed in the soils of the conventional system (Table 1). On the other hand, the phosphate fertilizer applied to the conventional system may inhibit phosphate solubilization (Naher et al. 2013).

Molecular identification of the isolates showed that they belong to the genera *Burkholderia*, *Enterobacter*, *Klebsiella* and *Leclercia*, the first three of which include many known PGPR (Rodríguez and Fraga1999; Madhaiyan et al. 2006). Many strains of *Burkholderia cepacia* have already been described as PGPR (Dawwam et al. 2013; Di Cello et al. 1997; Rodríguez et al. 2006; Pereg and McMillan2014), others are known as opportunistic pathogens in patients with cystic fibrosis and other immune depressive diseases (Richardson et al. 2009; Bhardwaj et al. 2014; Peeters et al. 2013). However, Bevivino et al. (1994) claimed that strains isolated from different sources show genetic differences and differ in their degree of pathogenicity. In fact, some strains of *B. cepacia* are already being used in the biological control of plant pathogens (Bevivino et al. 1994; Huang et al. 2013). Based on the identification of these 12 isolates, we could not determine any structural differences between the phosphorus-solubilizing communities of the two management systems' soils. One genus was only identified in soils from the conventional system (*Enterobacter*), while another was only detected in soils from the organic system (*Leclercia*, Figure 2) However most of the isolates from soils of both management systems belonged to the genus *Klebsiella*.

Table 1. Chemical properties of a *C. papaya* L. rhizospheric soil under both conventional and organic farming systems

Chemical characterization	Conventional	Organic	P value
pH	5.5	6.6*	0.041
P (mg dm^{-3})	355.3	$382.7^{n.s.}$	0.643
K (mg dm^{-3})	189.0	116.0*	0.032
Ca (cmolc dm^{-3})	4.4	$4.4^{n.s.}$	0.812
Mg (cmolc dm^{-3})	0.6	$0.9^{n.s.}$	0.073
Na (cmolc dm^{-3})	0.1	$0.2^{n.s.}$	0.056
C (%)	1.4	1.3	0.104
MO (g dm^{-3})	23.3	$22.4^{n.s.}$	0.138
Cu (mg dm^{-3})	2.8	8.6***	0.00014
Zn (mg dm^{-3})	22.6	14.0**	0.0085
Mn (mg dm^{-3})	10.0	23.0**	0.0012
S (mg dm^{-3})	127.5	35.3***	0.00001
B (mg dm^{-3})	0.9	1.4*	0.0193
Fe (mg dm^{-3})	26.8	21.2*	0.045
H + Al (cmolc dm^{-3})	2.3	0.8*	0.014
BS (cmolc dm^{-3})	5.6	$5.7^{n.s.}$	0.569
T (cmolc dm^{-3})	7.9	6.5*	0.027
t (cmolc dm^{-3})	5.6	$5.7^{n.s.}$	0.452
V (%)	70.7	87.3*	0.017

H + Al, potential acidity; BS, sum of bases; T, cation exchange capacity (CEC) at pH 7.0; t, effective cation exchange capacity (eCEC); V, saturation percentage for bases; $^{n.s}$, means are not significantly different by Student's t test at $P \leq 0.05$

* Means are statistically different by Student's t test at $P \leq 0.05$, ** $P \leq 0.01$ and *** $P \leq 0.001$

The capacity to solubilize inorganic phosphate is a necessary characteristic of a PGPR, however in most soils phytate, which is not assimilated by plants, is the largest soil P pool (Singh et al. 2014). In this study, all 12 bacterial isolates from soils of both systems showed a high capacity to solubilize phytate (Figure 3). Similar results were obtained in the rhizosphere of *Lolium perenne* L., *Trifolium repens* L., *Triticum aestivum* L., *Avena sativa* L. (Jorquera et al. 2008); *Lupinus luteus* L. (Unno et al. 2005) or *Lupinus albus* (Patel et al. 2010; Hayat et al. 2010). The capacity to mineralize phytate is important: in *B. amyloliquefaciens* it positively correlates with the promotion of corn seedlings' growth (Idriss et al. 2002). Thus, mineralizing microorganisms capable of promoting phytate use in the rhizosphere have been considered for application as biofertilizers (Unno et al. 2005; Jorquera et al. 2008; Singh et al. 2014). In contrast to phytate mineralization, isolates from the organic farming system were better able to solubilize tricalcium phosphate than those from the conventional system.

It is generally agreed that bacteria with a high capacity to solubilize phosphorus also exhibit diazotrophic function (Araújo et al. 2009). However, only 4 of the 12 isolates obtained in this study were diazotrophic: 3 from the organically-managed plot and 1 from the conventional management system (Table 2). The four isolated strains with high phosphorus solubiization capacity and the ability to fix nitrogen were *B. cepacia* O, *B. vietnamiensis* O, *Leclercia* sp O, and *B. vietnamiensis* C. As expected and reported by other authors, the *Enterobacter* isolate was not diazotrophic. The increase in plant nitrogen acquisition usually associated with *Enterobacter* PGPR may be related with changes in the plant hormonal balance induced by the bacteria (Martínez-Aguilar et al. 2008; Azadeh et al. 2010; Ferrara et al. 2012). The *Klebsiella* isolated did not display nitrogen fixation, although some species of these genus are diazotrophic (Rogers et al. 2011).

Table 2. Physiologic characteristics and antibiogram for the phosphate solubilizing bacteria isolated from rhizosphere of *C. papaya* L. in both conventional and organic farming systems

	Farming system											
	Conventional					Organic						
	Enterobacter sp C	*Klebsiella* sp C1	*Klebsiella* sp C2	*Klebsiella* sp C3	*B. vietnamiensis* C	*B.cepacia* O	*B. vietnamiensis* O	*Klebsiella* sp O1	*Klebsiella* sp O2	*Klebsiella* sp O3	*Klebsiella* sp O4	*Leclercia* sp O
Physiology												
N Fixation	*	*	*	*	++	++	++	*	*	*	*	++
Siderophore Production	+	+	+	+	+	+	+	+	+	+	+	+
Antibiotic Resistance												
Metacycline	r	r	r	r	r	r	r	r	r	r	r	nd
Ampicillin	r	r	r	r	r	r	r	r	r	r	r	nd
Chloramphenicol	s	s	s	s	s	s	s	s	s	s	s	nd
Erythromycin	s	s	s	s	s	s	s	s	s	s	s	nd
Kanamycin	s	s	s	s	s	s	s	s	s	s	s	nd
Tetracycline	s	s	s	s	s	s	s	s	s	s	s	nd

*, no nitrogen-fixing; ++, nitrogen-fixing; +, producing Siderophores; r, antibiotic resistant; s, susceptible to antibiotics; nd, not detectable

All strains isolated were able to produce siderophores, a frequent characteristic of PGPR (Richardson et al. 2009; Gupta et al. 2012; Loaces et al. 2011, Table 2). Siderophores are low molecular weight peptides or iron chelators which are synthesized by many microorganisms (Vassilev et al. 2006; Patel et al. 2010; Santos-Villalobos et al. 2012) and some plant species. The types of siderophores produced vary according to the microbial functional group (Lugtenberg and Kamilova 2009). Microorganisms capable of producing siderophores are beneficial to plants because they increase iron availability to the plant (Lugtenberg and Kamilova 2009; Ahmed and Holmström 2014), and perhaps because they are related with increased synthesis of antifungal compounds, thus providing protection against pathogens (Davidson 1988; Laslo et al. 2012).

So far no bacterial physiological role has been attributed to IAA. However many bacteria produce high amounts, especially when in the presence of tryptophan (Amin et al. 2015; Zúñiga et al. 2013). It has been suggested that IAA production by bacteria may be part of a signaling network between bacteria and non-bacterial partners. In the case of the interaction between PGPR and plants, this chemical crosstalk is very important, since IAA is a phytohormone controlling many important physiological processes in plants, such as cell division, tissue differentiation, root initiation, and growth (Khan et al. 2014). IAA accumulation was observed during the later stages of growth, after the stationary growth phase, which is characteristic of secondary metabolite production (Unno et al. 2005; Tsavkelova et al. 2005; Richardson et al. 2009; Ahemad and Khan 2011; Gupta et al. 2012; Figs. 4, 5). As expected (Tsavkelova et al. 2005), the highest production of IAA was observed in the presence of 200 mg L^{-1} of tryptophan (Figure 6). Considering that root exudates of many plants are rich in tryptophan (Tsavkelova et al. 2005; Singh et al. 2013), the chemical network between plants and PGPR which produce IAA is obvious.

All the isolates except *Leclercia* sp O inhibited the mycelial growth of the phytopathogenic fungi *Geotrichum candidum*, *Fusarium culmorum*, *Pestalotia* sp, *Alternaria* sp., *Phoma* sp and *Botrytis cinerea* (Figure 7). The antifungal activity of *B. cepacia* has also been observed by other authors and was correlated with the fungus' ability to produce siderophore peptides (Bevivino et al. 1998; Trujillo et al. 2007; Ahemad and Khan 2010). Some *Enterobacter* strains achieve suppression of mycelial growth of various phytopathogenic fungi through high activities of chitinolytic enzymes (Chernin et al. 1995).

Based on the sequencing of 16 s rRNA, all the isolates identified as belonging to the *Klebsiella* genus were genetically very similar. However, they presented distinct physiological profiles in relation to their capacity to solubilize rock phosphate, produce IAA and inhibit the growth of phytopathogenic fungi. These distinct phenotypes may correspond to distinct genotypes (not detected by the 16 s rRNA sequencing) or distinct gene expression due to environmental factors (Buckling et al. 2003; Bochner 2009).

CONCLUSIONS

The objective of this study was to compare the influence of two crop management (conventional and organic) in several important PGPB features.

In this study it was observed that the crop management regime under which *C. papaya* was grown influenced the physiological functions the strains isolated from their rhizospheres, most notably inorganic phosphate solubilization and nitrogen fixation. Although most strains isolated in this study showed high potential to improve plant growth and be used as PGPR, highlighting the importance of assessing the pathogenicity and opportunistic behavior of isolates. The isolated strains *B. vietnamiensis* C, *B. vietnamiensis* O, *Klebsiella* sp. O1 and *Enterobacter* sp. C clearly showed the capacity to solubilize phosphate, produce phytohormones and siderophores, and inhibit the growth of phytopathogenic fungi in vitro, and therefore can be considered as promising candidates to be used in biofertilizers. However, more studies are needed to understand how these isolates behave when used to inoculate plants, and evaluate the performance of plants inoculated under field conditions.

Although no substantial difference was found in several PGPR traits (IAA and siderophore production, nitrogen fixation and antifungal activity against plant pathogenic fungi) our results revealed that as a group, the isolates from the organic farming system, showed a higher potential efficiency to solubilize inorganic phosphates than the group from the conventional crop. However, more studies are needed to understand the effect of crop management on the structure of the bacterial community, including a metagenomic analysis using taxonomic and functional genetic markers.

METHODS

Sample description and chemical analysis of the rhizospheric soil

The study was carried out in an area where *C. papaya* L. (Golden variety) is grown under two different management regimes (conventional and organic), in Sooretama, in Espírito Santo state, Brazil, between 2012 and 2013. The organic area has quality certifications: in 2004 it obtained the "EUREPGAR Protocol for Fresh Fruits and Vegetables" certificate and in 2009 the organically-farmed area obtained the IBD organic certificate (Inspeções e Certificações Agropecuárias e Alimentícias). Local average annual temperature and precipitation are 22 °C and 700–800 mm. The experimental areas have a plain topography and a deep soil classified as moderate type A Dystrophic Yellow Podzolic with a sandy/medium texture. The plantation was carried out in planting holes (40 × 40 × 40 cm^3), with a volume of 64 L, with gaps of 3.0 × 2.0 m in a simple row system resulting in 1700 plants/ha^{-1}. Each planting hole was fertilized.

The conventional farming system (19°12′22.9″S 40°05′52.0″W) had been under natural vegetation until 1994, and was then used to grow conventional conillon coffee plants. Since 2004 *C. papaya* has been cultivated conventionally, using industrial chemicals for soil fertilization. Chemical pesticides are used for pest control and the soil irrigation is carried out by micro-sprinklers.

The organic farming system (19°14′13.40″S 40°4′38.73″W) was also under natural vegetation until 1994 and was also used to grow conventional conillon coffee plants, however, this crop was abandoned in 1999. During the year 2000, the plants in question were removed and the soil was then rested until 2004, allowing the grass to grow. In 2005, cultivation of *C. papaya* started with monthly soil fertilizations using liquid fertilizers with the amino acids Turbofil formulation: wood dirt, filter cake, natural phosphates, magnesium oxide, dolomite limestone, bone flour, fresh cow manure, molassed chicken manure, and ground ox liver. All the ingredients were mixed and left to rest for 40 days before being used. Double ventilated sulphur powder was used for the control of white mites, whilst cassava water was used every week to control other pests. Soil irrigation was carried out by micro-sprinklers.

Rhizospheric soil samples were collected at 0–20 cm depth, following EMBRAPA soil collection protocols (EMBRAPA 2005). Each sampled soil parcel contained 6 plants spaced at a distance of 20 cm. Four soil subsamples were collected from around each plant, at 8 cm from the plant neck. Soil samples were collected in labelled plastic bags and placed in polystyrene boxes that kept soil temperature constant, then transported to the laboratory, where they were immediately analysed.

For chemical analysis, a portion of each sample was dried in an incubator with forced air circulation at 60–70 °C for 24 h. After drying, samples were ground and sieved (2 mm) before storage in vacuum sealed plastic containers for further analysis (Sarruge and Haag 1974; Martins and Reissmann 2007). P was determined colorimetrically based on the ammonium vanadate-molybdate method. K, Al and Na were measured by flame photometry. Ca, Mg, S, Cu, Fe, Mn, B and Zn were determined by atomic absorption spectrophotometry (Table 1).

Isolation of total bacteria

Total culturable bacteria were isolated by serial dilution in a homogeneous suspension of 10 g soil in 90 mL of saline solution (0.85 % NaCl), shaken at 250 rpm for 30 min at 25 °C. Dilutions from 10^{-2} to 10^{-6} were plated on solid medium Nutrient Agar, and incubated for 5 days at 28 °C. Data were expressed as number of colony forming units (CFU) per g of dry soil (Grantina et al. 2011).

Isolation of phosphate-solubilizing bacteria

Phosphate-solubilizing bacterial strains were isolated by serial dilution as described for total bacteria, but the $10^{-2}$and 10^{-6} dilutions were smeared into solid medium (National Botanical Research Institute's phosphate (NBRIP) (Nautiyal 1999) containing 10 g of glucose, 5 g of $Ca_3(PO_4)_2$, 5 g of $MgCl_2{\cdot}6H_2O$, 0.25 g of $MgSO_4{\cdot}7H2O$, 0.2 g of KCl, 0.1 g of $(NH_4)_2SO_4$ and 1.5 % of agar in 1 L deionised water (pH 7.0). After incubation for 15 days at 28 °C, bacteria that presented a clear halo around the colony were considered capable of solubilizing tricalcium phosphate. Their numbers were expressed as colony forming units (CFU) per gram (g) of dried soil. Selected strains were replicated into new plates containing NBRIP solid medium. After the second passage, the isolated colonies were stored in DYGS culture media containing 2 g of glucose, 1.5 g of peptone, 2 g of yeast extract, 0.5 g of $KH_2PO_4{\cdot}7H_2O$ and 0.5 g of $MgSO_4{\cdot}7H_2O$ in 1 L deionised water (pH 6.8). They were subsequently stored in 50 % glycerol at −80 °C.

Molecular characterization of isolates

DNA extraction

Genomic DNA from all studied isolates was extracted following the method described by (Pitcher et al. 1989), with some modifications. Briefly, bacterial cells were collected by centrifugation and washed with TE buffer. The pellets were then re-suspended in 500 μL TE and digested with 250 μg lysozyme for 1 h at 37 °C. 500 μL of GES (5 M guanidium tiocyanate, 100 mM EDTA, 0.5 % (v/v) sarcosil) were added. The suspension was incubated on ice until cell lysis, then 250 μL of 10 M ammonium acetate were added and the suspension was re-incubated on ice for 10 min. Nucleic acids were extracted with chloroform:isoamilic alcohol (24:1) and precipitated with one volume of isopropanol. After centrifugation, DNA was washed with ethanol 70 % (v/v), resuspended in TE buffer with 50 μg ml^{-1} RNase, and incubated for 2 h at 37 °C. A second extraction with chloroform:isoamilic alcohol (24:1) was performed, in which DNA was precipitated with 1/10 volume of sodium acetate and 2.5 volumes of ethanol, centrifuged, washed with ethanol 70 % (v/v) and resuspended in 50–100 μL of sterile distilled water.

16S rDNA PCR amplification

PCR amplification of the complete 16S rRNA gene was carried out using the primers PA (5′-AGAGTTTGATCCTGGCTCAG-3′) and 907R (5′-CCGTCAATTCCTTTRAGTT 3′) (Muyzer et al. 1993; Massol-Deya et al. 1995). Reactions were performed in a final volume of 50 μL, containing 1X PCR buffer (Invitrogene), 200 μmol L^{-1} each of dATP, dCTP, dGTP and dTTP (Invitrogene), 2 mmol l^{-1} $MgCl_2$, 5 μmol L^{-1} of each primer, 100 ng of genomic DNA and 1U of Taq DNA polymerase (Invitrogene). A Biometra thermocycler was used for amplification. The reaction consisted of an initial denaturation step at 94 °C for 5 min, followed by 35 cycles of 1 min at 94 °C, 1 min at 50 °C and 1 min at 72 °C and a final extension step of 5 min at 72 °C.

16S rDNA sequencing and phylogenetic analysis

16S rDNA fragments amplified with primers PA and 907R from genomic DNA of isolates were purified with a JETquick spin column (Genomed, Germany) and sequenced with the same primers using a Beckman-Coulter automated DNA sequencer (model CEQ-2000) with dye terminators, following standard protocols and then identified using the BLAST tool of the National Center for Biotechnology Information (http://www.ncbi.nlm.nih.gov).

For DNA sequence analysis, alignments were made with ClustalX 2.1 and visually corrected with BioEdit Sequence Alignment Editor. The Neighbor Joining method was applied to estimate phylogenetic relationships among the isolates.

Nitrogen fixation

In order to assess the nitrogen fixing capacity of the isolates, an assay based on the formation of a layer in semi-solid JNFb medium (without N) was carried out (Döbereiner et al. 1995). Bacteria were first grown in DYGS liquid media, then washed with saline solution (NaCl 0.85 %), and 20 μl of the culture were inoculated in triplicate in 10 mL vials containing 5 mL of semi-solid JNFb medium. The vials were stored at 30 °C for 7 days. After this period of time it was possible to observe a layer forming, characteristic of diazotrophic bacteria when in culture media. As a positive control, a wild bacteria was used (*Gluconacetobacter diazotrophicus* PAL 5, kindly supplied by Professor Fábio Lopes Olivares from the Universidade Federal Norte Fluminense).

Assessment of antibiotic resistance by the colonies

Six antibiotics were used to test antibiotic resistance, at a concentration of 1000 μg/mL: chloramphenicol, ampicillin, metacicllin, erythromycin, canamycin and tetracycline. The colonies were inoculated into Nutrient Agar (NA) for 24 h at 28 °C. After growth, the colonies were transferred to a new agar plate with NA using a swab, and discs with 7 μl of each antibiotic were added. The plates were incubated at 28 °C for 24 h, after which the colonies were classified as either resistant or susceptible according to the halo formed around the antibiotic disc (Tawiah et al. 2012).

Production of siderophores

The production of siderophores isolated from the bacterial colonies was assayed using the chrome azurol S (CAS) method as described by (Schwyn and Neilands 1987), with some modifications. The siderophores were produced in soy trypticase liquid media (STL) diluted (1:10) in 1 litre of deionised water. The CAS solution was prepared in a 100 mL volumetric flask with 6 mL of a 10 mM HDTMA solution in deionised water. Slowly stirring, 1.5 mL of an iron solution (1 Mm $FeCl_3 \cdot 6H_2O$ and 0.01 N HCl) and 7.5 mL of an aqueous solution of CAS (2 mM) were also added. Separately, 4.307 g of anhydrous piperazine was solubilized in approximately 20 mL of deionised water and 6.25 mL of concentrated HCl were added. The prepared buffer (pH 6.5) was transferred to the aforementioned volumetric flask and the volume was completed with 100 mL of deionised water. Bacteria were grown in a 50 mL Erlenmeyer flask containing 10 mL of a 1:10 STL diluted medium. The bacteria were then incubated at 28 °C for 24 h with a rotation of 160 rpm. 2 mL of the suspension were taken from the suspension aseptically and centrifuged at 12,000*g* for 10 min. 1 mL of CAS was then added to 1 mL of the resulting solution. The bacterial colonies that transformed the colour of the CAS from blue to yellow within 15 min were considered producers of siderophores (Louden et al. 2011). This assessment was carried out by using a qualitative method, therefore only the presence or absence of production was detected.

Production of indole acetic acid (IAA)

The production of IAA was assessed using the colorimetric method described by (Gordon and Weber 1951), with a few modifications. Bacteria were grown in 10 mL of a nutritive broth at 28 °C for 24 h with constant agitation at 160 rpm before being transferred to 50 mL of the same medium to which tryptophan was added to make a final concentration of 2 and 200 µg/mL. At intervals of approximately 2 h (for a total period of 96 h), 3 mL (1 mL for the determination of the Optical Density (OD) and 2 mL for the determination of the production of IAA) of the broth were taken aseptically. In order to determine the production of IAA, samples were centrifuged for 10 min at 10,000×*g*. 1 ml of the supernatant of each sample was transferred to a test tube (10 mL) where 2 mL of Salkowski reagent (2 % $FeCl_3 \cdot 6H_2O$ and 37 % HCl_3) were added. The test tubes were kept in the dark at 28 °C for 30 min, after which the presence of the hormone was established by the pink coloration and quantified through the spectrophotometer reading at 530 nm. Each strain was assessed in triplicate, with the negative control being constituted by the medium alone. The hormone's production was predicted using a calibration curve obtained with 1, 2.5, 5, 10, 25, 50 and 100 µg mL^{-1} Acetic Indole Acid (Sigma®). Tryptophan was also tested at two doses, 2 (control) and 200 µg/mL, to evaluate the stimulation or inhibition of IAA production by isolates.

Pseudomonas fluorescens was included as a positive control PGPR strain known as producer of IAA (Oberhansli et al. 1991).

Capacity for solubilizing distinct phosphate sources

The ability of bacterial isolates to solubilize various phosphate sources in solid NBRIP supplemented with $Ca_3(PO_4)_2$ 5 gL^{-1} and phytic acid 1.6 gL^{-1} was evaluated. The plates containing the phosphate sources were inoculated with each bacterial isolate, by means of four equispaced bites on the surface of the medium, then further incubated at 28 °C for 15 days. The solubilizing capacity was evaluated based on the formation of halos around the colonies. The solubilization index (SI) was calculated using the ratio between the halo and colony diameters (Barraqueiro et al. 1976): IS = diameter halo/colony diameter.

Antifungal effect

Interactions tests between the bacterial isolates and pathogenic fungi were carried out for six species of the latter: *Botrytis cinerea*, *Pestalotia* sp., *Alternaria* sp., *Phoma* sp., *Fusarium culmorum* and *Geotrichum candidum.*

Since bacteria and fungi have different growth rates, this assessment began by calculating the growth rates of the fungi in YEPGA medium (10 g of peptone, 10 g of yeast extract, 50 g of glucose and 15 g of agar), and based on this rate the bacterial incubation time was also determined. The YEPGA medium was chosen to allow the growth of both bacteria and fungi. Over four days the diameter of each fungus' colony was measured (two perpendicular directions) (Broadbent et al. 1971). From the measurements, the expansion of the mycelium was calculated based on the linear equation diameter = (mycelium expansion x time) + y intercept, since fungi exhibit linear growth when on the surface of a petri dish with saturated nutritive conditions. After 2 days of fungal growth, the dishes received the inoculum of the bacteria, excepted for *B. cinerea* which demonstrated a slower growth rate; for this reason this bacteria were added after 7 days. Petri dishes were divided into 6 parts, with each of the fungi to be inoculated in the centre, and in each of the 6 parts a different strain of bacteria was added. A sector without any bacterial inoculation served as a negative control, and each bacteria was inoculated in triplicate. After the incubation period, the fungal growth was measured and photographed, then the resulting image was analysed using ImageJ, version 1.47®. This analysis allowed for determination of inhibition, by measuring the decrease in growth of each mycelium in the presence of bacteria in comparison with the negative control without bacteria.

Statistical analysis

The experimental design was completely randomized with four repetitions. All analyses were validated by convenient residual analyses that did not show departure from the normal distribution according to the Kolmogorov–Smirnov test and showed homocedasticity among factors. The data were analyzed using ANOVA and means comparison were estimated values of Tukey´s test, when factors and interactions were significant. To compare the organic farming system against the conventional, we applied a *t* test for two independent samples and computed the 95 % confidence intervals for the corresponding mean difference. All statistical tests were performed using the statistical program IBM-SPSS V.22.

DECLARATIONS

Authors' contributions

JM carried out all experiments with the contribution of other co-authors. MC contributed with siderophore and IAA determination. LC analysed statistically the data with assistance of TD and ACR. PC and AIM contributed in the phosphate solubilization analyses. SBS and MC contributed to the biocontrol experiments with phytopathogens. SC and RT contributed to the molecular sequencing. CC and ACR were the main responsible for writing the manuscript. All authors read and approved the final manuscript.

ACKNOWLEDGEMENTS

We would like to acknowledge Steve Houghton for help with the manuscript's preparation and for the helpful suggestions. This work was supported by CAPES and FAPES PhD fellowship conceded to JM and CNPq. Researcher fellowship conceded to ACR. ACR's laboratory is supported by Fundação de Amparo à Pesquisa do Estado do Rio de Janeiro - FAPERJ (#E-26/110.081/2014; #E-26/111.428/2014; #E-26/111.458/2014) and Conselho Nacional de Desenvolvimento Científico e Tecnológico (CNPq) (#475436/2010-5; #312399/2013-8) grants. CC is supported by Fundação para a Ciência e Tecnologia (FCT) grant PTDC/BIA-ECS/122214/2010.

COMPETING INTERESTS

The authors declare that they have no competing of interest.

REFERENCES

1. Empresa brasileira de pesquisa agropecuária (EMBRAPA) (2005). Procedimento para coleta de amostra de solos. Ministério de agricultura pecuária e

desenvolvimento. Agrobiologia 2005. Accessed from http://www.agencia.cnptia.embrapa.br/Repositorio/coleta_amostras_solo_000fhtbvqw702wyiv80v17a09ztd08zh.pdf. 15 Oct 2012

2. Ahemad M, Khan MS (2010) Plant growth promoting activities of phosphate solubilizing *Enterobacter asburiae* as influenced by Fungicides. EurAsian J BioSci 4:88–95.
3. Ahemad M, Khan MS (2011) Assessment of pesticide-tolerance and functional diversity of bacterial strains isolated from rhizospheres of different crops. Insight Microbiol 1:8–19**View ArticleGoogle Scholar**
4. Ahemad M, Kibret M (2014) Mechanisms and applications of plant growth promoting rhizobacteria: current perspective. J King Saud Univ Sci 26:1–20.
5. Ahemad F, Ahmad I, Khan MS (2008) Screening of free-living rhizospheric bacteria for their multiple plant growth promoting activities. Microbiol Res 163:173–181.
6. Ahmed E, Holmström SJM (2014) Siderophores in environmental research: roles and applications. Microb Biotechnol 7:196–208.
7. Amin SA, Hmelo LR, van Tol HM, Durham BP, Carlson LT, Heal KR, Morales RL, Berthiaume CT, Parker MS, Djunaedi B, Ingalls AE, Parsek MR, Moran MA, Armbrust EV (2015) Interaction and signalling between a cosmopolitan phytoplankton and associated bacteria. Nature 522:98–101.
8. Araújo ASF, Leite LFC, Santos VB, Carneiro RFV (2009) Soil microbial activity in conventional and organic agricultural systems. Sustainability 1:268–276.
9. Avis TJ, Gravel V, Antoun H, Tweddell RJ (2008) Multifaceted beneficial effects of rhizosphere microorganisms on plant health and productivity. Soil Biol Biochem 40:1733–1740.
10. Azadeh BF, Sariah M, Wong MY (2010) Characterization of *Burkholderia cepacia* genomovar as a potential biocontrol agent of *Ganoderma boninense* in oil palm. Afr J Biotechnol 9:3542–3548
11. Barea JM, Pozo MJ, Azcón R, Azcón-Aguilar C (2005) Microbial co-operation in the rhizosphere. J Exp Bot 56:1761–1778.
12. Barea AK, Pramanik K, Mandal B (2014) Response of biofertilizers and homo-brassinolide on growth, yield and oil content of sunflower (*Helianthus annuus* L. Afr J Agric Res 9:3494–3503.
13. Barraqueiro FR, Baya AM, Cormenzana AR (1976) Establecimiento de índices para el estudio de la solubilizacion de fosfatos por bacterias del suelo. ARS Pharm 17:399–406
14. Bashan Y, de-Bashan LE, Prabhu SR, Hernandez JP (2014) Advances in plant growth-promoting bacterial inoculant technology: formulations and practical perspectives (1998–2013). Plant Soil 378:1–33.
15. Beltrán-Esteve M, Reig-Martínez E (2014) Comparing conventional and organic citrus grower efficiency in Spain. Agric Syst 129:115–123.
16. Bevivino A, Tabacchioni S, Chiarini L, Carusi MV, Del Gallo M, Visca P (1994) Phenotypic comparison between rhizosphere and clinical isolates of *Burkholderia cepacia*. Microbiology 140:1069–1077.
17. Bevivino A, Sarrocco S, Dalmastri C, Tabacchioni S, Cantale C, Chiarini L (1998) Characterization of a free-living maize-rhizosphere population of *Burkholderia cepacia*: effect of seed treatment on disease suppression and growth promotion of maize. FEMS Microbiol Ecol 27:225–237. Bhardwaj D, Ansari MW, Sahoo RK, Tuteja N (2014) Biofertilizers function as key player in sustainable agriculture by improving soil fertility, plant tolerance and crop productivity. Microb Cell Fact 13:66.
18. Bochner BR (2009) Global phenotypic characterization of bacteria. FEMS Microbiol Rev 33:191–205.
19. Broadbent P, Baker KF, Waterworth Y (1971) Bacteria and actinomycetes antagonistic to fungal root pathogens in australian soils. Aust J Biol Sci 24:925–944
20. Buckling A, Wills MA, Colegrave N (2003) Adaptation limits diversification of experimental bacterial populations. Science 302:2107–2109.

21. Caballero-Mellado J, Onofre-Lemus J, Estrada-de los Santos P, Martınez-Aguilar L (2007) The tomato rhizosphere, an environment rich in nitrogen-fixing *Burkholderia* species with capabilities of interest for agriculture and bioremediation. Appl Environ Microbiol 73:5308–5319.
22. Chernin L, Ismailov Z, Haran S, Chet I (1995) Chitinolytic *Enterobacter agglomerans* antagonistic to fungal plant pathogens. Appl Environ Microbiol 61:1720–1726
23. Chiarini L, Bevivino A, Tabacchioni S, Dalmastri C (1998) Inoculation of *Burkholderia cepacia, Pseudomonas fluorescens* and *Enterobacter* sp. on *Sorghum bicolor:* root colonization and plant growth promotion of dual strain inocula. Soil Biol Biochem 30:81–87.
24. Dai M, Hamel C, Bainard LD, Arnaud MS, Grant CA, Lupwayi NZ, Malhi SS, Lemke R (2014) Negative and positive contributions of arbuscular mycorrhizal fungal taxa to wheat production and nutrient uptake efficiency in organic and conventional systems in the Canadian prairie. Soil Biol Biochem 74:156–166.
25. Davidson L (1988) Plant beneficial bacteria. Biotechnology 6:282–286
26. Dawwam GE, Elbeltagy A, Emara HM, Abbas IH, Hassan MM (2013) Beneficial effect of plant growth promoting bactéria isolated from the roots of potato plant. Ann Agric Sci 58:195–201.
27. Di Cello F, Bevivino A, Chiarini L, Fani R, Paffetti D, Tabacchioni S, Dalmastri C (1997) Biodiversity of a *Burkholderia cepacia* population isolated from the maize rhizosphere at different plant growth stages. Appl Environ Microbiol 63:4485–4493
28. Döbereiner J, Baldani VLD, Baldani JI (1995) Como isolar e identificar bactérias diazotróficas de plantas não-leguminosas. Embrapa-SPI, Brasília, p 60
29. Ferrara FIS, Oliveira ZM, Gonzales HHS, Floh EIS, Barbosa HR (2012) Endophytic and rhizospheric enterobacteria isolated from sugar cane have different potentials for producing plant growth-promoting substances. Plant Soil 353:409–417.
30. Food and Agriculture Organization (FAO) (2012) FAO. FAOSTAT. http://faostat.fao.org/site/291/default.aspx. Accessed 30 June 2014
31. Godfray HCJ, Beddington JR, Crute IR, Haddad L, Lawrence D, Muir JF, Pretty J, Robinson S, Thomas SM, Toulmin C (2010) Food security: the challenge of feeding 9 billion people. Science 327:812–818.
32. Gordon SA, Weber RP (1951) Colorimetric estimation of indoleacetic acid. Plant Physiol 26:192–195
33. Grantina L, Kenigsvalde K, Eze D, Petrina Z, Skrabule I, Rostoks N, Nikolajeva V (2011) Impact of six-year-long organic cropping on soil microorganisms and crop disease suppressiveness. Žemdirb Agricult 98:399–408
34. Gupta M, Kiran S, Gulatic A, Singh B, Tewari R (2012) Isolation and identification of phosphate solubilizing bacteria able to enhance the growth and aloin-A biosynthesis of *Aloe barbadensis* Miller. Microbiol Res 167:358–363.
35. Hayat R, Ali S, Amara U, Khalid Ahmed I (2010) Soil beneficial bacteria and their role in plant growth promotion: a review. Ann Microbiol 60:579–598.
36. Huang GH, Tian HH, Liu HY, Fan XW, Liang Y, Li YZ (2013) Characterization of plant-growth-promoting effects and concurrent promotion of heavy metal accumulation in the tissues of the plants grown in the polluted soil by *Burkholderia* Strain LD-11. Int J Phytoremediation 15:991–1009.
37. Idriss EE, Makarewicz O, Farouk A, Rosner K, Greiner R, Bochow H, Richter T, Borriss R (2002) Extracellular phytase activity of *Bacillus amyloliquefaciens* FZB45 contributes to its plant-growth-promoting effect. Microbiology 148:2097–2109.
38. Jackson LE, Burger M, Cavagnaro TR (2008) Roots, nitrogen transformations, and ecosystem services. Annu Rev Plant Biol 59:341–363.
39. Jorquera MA, Hernández MT, Rengel Z, Marschner P, Mora ML (2008) Isolation of culturable phosphobacteria with both phytate-mineralization and phosphate-solubilization activity from the rhizosphere of plants grown in a volcanic soil. Biol Fertil Soils 44:1025–1034.

40. Khan AA, Jilani G, Akhtar MS, Naqvi SMS, Rasheed M (2009) Phosphorus solubilizing bacteria: occurrence, mechanisms and their role in crop production. J Agric Biol Sci 1:48–58
41. Khan AL, Waqas M, Kang S-M, Al-Harrasi A, Hussain J, Al-Rawahi A, Al-Khiziri S, Ullah I, Ali L, Jung H-Y, Lee I-J (2014) Bacterial endophyte Sphingomonas sp. LK11 produces gibberellins and IAA and promotes tomato plant growth. J Microbiol 52(8):689–695
42. Laslo É, György É, Mara G, Tamás É, Ábrahám B, Lányi S (2012) Screening of plant growth promoting rhizobacteria as potential microbial inoculants. Crop Prot 40:43–48.
43. Loaces I, Ferrando L, Scavino AF (2011) Dynamics, diversity and function of endophytic siderophore-producing bacteria in rice. Microb Ecol 61:606–618.
44. Louden BC, Haarmann D, Lynne AM (2011) Use of blue agar CAS assay for siderophore detection. J Microbiol Biol Educ.
45. Lugtenberg B, Kamilova F (2009) Plant-growth-promoting rhizobacteria. Annu Rev Microbiol 63:541–556
46. Madhaiyan M, Poonguzhali S, Hari K, Saravanan VS, Sa T (2006) Influence of pesticides on the growth rate and plant-growth promoting traits of *Gluconacetobacter diazotrophicus*. Pestic Biochem Physiol 84:143–154.
47. Malboobi MA, Behbahani M, Madani H, Owlia P, Deljou A, Yakhchali B, Moradi M, Hassanabadi H (2009) Performance evaluation of potent phosphate solubilizing bacteria in potato rhizosphere. World J Microbiol Biotechnol 25:1479–1484.
48. Marinari S, Lagomarsino A, Moscatelli MC, Di Tizio A, Campiglia E (2010) Soil carbon and nitrogen mineralization kinetics in organic and conventional three-year cropping systems. Soil Tillage Res 109:61–168.
49. Marja R, Herzon I, Viik E, Elts J, Mänd M, Tscharntke T, Batáry P (2014) Environmentally friendly management as an intermediate strategy between organic and conventional agriculture to support biodiversity. Biol Conserv 178:146–154.
50. Martínez-Aguilar L, Díaz R, Peña-Cabriales JJ, Estrada-de los Santos P, Dunn MF, Caballero-Mellado J (2008) Multichromosomal genome structure and confirmation of diazotrophy in novel plant-associated *Burkholderia* Species. Appl Environ Microbiol 74:4574–4579.
51. Martins APL, Reissmann CB (2007) Material vegetal e as rotinas laboratoriais nos procedimentos químicoanalíticos. Scientia Agraria 8:1–17
52. Massol-Deya AA, Odelson DA, Hichey RP, Tiedje JM (1995) Bacterial community fingerprinting of amplified 16S and 16S–23S ribosomal DNA gene sequences and restriction endonuclease analysis (ARDRA). In: Akkermans ADL, van Elsas JD, de Bruijn FJ (eds) Molecular Microbial Ecology Manual. Kluwer Academic, Dordrecht, pp 3.3.2-1–3.3.2-8
53. Mendonça V, Pedrosa C, Feldberg NP, Abreu NAA, Brito APF, Ramos JD (2006) Doses of nitrogen and simple superphosphate on papaya Formosa plant growth. Ciênc. Agrotec. Lavras 30:1065–1070.
54. Moeskops B, Sukristiyonubowo Buchan D, Sleutel S, Herawaty L, Husen E, Saraswati R, Setyorini D, De Neve S (2010) Soil microbial communities and activities under intensive organic and conventional vegetable farming in West Java, Indonesia. Appl Soil Ecol 45:112–120.
55. Muyzer G, de Waal EC, Uitterlinden AG (1993) Profiling of complex microbial populations by denaturing gradient gel electrophoresis analysis of polymerase chain reaction-amplified genes coding for 16S rRNA. Appl Environ Microbiol 59:695–700
56. Naher UA, Othman R, Panhwar QA (2013) Culturable total and beneficial microbial occurrences in long-term nutrient deficit wetland rice soil. Aust J Crop Sci 7:1848–1853
57. Nautiyal CS (1999) An efficient microbiological growth medium for screening phosphate solubilizing microorganisms. FEMS Microbiol Lett 170:265–270.

58. Nery-Silva FA, Machado JC, Vilela de Resende ML, Lima LCO (2007) Inoculation methodology s of papaya fruits with fungi causing stem-end-rot. Ciênc. agrotec. Lavras 31:1374–1379.
59. Oberhansli T, Defago G, Haas D (1991) Indole-3-acetic acid (IAA) synthesis in the biocontrol strain CHA0 of Pseudomonas fluorescens: role of tryptophan side chain oxidase. J Gen Microbiol 137(10):2273–2279
60. Patel KJ, Singh AK, Nareshkumar G, Archana G (2010) Organic-acid-producing, phytate-mineralizing rhizobacteria and their effect on growth of pigeon pea (*Cajanus cajan*). Appl Soil Ecol 44:252–261.
61. Peeters C, Zlosnik JEA, Spilker T, Hird TJ, LiPuma JJ, Vandamme P (2013) *Burkholderia pseudomultivorans* sp. nov., a novel *Burkholderia cepacia* complex species from human respiratory samples and the rhizosphere. Syst Appl Microbiol 36:483–489.
62. Peix A, Rivas R, Mateos PF, Martinez-Molina E, Rodriguez-Barrueco CE, Valazquez E (2003) *Pseudomonas rhizosphaerae* sp. nov., a novel species that actively solubilizes phosphate in vitro. Int J Syst Evol Microbiol 53:2067–2072.
63. Pereg L, McMillan M (2014) Scoping the potential uses of beneficial microorganisms for increasing productivity in cotton cropping systems. Soil Biol Biochem.
64. Perez PG, Ye J, Wang S, Wang XL, Huang DF (2014) Analysis of the occurrence and activity of diazotrophic communities in organic and conventional horticultural soils. Appl Soil Ecol 79:37–48.
65. Pitcher DG, Saunders NA, Owen RJ (1989) Rapid extraction of bacterial genomic DNA with guanidium thiocyanate. Lett Appl Microbiol 8:151–156.
66. Radíc T, Likar M, Hancevíc K, Bogdanovíc I, Paskovíc I (2014) Occurrence of root endophytic fungi in organic versus conventional vineyards on the Croatian coast. Agric Ecosyst Environ 192:115–121.
67. Reed SC, Yang X, Thornton PE (2015) Incorporating phosphorus cycling into global modeling efforts: a worthwhile, tractable endeavor. New Phytol 208:324–329.
68. Richardson AE, Barea JM, McNeill AM, Prigent-Combaret C (2009) Acquisition of phosphorus and nitrogen in the rhizosphere and plant growth promotion by microorganisms. Plant Soil 321:305–339.
69. Rodríguez H, Fraga R (1999) Phosphate solubilizing bacteria and their role in plant growth promotion. Biotechnol Adv 17:319–339.
70. Rodríguez H, Fraga R, Gonzalez T, Bashan Y (2006) Genetics of phosphate solubilization and its potential applications for improving plant growth-promoting bacteria. Plant Soil 287:15–21.
71. Rogers A, McDonald K, Muehlbauer MF, Hoffman A, Koenig K, Newman L, Taghavi S, Lelie D (2011) Inoculation of hybrid poplar with the endophytic bacterium Enterobacter sp. 638 increases biomass but does not impact leaf level physiology. GCB Bioenergy 4:364–370.
72. Santos-Villalobos S, Barrera-Galicia GC, Miranda-Salcedo MA, Peña- Cabriales JJ (2012) Burkholderia cepacia XXVI siderophore with biocontrol capacity against *Colletotrichum gloeosporioides*. World J Microbiol Biotechnol 28:2615–2623.
73. Sarruge JR, Haag HP (1974) Análise química em plantas. ESALQ, Piracicaba, p 56
74. Schwyn B, Neilands JB (1987) Universal chemical assay for the detection and determination of siderophores. Anal Biochem 160:47–56
75. Shannon D, Sen AM, Johnson DB (2002) A comparative study of the microbiology of soils managed under organic and conventional regimes. Soil Use Manag 18:274–283.
76. Sharma SB, Sayyed RZ, Trivedi MH, Gobi TA (2013) Phosphate solubilizing microbes: sustainable approach for managing phosphorus deficiency in agricultural soils. Springer Plus 2:587.
77. Singh RK, Malik N, Singh S (2013) Improved nutrient use efficiency increases plant growth of rice with the use of IAA-overproducing strains of endophytic *Burkholderia cepacia* strain RRE25. Microb Ecol 66:375–384.
78. Singh P, Kumar V, Agrawal S (2014) Evaluation of phytase producing bacteria for their plant growth promoting activities. Int J Microbiol.

79. Srinivas T, Sridevi M, Mallaiah KV (2008) Effect of pesticides on *Rhizobium* and nodulation of green gram *Vigna Radita* (L.) Wilczek. IUP J Life Sci 2:36–44
80. Suzuki MS, Zambolim L, Liberato JR (2007) Progress of fungal diseases and correlation with climatic variables in papaya. Summa Phytopathol 33:167–177
81. Tawiah AA, Gbedema SY, Adu F, Boamah VE, Annan K (2012) Antibiotic producing microorganisms from River Wiwi, Lake Bosomtwe and the Gulf of Guinea at Doakor Sea Beach, Ghana. BMC Microbiol 12:1–8.
82. Trindade AV, Siqueira JO, Stürmer SL (2006) Arbuscular mycorrhizal fungi in papaya plantations of Espírito Santo and Bahia, Brazil. Braz J Microbiol 37:283–289.
83. Trujillo I, Díaz A, Hernández A, Heydrich M (2007) Antagonismo de cepas de *Pseudomonas fluorescens* Y *Burkholderia cepacia* contra hongos fitopatógenos delarroz y el maíz. Rev. Protección Veg. 22:41–46
84. Tsavkelova EA, Cherdyntseva TA, Netrusov AI (2005) Auxin production by bacteria associated with orchid roots. Microbiology 74:46–53
85. Unno Y, Okubo K, Wasaki J, Shinano T, Osaki M (2005) Plant growth promotion abilities and microscale bacterial dynamics in the rhizosphere of Lupin analysed by phytate utilization ability. Environ Microbiol 7:396–404.
86. van Diepeningen AD, de Vos OJ, Korthals GW, van Bruggen AHC (2006) Effects of organic versus conventional management on chemical and biological parameters in agricultural soils. Appl Soil Ecol 31:120–135.
87. Vassilev N, Vassileva M, Nikolaeva I (2006) Simultaneous P-solubilizing and biocontrol activity of microorganisms: potentials and future trends. Appl Microbiol Biotechnol 71:137–144. \
88. Zúñiga A, Poupin MJ, DonosoR Ledger T, Guiliani N, Gutiérrez RA, González B (2013) Quorum sensing and indole-3-acetic acid degradation play a role in colonization and plant growth promotion of arabidopsis thaliana by *Burkholderia phytofirmans* PsJN. Mol Plant Microbe Interact 26:546–553.

CHAPTER 21

Plant Growth Promotion by Streptomycetes: Ecophysiology, Mechanisms and Applications

Jucimara Anunciação de Jesus Sousa and Fabio Lopes Olivares*

Laboratorio de Biologia Celular e Tecidual (LBCT) and Núcleo de Desenvolvimento de Insumos Biológicos para Agricultura (NUDIBA), Universidade Estadual do Norte Fluminense Darcy Ribeiro (UENF), Av. Alberto Lamego, 2000, Campos dos Goytacazes, Rio de Janeiro 28013-602, Brazil

ABSTRACT

The genus *Streptomyces* comprises filamentous Gram-positive bacteria that are widely recognized for their ability to produce bioactive compounds such as antimicrobial, antiparasitic and immune-suppressing compounds via secondary metabolism. These bioactive compounds represent a third of all commercially available antibiotics. Streptomycetes have been found in beneficial associations with plants where they have improved plant growth and protected against pests, which have attracted the attention of researchers worldwide. This review focuses on the potential of streptomycetes as plant growth-promoting bacteria (PGPS) and considers features related to secondary metabolic pathways, interactions with host plants and recent advances in elucidating plant growth-promoting mechanisms. Such advances in basic knowledge have increased the prospects for streptomycetes to be used as bioinoculants for sustainable agriculture.

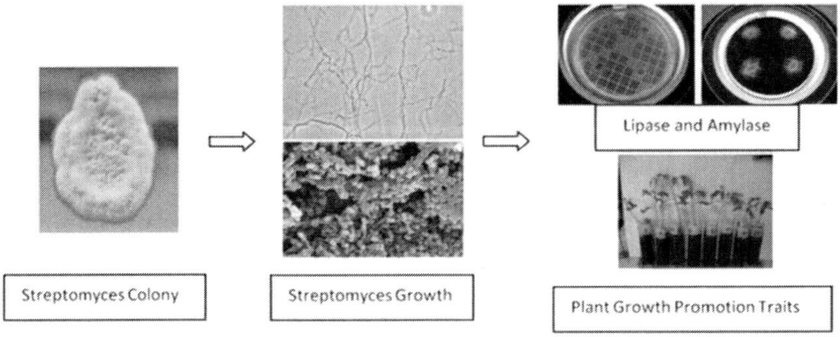

Keywords: *Streptomyces* Actinobacteria Biotechnological potential

INTRODUCTION

It has been estimated that the world's population will reach about nine billion by 2050 [1], which will require high levels of yield from agricultural systems. Centered on the Green Revolution model established in the last century, a range of research and technology transfer initiatives have been employed to meet the demand for food, fiber, and energy. However, it has become increasingly clear that the conventional systems of food production have many negative impacts on the environment [2].

To develop food production under environmentally and socially sustainable systems represents one of the twenty-first century's greatest challenges for agricultural researchers. One of the most promising initiatives for a new model of agriculture is based on converting the natural processes that occur in the soil–plant system into biological input technologies. In this context, microorganisms, their products and their processes are essential resources for a new generation of biotechnologies applicable to plant production and protection, and potentially a paradigm shift in agricultural practice.

Using plant growth-promoting bacteria (PGPB) for the benefit of agriculture has received increasing attention and acceptance [3, 4, 5]. Besides the well-studied Gram-negative plant-associated bacteria, Gram-positive bacteria can also have beneficial interactions with plants and promote plant growth [6, 7, 8]. *Streptomyces* is the most widely studied genus of Gram-positive PGPB and is the central subject of this review. This genus comprises a wide diversity of species that have a high G-C (guanine and cytosine) ratio in its DNA, up to 75 % of its genome [9]. This genus produces a wide variety of biologically active compounds; some with plant growth activity. It has been suggested by some authors that the increasingly intensive surveys of soil-borne microbes have resulted in a decreased frequency of newly discovered compounds [10]. However, the metabolic versatility and cosmopolitan behavior of *Streptomyces* species have enabled them to be isolated from different environments, some of which have not yet been explored, hence, presenting an opportunity to discover new bioactive compounds.

Morphogenetic and physiological aspects of the *Streptomyces* genus have been reviewed [11, 12]. Other reviews have focused on the biotechnological potential of plant-associated endophytic actinobacteria [10] and free-living plant growth-promoting actinobacteria [7]. This review presents an overview of the plant growth-promoting ability of various species from the genus *Streptomyces*; it considers historical aspects of their discovery, their physiological features, their plant growth mechanisms and their application as bioinoculants in agriculture.

REVIEW

Historical and taxonomic aspects

Nearly 200 years after Antony van Leeuwenhoek reported the first observation of bacteria in 1684 using his own handcrafted microscope, other pioneers, such as Ferdinand Cohn and Robert Koch, founded modern concepts about bacteriology as a science domain [13]. It took another 200 years to identify and reach the current understanding of actinobacteria. Further, only 204 years after van Leeuwenhoek studies, the first description of a microorganism that eventually became known as an actinobacterium was described, when Armauer Hansen discovered a microorganism in the tissues of leprosy patients, which was later described as the etiologic agent of this disease in 1874 [14].

In 1875, Cohn described the first Actinobacteria species, which he named *Strepthrotrix foersteri*. He isolated this microbe from samples of human tear ducts provided by R. Foerster, a medical friend. Cohn supposed that *Strepthrotrix foersteri* was not associated with any disease, but that it reached the patient's eye through airborne soil particles. Later, he observed that *Strepthrotrix foersteri* had morphological features of fungi and bacteria [14]. However, the proposed nomenclature for the bacterial genus was deemed invalid because *Strepthrotrix* had already been classified as a true fungus by Corda in 1839 [15]. Then in 1877, Carl Otto Harz described the etiologic agent of "lampy jaw". Harz observed structures similar to reproductive bodies and hyphae of fungi; therefore, he considered the microorganism to be a fungus and named it *Actinomyces bovis* [14]. In 1882, Robert Koch discovered another microorganism during his observations using light microscopy that is now recognized as an actinobacterium: the tuberculosis pathogen *Mycobacterium tuberculosis* [16]. Koch observed that the microbes presented morphological characteristics that were similar to those of microorganisms previously described by Hansen associated with leprosy disease [17].

Although there was a clear relationship between these microorganisms, it was not until 1916 that R.E. Buchanan suggested a nomenclature and classification for this group. Buchanan proposed the order Actinomycetales, containing the family *Actinomycetaceae* and the following genera: *Actinobacillus, Leptotrichia, Actinomyces,* and *Nocardia* [18]. In 1943, Waksman and Henrici proposed a new classification for the actinomycetes, which was based on their ability to form branching cells. Waksman and Henrici observed that one actinomycete group formed a condensed mat of interlinked branching hyphae that produced reproductive spores. The *Streptothrix* described by Cohn fell into this group, but due to the invalid genus name, Waksman and Henrici named it *Streptomyces,* which means "twisted fungus" [19].

In the fourth decade of the twentieth century, *Streptomyces* was recognized. Many studies were performed during this time to find chemotherapeutic treatments to control tuberculosis [17]. In 1943, Waksman again received attention, this time due to his greatest discovery: the antibiotic streptomycin isolated from *Streptomyces griseus* is effective against the tuberculosis pathogen [19]. About 600 validated species of *Streptomyces* have now been described following their isolation from many environmental sources. They are now the subject of research to discover new bioactive compounds for the pharmaceutical and agricultural industries [20].

The first proposed nomenclature of actinobacteria was based on sporulation patterns. Although morphological characteristics are typically important for *Streptomyces* identification, some studies have demonstrated that classification based on cell morphology, colony pigmentation, and physiological features do not always reflect the natural phylogenetic relationship between actinobacteria and related organisms [21]. Introduction of the polyphasic taxonomic approach combined molecular and biochemical analyses which elucidated streptomycetes systematics. In addition, increased availability of 16S rRNA sequence data has enabled accurate studies of taxonomic affiliations and phylogenetic relationships [15, 21, 22].

The *Streptomyces* genus belongs to the Streptomycetaceae family and the single-order *Streptomycetales* [23]. This order belongs to both phylum and class Actinobacteria [23]. The actinobacteria from the *Streptomyces* genus are the most extensively studied mycelial actinobacteria. They are aerobic,

Gram-positive bacteria that grow as branching filaments that consist of vegetative mycelia and aerial hyphae [15, 24]. Some morphological and physiological properties, along with the ability to produce a wide range of pigments, have been used not only to classify the *Streptomyces* genus [25, 26], but also to study its ecological distribution and biotechnological potential [27, 28].

Morphological differentiation and physiology

Streptomycetes have a markedly different cell envelope structure than Gram-negative bacteria, such that *Streptomyces* genus has been identified using cell wall composition [29, 30]. Similar to other actinobacteria, streptomycetes have no outer membrane and their cell walls have a thick peptidoglycan (or murein) layer [31]. The presence of LL-diaminopimelic (LL-DAP) in the cell wall confers a typical chemotaxonomic characteristic to all members of the *Streptomyces* genus [32, 33], and its presence together with glycine characterizes the cell wall as Type I [31, 34]. Teichoic acids (anionic glycopolymers) comprise another important cell wall component that confers a negative charge to the cell surface and contributes to physiological functioning and cell co-aggregation [35, 36].

The life cycle is initiated when favorable environmental conditions and nutrient availability promote spore germination [12] (Fig. 1). Next, germ tubes grow to form syncytial vegetative or substrate mycelia, which consist of interconnected feeding hyphae that are responsible for nutrient uptake [37]. When nutrients become scarce, or another stress condition occurs, programmed cell death of the substrate mycelia and cell differentiation at the center of the colony result in aerial hyphae [12, 38]. These aerial hyphae are subtly distinguishable from the feeding hyphae, as they are covered by a hydrophobic fibrous layer, perhaps to help the aerial hyphae break the surface tension on air pockets in the soil, whereas the feeding hyphae have a smooth hydrophilic surface [24].

The growth of *Streptomyces* involves hyphal tip extension and sub-apical branching [39]. Unlike the process in rod-shaped bacteria where cytokinesis is based on building a cross wall by depositing murein into lateral walls, *Streptomyces* growth occurs by hyphae production at the cell pole [37]. Although it is not clearly elucidated, this cell growth pattern is regulated by the apical protein complex DivIVA. In *Bacillus subtilis*, DivIVA interacts with the Min system to coordinate division at the middle of the cell. In contrast, in *Streptomyces*, the Min system is absent, thus DivIVA affects division at the cell tip. Another aspect of streptomycetes growth involves the conservation of two groups of proteins, the tubulin homolog FtsZ and several membrane proteins, which are both associated with cytokinetic Z-ring and septal peptidoglycan [11].

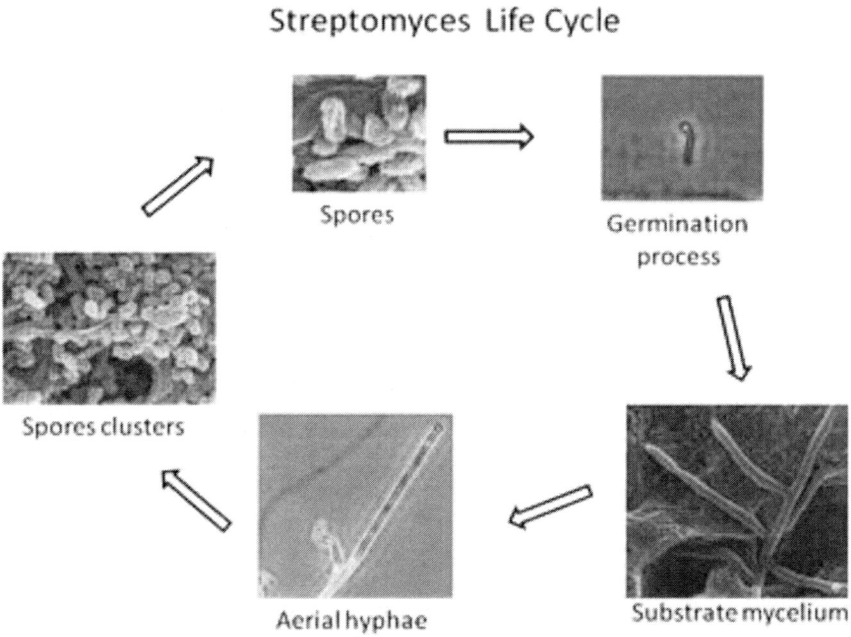

Fig. 1 Life cycle of *Streptomyces* [167] modified

The last phase of the *Streptomyces* life cycle consists of the apical cells of the aerial hyphae differentiating into a spore chain [12]. A differentiating apical compartment grows by tip extension and starts synchronous, multiple cell divisions into a developmentally controlled form [40]. Again, there is the participation of FtsZ, which leads to sporulation septa and then these pre-spores assemble thick spore walls by depositing actin [41]. The size of *Streptomyces* spores can range from 0.7 to 1.2 μm [42, 43]. These last two phases of the *Streptomyces* life cycle are closely related to antibiotic production [14]. During programmed cell death of the substrate mycelia, antibiotics are simultaneously produced, perhaps to protect the nutrient sources against competitor microorganisms [44, 45].

Among the 23,000 bioactive secondary metabolites produced by microorganisms, two-thirds are produced by actinobacteria, and *Streptomyces* spp. accounts for more than 70 % of these [46]. This production of secondary metabolites is attributed to the development of aerial hyphae as a result of nutrient limitation [47]. The biological activity of these compounds involves inhibitory or microbiocide activity against microorganisms (i.e., antibiotics), toxicity against metazoans, microbial hormone-like activity, and metal transport [48, 49, 50]. It has been well demonstrated that the secondary metabolites produced by streptomycetes increases adaptation to biological, physical, and chemical stresses; thus, they are recognized as 'stress metabolites' [36].

Structural genome studies tell us that most genes involved in regulating secondary metabolite pathways are arranged in clusters. These clusters might encode the highly phosphorylated guanosine nucleotide (p)ppGpp and some regulatory proteins involved in producing secondary metabolites [51]. In *Streptomyces* species under nutritional stress, alarmone ppGpp plays a role as a regulator of antibiotic production [52, 53]. Members of both the SARP and LAL families of regulatory proteins appear to be confined to actinobacteria, mainly genus *Streptomyces*, and have shown species-specific controls for secondary metabolism pathways [54, 55, 56]. Cell-to-cell communication is a determining factor for modulating antibiotic production, and γ-butyrolactones (GBLs) are the main intercellular signaling compounds [48].

The overall ecophysiological traits of the *Streptomyces* genus support the concept of cosmopolitan biogeographical behavior. The traits include the ability to form spores under unfavorable abiotic conditions, a competitive ability related to antibiotic production, a broad pH range that is favorable for growth, and a wide pH range that is optimal for growth between different *Streptomyces* species (such as pH 4.3 for the acidophilic *S. yeochonensis* [57], pH 7.0 for the neutrophilic *S. roseus* [58], and pH 10 for the alkaliphilic *S. alkalithermotolerans* [59]). Streptomycetes are typically chemoorganotrophic with a great versatility for metabolizing a wide range of carbon sources including mono- and disaccharides, polyol, organic acids (glucose, dextrose, fructose, lactose, maltose, mannitol, rhamnose, sucrose, glycerol, and glycolic acid), polysaccharides (including cellulose and starch), and more complex and recalcitrant C-sources (such as humic and fulvic acids) [60, 61, 62].

Representatives of the genus *Streptomyces* are well recognized as soil-dwelling bacteria. Nevertheless, many reports have shown that they are widely distributed in both aquatic and terrestrial environments [60]. This cosmopolitan distribution of streptomycetes might be attributed to their production of spores, which are readily spread and thus could explain its presence in different environments (see Table 1). Several members of the genus *Streptomyces* have been isolated from various vegetative and reproductive plant parts, such as roots, tubers, stems, leaves, and seeds. There are many questions concerning the ecological and physiological significance of this interaction that would potentially converge for successful practical applications related to plant growth-promoting and plant-protection properties [63, 64].

Table 1. Biotechnological potential and cosmopolitan features of the *Streptomyces* species

Species	Strain	Isolated from	Major characteristics	References
S. coelicolor	A3(2)	Soil	The best genetically known representative of the genus	[145]
S. xiamenensis	DSM 41903^T	Mangrove sediment	Not described	[62]
S. axinellae	DSM 41948^T	Mediterranean sponge *Axinella polypoides* (Porifera)	Not described	[42]
S. griseus	DSM 40236^T	Soil	Producer of streptomycin antibiotic	[19]
S. chumphonensis	KK1-2^T	Marine sediments	Not described	[146]
S. rochei	SM3	Decomposed cow dung	Alleviates the stresses caused by *Sc. sclerotiorum* and salt in chickpea	[147]
S. fildesensis	GW25-5^T	Antarctic soil	Not described	[148]
S. scabies	ATCC 49173	Potato scab	Pathogen of potato scab	[149]
S. pseudovenezuelae	ACTA 1383	Rhizosphere of *Ebenus sibthorpii*	Antagonist activity against *Rhizoctonia solani*	[122]
S. oryzae	S16-07^T	Surface-sterilized stems of rice	Not described	[150]
S. wadayamensis	A23	Plant tissue of *Citrus reticulata*	Antagonist activity against *Xylella fastidiosa*	[151]
S. kebangsaanensis	SUK12^T	Inner tissue of *Portulaca oleracea* L. stems	Phenazine-1-carboxylic acid producer	[152]
S. phytohabitans	KLBMP4601^T	Surface-sterilized roots of *Curcuma phaeocaulis*	Not described	[153]
S. diastaticus	UENF AC01	Vermicompost of sunflower cake	Plant growth-promoting streptomycete. AIA and catalase producer	[154, 155]
S. variabilis	UENF AC31	Vermicompost of sunflower cake	Phosphate solubilizer	[154, 155]

^{T} Type strain

Ecology of Streptomyces–plant host interaction

To acquire a better understanding and to manipulate the interactions between plant growth-promoting streptomyces (PGPS) and their hosts, it is necessary to elucidate those biochemical mechanisms that lead to compatible relationships. As shown above (Table 1), most of these streptomycetes are soil-dwelling bacteria with a free-living life cycle in the soil (i.e., saprophytic competence) and they are able to efficiently colonize the rhizosphere and rhizoplane compartments. Eventually, some PGPS might become endophytic and colonize the inner tissues of the host plant and partly or fully conduct their life cycle within them [65]. Therefore, before discussing plant responses to PGPS, it is necessary to describe the rhizosphere and rhizoplane colonization process.

The rhizosphere is the soil volume under the influence of plant root exudates, secretions, and loose cell deposition [66]. The rhizoplane plays a crucial modulatory role on the microbial community structure and microbial diversity that will ultimately influence plant growth and performance in the soil–plant system [67]. Additionally, microbial activity in the rhizosphere soil is markedly influenced by carbon-containing metabolites released from roots via the rhizodeposition process [68]. Rhizodeposition consists of numerous compounds such as ionic secretions, free oxygen and water, enzymes, proteins, mucilage, amino acids, organic acids, sugars, and phenolics [69]. These compounds act as nutritional resources, chemoattractants, chemorepulsants, and signaling compounds that shape the microbial community structure and activity of different groups of microorganisms and have a significant influence on plant root–microorganism interactions [70].

The PGPS are widely recognized to be good rhizosphere colonizers and their rhizosphere competence may be partially explained by several chemotaxic features, such as bacterial rate multiplication, quorum sensing-controlled gene expression, amino acids, antibiotics, and siderophore synthesis [71]. Initially, soil-borne or introduced streptomycete cells that respond to released compounds are actively attracted to the rhizosphere by chemotaxis [69, 72]. At this point, the versatile nutritional requirements and increased population rates coupled with bioactive compound production and detoxification mechanisms are determinants for successful rhizosphere establishment. The former properties play a pivotal role in overcoming complex competition and can be partially explained by the ability to produce antibiotics and other bioactive compounds that confer an ecological advantage and allow PGPS to colonize niches. Several biomolecules that are secondary metabolites of *Streptomyces* contribute to biocontrol and successful colonization; e.g., siderophores from *S. coelicolor* [73], the antifungal nigericin and antibiotic geldanamycin from *S. violaceusniger* YCED-9 [74], and chitinase from *S. violaceusniger* YH27A [75].

Some secreted proteins are important for successful root colonization by *Streptomyces* [76]. Differential protein expression was exhibited by *S. coelicolor* when it was cultivated in minimal medium with and without *Lemna minor* fronds. Bacterial enzymes involved in degrading cellulose, alkenes, and amino acids were induced in the presence of plant extracts, suggesting that the carbon and energy were acquired through degrading compounds present in *L. minor* exudates [77].

Isolates of PGPS can be screened for the attribute that only a minority of species are recognized as phytopathogenic agents; namely *Streptomyces scabies, S. acidiscabies, S. turgidiscabies,* and *S. ipomoeae,* which are all etiologic agents of common scab diseases [78]. *Streptomyces scabies* is a model organism for investigating plant host–pathogen interactions in Gram-positive bacteria in which a protein secretion system is essential for its pathogenesis [76]. The effector proteins can be secreted by the secretory (Sec) and twin-arginine translocation (Tat) pathway or, additionally, by the specialized Type VII secretion system (T7SS) [79]. Thaxtomin, a family of nitrated dipeptide phytotoxins, is an important pathogenicity factor secreted by *S. scabies* and plays a role in cellulose biosynthesis inhibition [80]. Nec1 is a protein that is required for root colonization by *S. scabies* [81]. Another important intercellular signal attributed to pathogenic streptomycetes and its host is nitric oxide (NO) [82]. Synthase-derived NO is produced by plant-pathogenic *Streptomyces* in response to cellobiose production, which expands plant tissues and appears to be involved in the nitration of thaxtomin [83].

In contrast, a wide variety of *Streptomyces* species establish beneficial plant–microbe interactions [84, 85, 86]. The ability of some *Streptomyces* species to gain access into root tissues and to establish an endophytic lifestyle without causing visible harm or symptoms in the host plant has been reported (Table 1). These species can be found mainly in the apoplastic compartments that comprise the intercellular spaces and lumen of differentiated dead cells (sclerenchyma and xylem cells) of the host plant organs (roots, stems, leaves, flowers, fruits, and seeds); intracellular occurrences seem to be less frequent [10]. Using a *Streptomyces* sp. strain EN27 tagged with green fluorescent protein, the endophytic colonization of embryos, endosperm, and emerging radicles of wheat seeds has been reported [87].

Unlike Gram-negative bacteria, *Streptomyces* and other filamentous actinobacteria possess active penetration structures that grow on the plant surface and infect intact cells [88]. The ability to attach and develop an infection point through the cell wall of sweet potato (*Ipomoea batatas* [L.] Lam.) has been reported for *S. ipomoea*. Subsequently, the tip-growth hyphae colonize the interior of parenchyma cells and establish an endophytic interaction [89]. Similarly, using short branches that emerged from the main hyphae, *S. scabies* was observed to penetrate the cell walls of potato plants [90]. Based on the described cell wall penetration process by these short hyphae within a short and uniform distance from the branching point, the authors suggested a specialized penetration function for this structure. Furthermore, *Streptomyces* might enter plant tissues by natural openings such as stomatal apertures, lenticels, hydathodes, wounds, broken trichomes, and root hair cracks formed by lateral root emergence zones [88]. A transmission electron microscope analysis demonstrated that *S. galbus* MBR-5 entered the leaves of *Rhododendron* sp. seedlings through stomatal openings [91].

Plants exhibit defense responses during infection by PGPS, but they are less aggressive than those expressed during pathogenic interactions [92]. Although the biochemical mechanisms involved in these plant responses are not clearly elucidated, the pivotal role of phytohormones such as salicylic acid (SA), jasmonic acid (JA), and ethylene (ET) has been widely reported in the literature [93, 94, 95, 96]. *Arabidopsis thaliana* inoculated with the endophytic *Streptomyces* sp. strain EN27 induced low levels of gene expression related to the JA and ET pathways [93]. Furthermore, plant response to PGPS infection involves generating reactive oxygen species, which are cytotoxic [83]. This plant process can be subverted by enzymatic machinery related to antioxidant activity; e.g., catalase and superoxide dismutase by *S. coelicolor*, and aconitase by *S. viridochromogenes* strain

Tü494. This process plays an important role in alleviating the cytotoxic effects induced by reactive oxygen species and in allowing successful plant tissue colonization [97, 98, 99].

Endophytic colonization usually represents an important ecological advantage to the microorganism, because the protected environment is less susceptible to abiotic stress (pH, redox, osmotic, and hydraulic variations) and microbial competition compared with the rhizosphere–soil system. Moreover, the intimate contact between plant host cells and microbial cells might be more effective for the bidirectional exchange of signals, functional metabolites, and nutritional sources for a successful beneficial interaction, which may promote growth of the host plant [88, 100]. The mechanisms involved in plant growth-promoting activity by both endophytic and rhizosphere streptomycetes will be discussed below.

Plant growth promotion by streptomycetes

Most of the fundamental and applied studies of beneficial plant–microbe interactions relate to Gram-negative bacteria [101, 102, 103]. Although less often studied, many representative groups of Gram-positive bacteria, particularly those belonging to genus *Streptomyces*, exhibit a range of traits that may improve plant growth by using different mechanisms [7]. As quoted for other beneficial interactions, the plant growth-promoting effects related to *Streptomyces*–plant interactions can be divided into biofertilization, biostimulation, and bioprotection [104].

Biofertilization effects

Due to the mineralogical and electrochemical properties of many soils, some essential mineral nutrients are unavailable to plants because they are present in insoluble forms [105]. Biofertilization consists of a direct mechanism that improves the macro or micronutrient acquisition (uptake and assimilation) by plants. Nutrients sequestered in the crystalline lattice of the mineral fraction of soil can be solubilized and released into solution by organic acids (such as gluconic acid, citric acid, succinic acid, and oxalic acid) that are secreted by different microorganisms [106]. Jog and colleagues [64] reported the release of free phosphate by acidification resulting from the release of malic acid and gluconic acids by *Streptomyces* mhcr0816 and *Streptomyces* mhce0811, respectively. These authors also observed an increase in the number of branches and lateral roots, shoot length, and mineral content of Fe, Mn, and P in wheat plants inoculated with these streptomycetes. Soil treated with a phosphate-solubilizing strain of *Streptomyces* that was isolated from a wheat field increased N, Fe, P, and Mn content in shoots of wheat [107]. In addition, solubilization of rock phosphate by *S. youssoufiensis* has been reported [108, 109]. Streptomycetes can further promote mineral supply by synthesizing siderophores and siderophore uptake systems [107]. *Streptomyces* sp. GMKU 3100, a siderophore-producing endophytic streptomycete, was capable of promoting the growth of rice and mungbean, whereas its siderophore-deficient mutant did not differ from the uninoculated control [86]. Until now, there is no convincing evidence for free-living or endophytic *Streptomyces* species able to fix nitrogen, since the controversial report related to *S. thermoautotrophicus* by Ribbe and colleges [110] was recently refuted [111].

Biostimulation effects

Plant growth can be directly improved by the microbial production of metabolic compounds with phytohormonal activity at micromolar to nanomolar concentrations [64]. Auxin and auxin-like compounds regulate many aspects of plant growth and development including cell plasticity, tissue elongation, embryogenesis, tip dominance, and emergence of lateral roots [112]. The production of indole-3-acetic acid (IAA) by *Streptomyces* spp. has been quoted in several reports [107, 113, 114]. For example, *S. atrovirens* ASU14 utilized tryptophan and produced 22 μg/mL of the IAA [115]. An auxin-like activity due to pteridic acid A was produced by *S. hygroscopicus* TP-A0451 isolated from *Pteridium aquilinum* (bracken) stems [116]. This compound stimulated root elongation and induced the formation of adventitious roots in *Phaseolus vulgaris* (kidney bean) hypocotyls. Another class of phytohormones that have an important function in plant growth is gibberellin. This phytohormone is involved in physiological process that includes seed germination, growth of stems and leaves, floral induction, and growth of flowers and fruits [117, 118]. *Streptomyces* species isolated from a marine environment exhibited the ability to produce a range of phytohormones, including gibberellic acid, and enhanced the agronomic performance of eggplant (*Solanum melongena*) by influencing its growth parameters, including root length and fresh or dry root weight [119].

Bioprotection

Plant-associated microbes might have the ability to minimize the challenges imposed by phytopathogens. This biocontrol activity can manifest as antibiosis, nutrient and space competition, parasitism, predation, hypovirulence, and induced systemic resistance [69, 120], which are indirect mechanisms of plant growth promotion. The versatile production of secondary metabolites involved in biocontrol is primarily a competitive strategy to successfully colonize the root zone [71]. The genus *Streptomyces* is widely recognized as being able to synthesize several bioactive metabolites that act to control phytopathogen and to confer an advantage to rhizosphere or endophytic colonization [121]. It has been reported that antifungal metabolites produced by 213 *Streptomyces* strains isolated from different habitats exhibited in vitro antagonistic activity against *Rhizoctonia solani* [122]. An investigation into the nematicidal property of *Streptomyces roseoverticillatus* CMU-MH021 revealed the production of secondary metabolites that acted against the root-knot nematode *Meloidogyne incognita* [123]. *Streptomyces roseoverticillatus* CMU-MH021 produced fervenulin, which decreased the percentage of hatched eggs and increased the percentage mortality of second-stage juveniles; it produced isocoumarin, which also increased the percentage mortality of second-stage juveniles. A recently released draft genome sequence of the mushroom mycoparasite antagonist *Streptomyces* sp. strain 150FB, with close correspondence to the *S. avermitilis* MA-4680 genome, revealed possible factors related to disease suppression, namely, a set of genes encoding extracellular enzymes involved in degrading fungal cell wall polysaccharides, disrupting membranes and proteins, and also peroxidases and ribonucleases [124]. In addition, two terpenes and two siderophore biosynthetic gene clusters were detected. Streptomycetes can also stimulate plant defense by the *priming* phenomenon, resulting in induced systemic resistance [125]. *Streptomyces* sp. strain AcH505 suppressed oak powdery mildew infection [126]. This streptomycete strain elicited a systemic defense response in oak (*Quercus robur*), inducing the jasmonic acid/ethylene-dependent pathway and the salicylic acid-dependent pathway.

Secondary metabolites from streptomyces species include not only plant bioprotection effects (scope of the present review), but also other bioactivity properties related to application in medical science. Table 2 summarizes the examples of some of these compounds and their genes and operon structures. Advances in the genetic basis of secondary metabolite pathways enhance the perspectives for discovering new compounds in ecological surveys under distinct environmental conditions.

Plant-associated streptomycetes can also benefit the host plant by mitigating abiotic stress such as heat, cold, drought, and nutrient depletion, thus reducing their negative impacts and consequently increasing plant growth [127]. The application of *Streptomyces filipinensis* no. 15, a 1-aminocyclopropane-1-carboxylic acid (ACC) deaminase and IAA producer, reduced the endogenous levels of ACC, the immediate precursor of ethylene, in both roots and shoots and subsequently enhanced plant fitness to the environment [128]. Plant growth-promoting and stress-alleviating activities have been demonstrated for a halo-tolerant and ACC deaminase-producing *Streptomycete* sp. strain PGPA39 applied to tomato (*Solanum lycopersicum*) plants under salinity stress [129].

Table 2. Secondary metabolites produced by *Streptomyces* species, main bioactivity property, and their genes and structural organization

Streptomyces species	Secondary metabolite	Bioactivity	Genetic basis	References
S. clavuligerus	Clavulanic acid	Antibiotic	*herABCDEFG* genes	[156]
S. coelicolor	Actinorhodin	Antibiotic	*abeABCD*; *α-abeA* genes	[157]
S. griseus	Streptomycin	Antibiotic	*strFGHIK* genes	[158]
S. hygroscopicus	Geldanamycin	Antitumor	*ahba-B* locus	[159]
S. hygroscopicus var. ascomyceticus	Ascomycin	Immunosuppressant; neurotrophic; antifungal	FK520 gene cluster	[160]
S. thioluteus	Aureothin	Antitumor; antifungal; insecticidal	*aur* operon (*aurA* through *aurI*)	[161]
S. venezuelae	Chloramphenicol	Antibiotic	*sven0916-sven0928* gene cluster	[162]
S. violaceusniger	Meridamycin	Neuroprotectant	*merA-D*; *mere*; *merP* genes	[163]
S. viridochromogenes Tu¨494	Phosphinothricin tripeptide	Herbicide	*phsB*; *orfM*; *phsC*; *pmi*; *ppm*; *ppd*, *phsA*; *pat*, *dea*; *prpA*	[164]
S. griseus	Griseobactin	Siderophore	*dhbACEBG* operon	[165]
S. scabiei	Indole-3-acetic acid	Auxin	*iaaM*; *iaaH*	[166]

Furthermore, a different approach that is based on the indirect mechanisms of plant growth by PGPS has received particular attention. In a study involving a tripartite culture system [130], streptomycete application seemed to act as a modulator in both mycorrhizal and nitrogen-fixing symbiosis and raised their ability to induce plant growth [131, 132]. An increased number, size, and vigor of root nodules were observed on young pea (*Pisum sativum*) seedlings cultivated in soil with naturally abundant *Rhizobium* and that was inoculated with *S. lydicus* WYEC108 [133]. *Streptomyces* sp. AcH505 isolated from the hyphosphere of a spruce (*Picea abies*) promoted mycelial growth and the mycorrhization rate by *Amanita muscaria* and *Suillus bovinus*, while it suppressed the mycelial extension of the plant pathogens *Armillaria obscura* and *Heterobasidion annosum* [134, 135].

Our group has recently proposed some different methods to enhance the plant growth-promoting ability of the endophytic diazotrophic bacteria *Herbaspirillum seropedicae* strain HRC54 (unpublished results). One method involved axenic studies combining *H. seropedicae* with the *Streptomyces bellus* strain UENF AC06 that was isolated from mature vermicompost. This mixed cultivation increased the population growth and nitrogen fixation rates of *Herbaspirillum* when measured by acetylene reduction activity. Therefore, *S. bellus* strain AC06 may be used to improve the plant growth-promotion response by *H. seropedicae* strain HRC54, and we suggest that this mixed cultivation is a potential bioinoculant for agricultural systems.

Biotechnological application of PGPS in agriculture

Using plant growth-promoting bacteria to improve nutrient availability to plants is an important practice for sustainable agriculture and is an alternative to chemical pesticides and fertilizers [136]. PGPS can enhance nutrient availability to plants by biosynthesizing metal chelators and phosphorus solubilizers, producing phytohormones, controlling phytopathogen, and alleviating abiotic stress. Although few *Streptomyces* strains are used as biofertilizers, commercial bioinoculants (microbe-based products) have been developed to improve plant growth promotion (Table 3) [63].

Table 3. Commercially developed plant growth-promotion products from streptomycetes

Streptomycete active ingredients or their metabolites	Comercialized product	Target crop	Plant growth-promotion effects
Streptomyces lydicus WYEC 108	Actinovate® SP/Actinovate® AG (seed application)	Grass, ornamentals, vegetables, and forest species in greenhouses, nurseries, and more	Biocontrol by soil-borne plant pathogens and foliar diseases
	Micro108® soluble/Micro108® Seed Inoculant	Food and fiber crops, ornamentals, turf grass, landscape plants, including tree seedlings for transplanting into the forest	Enhances plant vitality and encourages more vigorous root systems
	Actino-iron®	For indoor/outdoor greenhouse, nursery, turf grass, ornamental plant, and field uses	Biocontrol of soil-borne plant pathogen
Streptomyces griseoviridis	Mycostop®	Seedling production, vegetables, herbs, and ornamentals	Biocontrol by soil and seed-borne pathogens
Streptomyces avermitilis	AVID® 0.15EC	Shadehouse, greenhouse, field-grown ornamentals, foliage plants, Christmas trees, and other woody ornamentals	Biocontrol of leaf miners, mites, and suppression of aphids, whiteflies, and thrips
	VERMITEC® (additional brands AGRI-MEK®)	Cotton, citrus, pome fruit, nuts and vegetables	Biocontrol of mites and insects
	PROCLAIM® (additional brands AFFIRM®, DENIM ™)	Vegetables (*Brassica*, leafy and fruiting vegetables) and cotton	Lepidoptera. Side effects on mites, leaf miners, and thrips
	AVICTA®	Cotton, corn, and soybean	Biocontrol of nematodes

Adapted from Hamedi and Mohammadipanah [63]

The first challenge to produce a bioinoculant using PGPS is to find the best strain of streptomycete (single formulation) or its combination with other microbes (mixed formulation). Screening studies involve different experimental assays in the laboratory and greenhouse. Elite strains must be tested under different environmental conditions and for different plant species and different genotypes of a target crop [4]. Following these steps, it is necessary to develop specific bioinoculant formulations, which involve determining the required physical and chemical characteristics of the carrier, and ascertaining which additives and metabolites would increase the viability, activity, and performance of the microorganism when introduced under field conditions [4]. Lastly, research to develop methods of bioinoculant delivery and timing must be carried out in the greenhouse and field to maximize the plant response to the applied bioinoculant. It has been found that applying streptomycetes is different to applying Gram-negative bacteria because better plant responses have generally been obtained when the streptomycete application to the soil or substrate occurred before sowing the seeds, thus allowing the selected strain to colonize and establish [6, 137, 138].

The viability and activity of the population inoculum is often affected by abiotic stress. Since the endophytic compartment represent a more protective niche for plant–bacteria interaction, formulations based on endophytic streptomycete strains had advantages over non-endophytic streptomycetes that would guarantee successful positive plant host response [139]. Representative streptomycetes possess the ability to colonize the rhizosphere or endophytic tissues of plants [6, 87], to produce spores resistant to irradiation, heat, and drought [12], and to have an impressive range of metabolites with a variety of biological activities [140]. *Streptomyces* may also convert plant exudates or macromolecules/supramolecules present in the rhizosphere into a form that can be used by other plant growth-promoting microbes. This proposition originated from our research group and represents a technological conversion of the fundamental concept of ecological succession of community members and metabolic cooperation. In this line of research, we combined the non-starch-degrading diazotrophic bacterium *H. seropedicae* strain HRC54 with *S. bellus* strain UENF AC 06, and verified that an increased population of the bacteria was probably sustained by C-sources from a starch degradation by-product. It is noteworthy that *Herbaspirillum* did not grow when *S. bellus* was absent in the medium and when starch was the sole C-source (unpublished results). It is another example of a simple idea that can be converted into a biotechnological product with a range of applications for sustainable crop production.

CONCLUDING REMARKS AND PERSPECTIVES

The greatest challenge for agriculture in the current century is to produce food for the increasing world population and to reduce the dependence on non-renewable resources and environmental impact. The use of plant growth-promoting microbes to enhance crop production has emerged as a sustainable and alternative tool to meet this challenge [141]. The most studied and technologically developed plant growth-promoting microbes mainly include rhizobia and other Gram-negative bacteria. However, besides their widely known ability to produce antibiotics, representatives of the Gram-positive genus *Streptomyces* have been found in beneficial associations with plants, including those of agronomic importance

[64, 142, 143, 144]. The diversity of bioactive compounds produced by *Streptomyces* spp. includes substances capable of improving plant growth; hence, they are recognized as PGPS. Increased knowledge of secondary metabolic pathways involving the production of bioactive compounds and mechanisms of plant growth promotion by *Streptomyces* will be assisted by the rapid development of functional genomics and bioinformatics over the coming years. PGPS can promote plant growth by colonizing the rhizosphere or the endophytic plant environment. However, the interactions between rhizosphere PGPS and indigenous microbiota as well as the infection process performed by endophytic PGPS are still not clearly elucidated. Metagenomic approaches and the use of molecular markers such as fluorescent proteins will certainly contribute to the study of dynamic microbial populations in the plant rhizosphere, PGPS inoculation, and streptomycetes endophytic plant colonization. In addition, the use of PGPS as commercial biofertilizers is developing, but research into the design of bioinoculant formulations (such as additives, carriers, and delivery methods) will increase the plant growth and yields and acceptance by farmers around the world.

AUTHORS' CONTRIBUTION

JAJS and FLO contributed equally to this review. Both authors read and approved the final manuscript.

COMPETING INTERESTS

The authors declare that they have no competing interests.

REFERENCES

1. Godfray HCJ, et al. Food security: the challenge of feeding 9 billion people. Science. 2010;327(5967):812–8.
2. Tilman D, et al. Global food demand and the sustainable intensification of agriculture. Proc Natl Acad Sci USA. 2011;108(50):20260–4.
3. Ahemad M, Kibret M. Mechanisms and applications of plant growth promoting rhizobacteria: current perspective. J King Saud Univ Sci. 2014;26(1):1–20.
4. Bashan Y, et al. Advances in plant growth-promoting bacterial inoculant technology: formulations and practical perspectives (1998–2013). Plant Soil. 2014;378(1–2):1–33.
5. Abbamondi GR, et al. Plant growth-promoting effects of rhizospheric and endophytic bacteria associated with different tomato cultivars and new tomato hybrids. Chem Biol Technol Agric. 2016;3(1):1–10.
6. Bonaldi M, et al. Colonization of lettuce rhizosphere and roots by tagged Streptomyces. Front Microbiol. 2015;6:25.
7. Francis I, Holsters M, Vereecke D. The Gram-positive side of plant-microbe interactions. Environ Microbiol. 2010;12(1):1–12.
8. Nebbioso A, et al. Phytochemical profiling of tomato roots following treatments with different microbial inoculants as revealed by IT-TOF mass spectrometry. Chem Biol Technol Agric. 2016;3(1):1–8.
9. Lightfield J, Fram NR, Ely B. Across bacterial phyla, distantly-related genomes with similar genomic GC content have similar patterns of amino acid usage. PLoS One. 2011;6(3):e17677.
10. Qin S, et al. Biodiversity, bioactive natural products and biotechnological potential of plant-associated endophytic actinobacteria. Appl Microbiol Biotechnol. 2011;89(3):457–73.
11. Flärdh K. Growth polarity and cell division in Streptomyces. Curr Opin Microbiol. 2003;6(6):564–71.
12. Flardh K, Buttner MJ. Streptomyces morphogenetics: dissecting differentiation in a filamentous bacterium. Nat Rev Microbiol. 2009;7(1):36–49.
13. Porter JR. Antony van Leeuwenhoek: tercentenary of his discovery of bacteria. Bacteriol Rev. 1976;40(2):260–9.
14. Hopwood DA. Streptomyces in nature and medicine: the antibiotic makers. Oxford: Oxford University Press; 2007.
15. Dworkin M, Falkow S. The prokaryotes: vol. 4: bacteria: firmicutes, cyanobacteria, vol. 4. Berlin: Springer Science & Business Media; 2006.
16. Cambau E, Drancourt M. Steps towards the discovery of Mycobacterium tuberculosis by Robert Koch, 1882. Clin Microbiol Infect. 2014;20(3):196–201.
17. Daniel TM. The history of tuberculosis. Respir Med. 2006;100(11):1862–70.
18. Buchanan RE. Studies in the nomenclature and classification of the bacteria: VIII. The subgroups and genera of the Actinomycetales. J Bacteriol. 1918;3(4):403–6.
19. Waksman SA. Streptomycin: background, isolation, properties, and utilization. Science. 1953;118(3062):259–66.
20. Kämpfer P, et al. The family Streptomycetaceae, in the prokaryotes. Berlin: Springer; 2014. p. 889–1010.
21. Chandra G, Chater KF. Developmental biology of Streptomyces from the perspective of 100 actinobacterial genome sequences. FEMS Microbiol Rev. 2014;38(3):345–79.
22. Ochi K. A taxonomic study of the genus Streptomyces by analysis of ribosomal-protein at-L30. Int J Syst Bacteriol. 1995;45(3):507–14.
23. Ludwig W, et al. Road map of the phylum Actinobacteria, in Bergey's manual® of systematic bacteriology. Berlin: Springer; 2012. p. 1–28.

24. Petrus MLC, Claessen D. Pivotal roles for Streptomyces cell surface polymers in morphological differentiation, attachment and mycelial architecture. Antonie Van Leeuwenhoek Int J Gen Mol Microbiol. 2014;106(1):127–39.
25. Pridham TG, Hesseltine CW, Benedict RG. A guide for the classification of Streptomycetes according to selected groups—placement of strains in morphological sections. Appl Microbiol. 1958;6(1):52–79.
26. Tresner HD, Davies MC, Backus EJ. Electron microscopy of Streptomyces spore morphology and its role in species differentiation. J Bacteriol. 1961;81:70–80.
27. Atalan E, et al. Biosystematic studies on novel streptomycetes from soil. Antonie Van Leeuwenhoek Int J Gen Mol Microbiol. 2000;77(4):337–53.
28. Laidi RF, et al. Taxonomy, identification and biological activities of a novel isolate of Streptomyces tendae. Arab J Biotechnol. 2006;9:427–36.
29. Schleifer KH, Kandler O. Peptidoglycan types of bacterial cell walls and their taxonomic implications. Bacteriol Rev. 1972;36(4):407–77.
30. Yamaguchi T. Comparison of the cell-wall composition of morphologically distinct actinomycetes. J Bacteriol. 1965;89:444–53.
31. Shashkov AS, et al. A polymer with a backbone of 3-deoxy-D-glycero-D-galacto-non-2-ulopyranosonic acid, a teichuronic acid, and a beta-glucosylated ribitol teichoic acid in the cell wall of plant pathogenic Streptomyces sp. VKM Ac-2124. Eur J Biochem. 2002;269(24):6020–5.
32. Tatar D, et al. Streptomyces iconiensis sp. nov. and Streptomyces smyrnaeus sp. nov., two halotolerant actinomycetes isolated from a salt lake and saltern. Int J Syst Evol Microbiol. 2014;64(Pt 9):3126–33.
33. Weidenmaier C, Peschel A. Teichoic acids and related cell-wall glycopolymers in gram-positive physiology and host interactions. Nat Rev Microbiol. 2008;6(4):276–87.
34. Acharyabhatta A, Kandula SK, Terli R. Taxonomy and polyphasic characterization of alkaline amylase producing marine actinomycete Streptomyces rochei BTSS 1001. Int J Microbiol. 2013;2013:276921.
35. Hamid M. The use of morphological and cell wall chemical markers in the identification of Streptomyces species associated with Actinomycetoma. Afr J Clin Exp Microbiol. 2013;14(2):45–50.
36. Lechevalier MP, Lechevalier H. Chemical composition as a criterion in the classification of aerobic actinomycetes. Int J Syst Bacteriol. 1970;20(4):435–43.
37. Dyson P. Streptomyces, in encyclopedia of microbiology. 3rd ed. In: Schaechter M, editor. Oxford: Academic Press; 2009. p. 318–332.
38. Hasani A, Kariminik A, Issazadeh K. Streptomycetes: characteristics and their antimicrobial activities. Int J Adv Biol Biomed Res. 2014;2(1):63–75.
39. Mistry BV, et al. FtsW is a dispensable cell division protein required for Z-ring stabilization during sporulation septation in Streptomyces coelicolor. J Bacteriol. 2008;190(16):5555–66.
40. Willemse J, et al. Positive control of cell division: FtsZ is recruited by SsgB during sporulation of Streptomyces. Genes Dev. 2011;25(1):89–99.
41. Flardh K, et al. Generation of a non-sporulating strain of Streptomyces coelicolor A3(2) by the manipulation of a developmentally controlled ftsZ promoter. Mol Microbiol. 2000;38(4):737–49.
42. Pimentel-Elardo SM, et al. Streptomyces axinellae sp. nov., isolated from the mediterranean sponge Axinella polypoides (Porifera). Int J Syst Evol Microbiol. 2009;59(Pt 6):1433–7.
43. Reponen TA, et al. Characteristics of airborne actinomycete spores. Appl Environ Microbiol. 1998;64(10):3807–12.
44. Claessen D, et al. Bacterial solutions to multicellularity: a tale of biofilms, filaments and fruiting bodies. Nat Rev Microbiol. 2014;12(2):115–24.
45. van Wezel GP, McDowall KJ. The regulation of the secondary metabolism of Streptomyces: new links and experimental advances. Nat Prod Rep. 2011;28(7):1311–33.

46. Olano C, et al. Improving production of bioactive secondary metabolites in actinomycetes by metabolic engineering. Metab Eng. 2008;10(5):281–92.
47. Bibb MJ. Regulation of secondary metabolism in streptomycetes. Curr Opin Microbiol. 2005;8(2):208–15.
48. Biarnes-Carrera M, Breitling R, Takano E. Butyrolactone signalling circuits for synthetic biology. Curr Opin Chem Biol. 2015;28:91–8.
49. Morgenstern A, et al. Divalent transition-metal-ion stress induces prodigiosin biosynthesis in Streptomyces coelicolor M145: formation of coeligiosins. Chemistry. 2015;21(16):6027–32.
50. Yekkour A, et al. A novel hydroxamic acid-containing antibiotic produced by a Saharan soil-living Streptomyces strain. Lett Appl Microbiol. 2015;60(6):589–96.
51. Rutledge PJ, Challis GL. Discovery of microbial natural products by activation of silent biosynthetic gene clusters. Nat Rev Microbiol. 2015;13(8):509–23.
52. Chakraburtty R, Bibb M. The ppGpp synthetase gene (relA) of Streptomyces coelicolor A3(2) play's a conditional role in antibiotic production and morphological differentiation. J Bacteriol. 1997;179(18):5854–61.
53. Gomez-Escribano JP, et al. Streptomyces clavuligerus relA-null mutants overproduce clavulanic acid and cephamycin C: negative regulation of secondary metabolism by (p)ppGpp. Microbiology-Sgm. 2008;154:744–55.
54. Chen C, et al. Effect of overexpression of endogenous and exogenous Streptomyces antibiotic regulatory proteins on c (FK506) production in Streptomyces sp KCCM11116P. Rsc Adv. 2015;5(21):15756–62.
55. Kurniawan YN, et al. Differential contributions of two SARP family regulatory genes to indigoidine biosynthesis in Streptomyces lavendulae FRI-5. Appl Microbiol Biotechnol. 2014;98(23):9713–21.
56. Xie C, Deng JJ, Wang HX. Identification of AstG1, A LAL family regulator that positively controls ansatrienins production in Streptomyces sp. XZQH13. Curr Microbiol. 2015;70(6):859–64.
57. Kim SB, et al. Taxonomic study of neutrotolerant acidophilic actinomycetes isolated from soil and description of Streptomyces yeochonensis sp. nov. Int J Syst Evol Microbiol. 2004;54(Pt 1):211–4.
58. Zakalyukina YV, Zenova G, Zvyagintsev D. Peculiarities of growth and morphological differentiation of acidophilic and neutrophilic soil streptomycetes. Microbiology. 2004;73(1):74–8.
59. Sultanpuram VR, Mothe T, Mohammed F. Streptomyces alkalithermotolerans sp. nov., a novel alkaliphilic and thermotolerant actinomycete isolated from a soda lake. Antonie Van Leeuwenhoek. 2015;107(2):337–44.
60. Antony-Babu S, Stach JE, Goodfellow M. Genetic and phenotypic evidence for Streptomyces griseus ecovars isolated from a beach and dune sand system. Antonie Van Leeuwenhoek. 2008;94(1):63–74.
61. Noda S, et al. 4-Vinylphenol biosynthesis from cellulose as the sole carbon source using phenolic acid decarboxylase- and tyrosine ammonia lyase-expressing Streptomyces lividans. Bioresour Technol. 2015;180:59–65.
62. Xu J, et al. Streptomyces xiamenensis sp. nov., isolated from mangrove sediment. Int J Syst Evol Microbiol. 2009;59(Pt 3):472–6.
63. Hamedi J, Mohammadipanah F. Biotechnological application and taxonomical distribution of plant growth promoting actinobacteria. J Ind Microbiol Biotechnol. 2015;42(2):157–71.
64. Jog R, et al. Mechanism of phosphate solubilization and antifungal activity of Streptomyces spp. isolated from wheat roots and rhizosphere and their application in improving plant growth. Microbiology. 2014;160(Pt 4):778–88.
65. Duncan KR, et al. Exploring the diversity and metabolic potential of actinomycetes from temperate marine sediments from Newfoundland, Canada. J Ind Microbiol Biotechnol. 2015;42(1):57–72.
66. Hartmann A, Rothballer M, Schmid M. Lorenz Hiltner, a pioneer in rhizosphere microbial ecology and soil bacteriology research. Plant Soil. 2008;312(1–2):7–14.

67. Oh YM, et al. Distinctive bacterial communities in the rhizoplane of four tropical tree species. Microb Ecol. 2012;64(4):1018–27.
68. Doornbos RF, van Loon LC, Bakker PA. Impact of root exudates and plant defense signaling on bacterial communities in the rhizosphere. A review. Agron Sustain Dev. 2012;32(1):227–43.
69. Bais HP, et al. The role of root exudates in rhizosphere iterations with plants and other organisms. Annu Rev Plant Biol. 2006;57:233–66.
70. Bever JD, Platt TG, Morton ER. Microbial population and community dynamics on plant roots and their feedbacks on plant communities. Annu Rev Microbiol. 2012;66:265–83.
71. Compant S, Clement C, Sessitsch A. Plant growth-promoting bacteria in the rhizo- and endosphere of plants: their role, colonization, mechanisms involved and prospects for utilization. Soil Biol Biochem. 2010;42(5):669–78.
72. Berg G. Plant-microbe interactions promoting plant growth and health: perspectives for controlled use of microorganisms in agriculture. Appl Microbiol Biotechnol. 2009;84(1):11–8.
73. Challis GL, Ravel J. Coelichelin, a new peptide siderophore encoded by the Streptomyces coelicolor genome: structure prediction from the sequence of its non-ribosomal peptide synthetase. FEMS Microbiol Lett. 2000;187(2):111–4.
74. Trejo-Estrada S, Paszczynski A, Crawford D. Antibiotics and enzymes produced by the biocontrol agent Streptomyces violaceusniger YCED-9. J Ind Microbiol Biotechnol. 1998;21(1–2):81–90.
75. Gherbawy Y, et al. Molecular screening of Streptomyces isolates for antifungal activity and family 19 chitinase enzymes. J Microbiol. 2012;50(3):459–68.
76. Fyans JK, et al. The ESX/type VII secretion system modulates development, but not virulence, of the plant pathogen Streptomyces scabies. Mol Plant Pathol. 2013;14(2):119–30.
77. Langlois P, et al. Identification of Streptomyces coelicolor proteins that are differentially expressed in the presence of plant material. Appl Environ Microbiol. 2003;69(4):1884–9.
78. Lin L, et al. Thaxtomin A-deficient endophytic Streptomyces sp. enhances plant disease resistance to pathogenic Streptomyces scabies. Planta. 2012;236(6):1849–61.
79. Anne J, et al. Protein secretion biotechnology in gram-positive bacteria with special emphasis on Streptomyces lividans. Biochim Biophys Acta. 2014;1843(8):1750–61.
80. Bischoff V, et al. Thaxtomin A affects CESA-complex density, expression of cell wall genes, cell wall composition, and causes ectopic lignification in Arabidopsis thaliana seedlings. J Exp Bot. 2009;60(3):955–65.
81. Joshi M, et al. Streptomyces turgidiscabies secretes a novel virulence protein, Nec1, which facilitates infection. Mol Plant Microbe Interact. 2007;20(6):599–608.
82. Loria R, et al. Thaxtomin biosynthesis: the path to plant pathogenicity in the genus Streptomyces. Antonie Van Leeuwenhoek. 2008;94(1):3–10.
83. Johnson EG, et al. Plant-pathogenic Streptomyces species produce nitric oxide synthase-derived nitric oxide in response to host signals. Chem Biol. 2008;15(1):43–50.
84. Meschke H, Schrempf H. Streptomyces lividans inhibits the proliferation of the fungus Verticillium dahliae on seeds and roots of Arabidopsis thaliana. Microb Biotechnol. 2010;3(4):428–43.
85. Palaniyandi SA, et al. Genetic and functional characterization of culturable plant-beneficial actinobacteria associated with yam rhizosphere. J Basic Microbiol. 2013;53(12):985–95.
86. Rungin S, et al. Plant growth enhancing effects by a siderophore-producing endophytic streptomycete isolated from a Thai jasmine rice plant (Oryza sativa L. cv. KDML105). Antonie Van Leeuwenhoek. 2012;102(3):463–72.
87. Coombs JT, Franco CM. Visualization of an endophytic Streptomyces species in wheat seed. Appl Environ Microbiol. 2003;69(7):4260–2.

88. Shimizu M. Endophytic actinomycetes: biocontrol agents and growth promoters, in bacteria in agrobiology: plant growth responses. Berlin: Springer; 2011. p. 201–20.
89. Clark C, Matthews SW. Histopathology of sweet potato root infection by Streptomyces ipomoea. Phytopathology. 1987;77(10):1418–23.
90. Loria R, et al. A paucity of bacterial root diseases: Streptomyces succeeds where others fail. Physiol Mol Plant Pathol. 2003;62(2):65–72.
91. Suzuki T, et al. Visualization of infection of an endophytic actinomycete Streptomyces galbus in leaves of tissue-cultured rhododendron. 日本放線菌学会誌. 2005;19(1):7–12.
92. Compant S, et al. Endophytic colonization of Vitis vinifera L. by plant growth-promoting bacterium Burkholderia sp. strain PsJN. Appl Environ Microbiol. 2005;71(4):1685–93.
93. Conn VM, Walker AR, Franco CM. Endophytic actinobacteria induce defense pathways in Arabidopsis thaliana. Mol Plant Microbe Interact. 2008;21(2):208–18.
94. Miche L, et al. Upregulation of jasmonate-inducible defense proteins and differential colonization of roots of Oryza sativa cultivars with the endophyte Azoarcus sp. Mol Plant Microbe Interact. 2006;19(5):502–11.
95. Shah J. Plants under attack: systemic signals in defence. Curr Opin Plant Biol. 2009;12(4):459–64.
96. Yasuda M, et al. Effects of colonization of a bacterial endophyte, Azospirillum sp. B510, on disease resistance in rice. Biosci Biotechnol Biochem. 2009;73(12):2595–9.
97. Cho YH, Lee EJ, Roe JH. A developmentally regulated catalase required for proper differentiation and osmoprotection of Streptomyces coelicolor. Mol Microbiol. 2000;35(1):150–60.
98. Kim FJ, et al. Differential expression of superoxide dismutases containing Ni and Fe/Zn in Streptomyces coelicolor. Eur J Biochem. 1996;241(1):178–85.
99. Michta E, et al. Proteomic approach to reveal the regulatory function of aconitase AcnA in oxidative stress response in the antibiotic producer Streptomyces viridochromogenes Tu494. PLoS One. 2014;9(2):e87905.
100. Johnston-Monje D, Raizada MN. Plant and endophyte relationships: nutrient management. Compr Biotechnol. 2011;2:713–27.
101. Balsanelli E, et al. Molecular adaptations of Herbaspirillum seropedicae during colonization of the maize rhizosphere. Environ Microbiol. 2015.
102. James EK, et al. Infection and colonization of rice seedlings by the plant growth-promoting bacterium Herbaspirillum seropedicae Z67. Mol Plant Microbe Interact. 2002;15(9):894–906.
103. Reinhold-Hurek B, et al. An endoglucanase is involved in infection of rice roots by the not-cellulose-metabolizing endophyte Azoarcus sp. strain BH72. Mol Plant Microbe Interact. 2006;19(2):181–8.
104. Saharan B, Nehra V. Plant growth promoting rhizobacteria: a critical review. Life Sci Med Res. 2011;21:1–30.
105. Devau N, et al. Soil pH controls the environmental availability of phosphorus: experimental and mechanistic modelling approaches. Appl Geochem. 2009;24(11):2163–74.
106. Rajput MS, Kumar G, Rajkumar S. Repression of oxalic acid-mediated mineral phosphate solubilization in rhizospheric isolates of Klebsiella pneumoniae by succinate. Arch Microbiol. 2013;195(2):81–8.
107. Sadeghi A, et al. Plant growth promoting activity of an auxin and siderophore producing isolate of Streptomyces under saline soil conditions. World J Microbiol Biotechnol. 2012;28(4):1503–9.
108. Hamdali H, et al. Screening for rock phosphate solubilizing Actinomycetes from Moroccan phosphate mines. Appl Soil Ecol. 2008;38(1):12–9.
109. Hamdali H, et al. Streptomyces youssoufiensis sp nov., isolated from a Moroccan phosphate mine. Int J Syst Evol Microbiol. 2011;61:1104–8.

110. Ribbe M, Gadkari D, Meyer O. N2 fixation by Streptomyces thermoautotrophicus involves a molybdenum-dinitrogenase and a manganese-superoxide oxidoreductase that couple N2 reduction to the oxidation of superoxide produced from O2 by a molybdenum-CO dehydrogenase. J Biol Chem. 1997;272(42):26627–33.
111. MacKellar D, et al. Streptomyces thermoautotrophicus does not fix nitrogen. Sci Rep. 2016;6:20086.
112. Teale WD, Paponov IA, Palme K. Auxin in action: signalling, transport and the control of plant growth and development. Nat Rev Mol Cell Biol. 2006;7(11):847–59.
113. Abd-Alla MH, El-Sayed E-SA, Rasmey A-HM. Indole-3-acetic acid (IAA) production by Streptomyces atrovirens isolated from rhizospheric soil in Egypt. J Biol Earth Sci. 2013;3(2):B182–93.
114. Manulis S, et al. Biosynthesis of indole-3-acetic acid via the indole-3-acetamide pathway in Streptomyces spp. Microbiology. 1994;140(Pt 5):1045–50.
115. Lin L, Xu XD. Indole-3-acetic acid production by endophytic Streptomyces sp. En-1 isolated from medicinal plants. Curr Microbiol. 2013;67(2):209–17.
116. Igarashi Y. Screening of novel bioactive compounds from plant-associated actinomycetes. Actinomycetologica. 2004;18(2):63–6.
117. King RW, Evans LT. Gibberellins and flowering of grasses and cereals: prizing open the lid of the "florigen" black box. Annu Rev Plant Biol. 2003;54:307–28.
118. Pharis RP, King RW. Gibberellins and reproductive development in seed plants. Ann Rev Plant Physiol Plant Mol Biol. 1985;36:517–68.
119. Rashad FM, et al. Isolation and characterization of multifunctional Streptomyces species with antimicrobial, nematicidal and phytohormone activities from marine environments in Egypt. Microbiol Res. 2015;175:34–47.
120. Qiao J-Q, et al. Stimulation of plant growth and biocontrol by Bacillus amyloliquefaciens subsp. plantarum FZB42 engineered for improved action. Chem Biol Technol Agric. 2014;1(1):1–14.
121. Golinska P, Dahm H. Antagonistic properties of Streptomyces isolated from forest soils against fungal pathogens of pine seedlings. Dendrobiology. 2013;69.
122. Kanini GS, et al. Greek indigenous streptomycetes as biocontrol agents against the soil-borne fungal plant pathogen Rhizoctonia solani. J Appl Microbiol. 2013;114(5):1468–79.
123. Ruanpanun P, et al. Nematicidal activity of fervenulin isolated from a nematicidal actinomycete, Streptomyces sp. CMU-MH021, on Meloidogyne incognita. World J Microbiol Biotechnol. 2011;27(6):1373–80.
124. Tarkka MT, et al. Draft genome sequence of Streptomyces sp. strain 150FB, a mushroom mycoparasite antagonist. Genome Announc. 2015;3(2):e01441.
125. Berg G, Smalla K. Plant species and soil type cooperatively shape the structure and function of microbial communities in the rhizosphere. FEMS Microbiol Ecol. 2009;68(1):1–13.
126. Kurth F, et al. Streptomyces-induced resistance against oak powdery mildew involves host plant responses in defense, photosynthesis, and secondary metabolism pathways. Mol Plant Microbe Interact. 2014;27(9):891–900.
127. Atkinson NJ, Urwin PE. The interaction of plant biotic and abiotic stresses: from genes to the field. J Exp Bot. 2012;63(10):3523–43.
128. El-Tarabily KA. Promotion of tomato (Lycopersicon esculentum Mill.) plant growth by rhizosphere competent 1-aminocyclopropane-1-carboxylic acid deaminase-producing Streptomycete actinomycetes. Plant Soil. 2008;308(1–2):161–74.
129. Palaniyandi SA, et al. Streptomyces sp. strain PGPA39 alleviates salt stress and promotes growth of 'Micro Tom' tomato plants. J Appl Microbiol. 2014;117(3):766–73.
130. Schrey SD, et al. Production of fungal and bacterial growth modulating secondary metabolites is widespread among mycorrhiza-associated streptomycetes. BMC Microbiol. 2012;12:164.
131. Bonfante P, Anca IA. Plants, mycorrhizal fungi, and bacteria: a network of interactions. Annu Rev Microbiol. 2009;63:363–83.

132. Schrey SD, Tarkka MT. Friends and foes: streptomycetes as modulators of plant disease and symbiosis. Antonie Van Leeuwenhoek Int J Gen Mol Microbiol. 2008;94(1):11–9.
133. Tokala RK, et al. Novel plant-microbe rhizosphere interaction involving Streptomyces lydicus WYEC108 and the pea plant (Pisum sativum). Appl Environ Microbiol. 2002;68(5):2161–71.
134. Maier A, et al. Actinomycetales bacteria from a spruce stand: characterization and effects on growth of root symbiotic and plant parasitic soil fungi in dual culture. Mycol Prog. 2004;3(2):129–36.
135. Schrey SD, et al. Mycorrhiza helper bacterium Streptomyces AcH 505 induces differential gene expression in the ectomycorrhizal fungus Amanita muscaria. New Phytol. 2005;168(1):205–16.
136. Figueiredo MDVB, et al. Plant growth promoting rhizobacteria: fundamentals and applications, in plant growth and health promoting bacteria. Berlin: Springer; 2011. p. 21–43.
137. Gopalakrishnan S, et al. Plant growth-promoting traits of Streptomyces with biocontrol potential isolated from herbal vermicompost. Biocontrol Sci Tech. 2012;22(10):1199–210.
138. Gopalakrishnan S, et al. Evaluation of Streptomyces strains isolated from herbal vermicompost for their plant growth-promotion traits in rice. Microbiol Res. 2014;169(1):40–8.
139. Kinkel LL, et al. Streptomyces competition and co-evolution in relation to plant disease suppression. Res Microbiol. 2012;163(8):490–9.
140. Chater KF. Streptomyces inside-out: a new perspective on the bacteria that provide us with antibiotics. Philos Trans R Soc Lond B Biol Sci. 2006;361(1469):761–8.
141. Canellas LP, Olivares FL. Physiological responses to humic substances as plant growth promoter. Chem Biol Technol Agric. 2014;1(1):1–11.
142. Araújo JMD, Silva ACD, Azevedo JL. Isolation of endophytic actinomycetes from roots and leaves of maize (Zea mays L.). Braz Arch Biol Technol. 2000;43(4):0–0.
143. Hu D, et al. Genome sequence of Streptomyces sp. strain TOR3209, a rhizosphere microecology regulator isolated from tomato rhizosphere. J Bacteriol. 2012;194(6):1627.
144. Kuramata M, et al. Arsenic biotransformation by Streptomyces sp. isolated from rice rhizosphere. Environ Microbiol. 2015;17(6):1897–909.
145. Bentley SD, et al. Complete genome sequence of the model actinomycete Streptomyces coelicolor A3(2). Nature. 2002;417(6885):141–7.
146. Phongsopitanun W, et al. Streptomyces chumphonensis sp. nov., isolated from marine sediments. Int J Syst Evol Microbiol. 2014;64(Pt 8):2605–10.
147. Srivastava S, et al. Streptomyces rochei SM3 induces stress tolerance in Chickpea against Sclerotinia sclerotiorum and NaCl. J Phytopathol. 2014;163(7–8):583–92.
148. Li J, et al. Streptomyces fildesensis sp. nov., a novel streptomycete isolated from Antarctic soil. Antonie Van Leeuwenhoek. 2011;100(4):537–43.
149. Lambert D, Loria R. Streptomyces scabies sp. nov., nom. rev. Int J Syst Bacteriol. 1989;39(4):387–92.
150. Mingma R, et al. Streptomyces oryzae sp. nov., an endophytic actinomycete isolated from stems of rice plant. J Antibiot (Tokyo). 2015;68(6):368–72.
151. Oliveira LG, et al. Genome sequence of Streptomyces wadayamensis strain A23, an endophytic actinobacterium from Citrus reticulata. Genome Announc. 2014;2(4):e00625.
152. Sarmin NI, et al. Streptomyces kebangsaanensis sp. nov., an endophytic actinomycete isolated from an ethnomedicinal plant, which produces phenazine-1-carboxylic acid. Int J Syst Evol Microbiol. 2013;63(Pt 10):3733–8.
153. Bian GK, et al. Streptomyces phytohabitans sp. nov., a novel endophytic actinomycete isolated from medicinal plant Curcuma phaeocaulis. Antonie Van Leeuwenhoek. 2012;102(2):289–96.

154. Aguiar KP. Prospecção de bactérias promotoras do crescimento vegetalassociadas a vermicompostos., in CCTA. Campos dos Goytacazes: Universidade Estadual do Norte Fluminense Darcy Ribeiro; 2012. p. 101.
155. Jesus JA. Potencial Biotecnológico de actinobactérias e bactérias diazotróficas para o crescimento de plantas, in CCTA. Campos dos Goytacazes: Universidade Estadual do Norte Fluminense Darcy Ribeiro; 2013. p. 107.
156. Shao L, et al. Identification of the Herboxidiene biosynthetic gene cluster in Streptomyces chromofuscus ATCC 49982. Appl Environ Microbiol. 2012;78(6):2034–8.
157. Hindra, Pak P, Elliot MA. Regulation of a novel gene cluster involved in secondary metabolite production in Streptomyces coelicolor. J Bacteriol. 2010;192(19):4973–82.
158. Mansouri K, Piepersberg W. Genetics of Streptomycin production in Streptomyces-griseus—nucleotide-sequence of 5 genes, Strfghik, including a phosphatase gene. Mol Gen Genet. 1991;228(3):459–69.
159. Rascher A, et al. Insights into the biosynthesis of the benzoquinone ansamycins geldanamycin and herbimycin, obtained by gene sequencing and disruption. Appl Environ Microbiol. 2005;71(8):4862–71.
160. Wu K, et al. The FK520 gene cluster of Streptomyces hygroscopicus var. ascomyceticus (ATCC 14891) contains genes for biosynthesis of unusual polyketide extender units. Gene. 2000;251(1):81–90.
161. He J, Hertweck C. Iteration as programmed event during polyketide assembly; molecular analysis of the aureothin biosynthesis gene cluster. Chem Biol. 2003;10(12):1225–32.
162. Fernandez-Martinez LT, et al. New insights into chloramphenicol biosynthesis in Streptomyces venezuelae ATCC 10712. Antimicrob Agents Chemother. 2014;58(12):7441–50.
163. He M, et al. Isolation and characterization of meridamycin biosynthetic gene cluster from Streptomyces sp. NRRL 30748. Gene. 2006;377:109–18.
164. Schwartz D, et al. Biosynthetic gene cluster of the herbicide phosphinothricin tripeptide from Streptomyces viridochromogenes Tu494. Appl Environ Microbiol. 2004;70(12):7093–102.
165. Patzer SI, Braun V. Gene cluster involved in the biosynthesis of griseobactin, a catechol-peptide siderophore of Streptomyces sp. ATCC 700974. J Bacteriol. 2010;192(2):426–35.
166. Legault GS, et al. Tryptophan regulates thaxtomin A and indole-3-acetic acid production in Streptomyces scabiei and modifies its interactions with radish seedlings. Phytopathology. 2011;101(9):1045–51.
167. Kieser T, et al. Practical Streptomyces genetics. Norwich: The John Innes Foundation; 2000.

CHAPTER 22

Key Knowledge and Data Gaps in Modelling The Influence of CO_2 Concentration on The Terrestrial Carbon Sink

T.A.M. Pugh[a,b,*], C. Müller [c], A. Arneth[b], V. Haverd[d], B. Smith[e]

[a] *School of Geography, Earth & Environmental Sciences and Birmingham Institute of Forest Research, University of Birmingham, Birmingham, B15 2TT, United Kingdom*

[b] *Karlsruhe Institute of Technology, Institute of Meteorology and Climate Research-Atmospheric Environmental Research (IMK-IFU), Kreuzeckbahnstraße 19, 82467 Garmisch-Partenkirchen, Germany*

[c] *Potsdam Institute for Climate Impact Research, Potsdam, Germany*

[d] *CSIRO Oceans and Atmosphere, P.O. Box 3023, Canberra ACT 2601, Australia*

[e] *Department of Physical Geography and Ecosystem Science, Lund University, SE-223 62 Lund, Sweden*

ABSTRACT

Primary productivity of terrestrial vegetation is expected to increase under the influence of increasing atmospheric carbon dioxide concentrations ([CO_2]). Depending on the fate of such additionally fixed carbon, this could lead to an increase in terrestrial carbon storage, and thus a net terrestrial sink of atmospheric carbon. Such a mechanism is generally believed to be the primary global driver behind the observed large net uptake of anthropogenic CO_2 emissions by the biosphere. Mechanisms driving CO_2 uptake in the Terrestrial Biosphere Models (TBMs) used to attribute and project terrestrial carbon sinks, including that from increased [CO_2], remain in large parts unchanged since those models were conceived two decades ago. However, there exists a large body of new data and understanding providing an opportunity to update these models, and directing towards important topics for further research. In this review we highlight recent developments in understanding of the effects of elevated [CO_2] on photosynthesis, and in particular on the fate of additionally fixed carbon within the plant with its implications for carbon turnover rates, on the regulation of photosynthesis in response to environmental limitations on in-plant carbon sinks, and on emergent ecosystem responses. We recommend possible avenues for model improvement and identify requirements for better data on core processes relevant to the understanding and modelling of the effect of increasing [CO_2] on the global terrestrial carbon sink.

1. INTRODUCTION

The capacity of terrestrial ecosystems to act as a net source or sink of carbon over decadal or centennial timescales is one of the largest uncertainties in estimates of the present and future global carbon budget, and its role in the radiative forcing of climate (Le Quéré et al., 2014, 2009; Friedlingstein et al., 2014). It is by now without dispute that a substantial amount of anthropogenic CO_2 emissions are removed from the atmosphere annually due to a net global CO_2 uptake by terrestrial ecosystems. A combination of global constraints based on mass balance (Le Quéré et al., 2014), atmospheric CO_2 inversions (Stephens et al., 2007), forest inventories (Pan et al., 2011) and process-based terrestrial carbon cycle models (Sitch et al., 2015) suggests that increases in terrestrial carbon storage resulting from the effects of CO_2fertilisation, here defined as enhancements in photosynthesis under elevated atmospheric CO_2 concentration ([CO_2]), may explain as much as 60% of the contemporary terrestrial carbon sink (Schimel et al., 2015). However, the contributions of different regional ecosystems to the global sink (Ahlström et al., 2015), and the relative importance of different mechanisms and historical factors in driving it, are still being debated; particularly the roles of CO_2 fertilisation, high-latitude climate warming and secondary forest regrowth (e.g. Smith et al., 2016). Estimates may be confounded by uncertainties in land-use change emissions (Pugh et al., 2015). In the absence of consensus on the origins and behaviour of the present-day carbon sink, future projections remain highly tenuous, often showing striking divergence among alternative models (Friedlingstein et al., 2006; Friedlingstein et al., 2014; Sitch et al., 2008; Friend et al., 2014; Jones et al., 2013). The simulated increases in vegetation biomass over the 21st century may be high, typically amounting to 100–400 Pg C for scenarios with strong

$[CO_2]$ (and therefore warming) increases (Friend et al., 2014), in the mean equivalent to about a 50% increase in global vegetation carbon pools relative to the present day. However, there has been a great deal of discussion over whether such projections are realistic, or whether the models used miss critical plant and ecosystem processes (e.g. Hungate et al., 2003; Körner, 2009; Fatichi et al., 2014; Hickler et al., 2015; Bugmann and Bigler 2011).

Calculations of the current terrestrial carbon sink, including the effects of elevated $[CO_2]$, along with projections of the carbon cycle in the late 21st century and beyond, are available from a variety of bottom-up modelling approaches that we here collectively term Terrestrial Biosphere Models (TBMs). These models include biogeochemical Land Surface Models (LSMs) and Dynamic Global Vegetation Models (DGVMs) configured for 'off-line' simulations, as well as interactive versions of the same models coupled to the physical climate component of Earth System Models (ESMs). The models rely on bottom-up principles whereby component processes affecting carbon fluxes into, out of, and between ecosystem compartments are represented explicitly and linked via dependencies on external drivers and evolving system state. This bottom-up approach is argued to be necessary because carbon cycle dynamics emerge from the joint behaviour of a multitude of processes linked across widely-ranging time (and space) scales, leading to legacy effects with time-signatures from seasons to centuries. There is thus a strong imperative to ensure these models build on process representations that reflect observational constraints, and formalise biological and ecological understanding.

The expectation of the existence of a CO_2 fertilisation effect stems from a mechanistic understanding of photosynthesis, whereby elevated $[CO_2]$ leads to increased concentrations of CO_2 within the leaf, thereby enhancing the carboxylation rate of Rubisco and reducing photorespiration, and thus increasing gross primary production (Drake et al., 1997). An indirect effect of $[CO_2]$ on photosynthesis is also widely observed, whereby stomatal conductance is reduced, reducing water loss for a given assimilation rate (Ainsworth and Rogers 2007). The resulting reduction in water stress may be particularly important in dry-climate ecosystems. These direct and indirect responses of photosynthesis have been well established in laboratory and open-top chamber experiments, where they have also been found to translate into enhanced plant growth (e.g. Drake et al., 1997; Norby et al., 1999). This increase in growth provides the basis for hypothesising the potential existence of an on-going net flux, or "sink", of carbon to the terrestrial biosphere resulting from CO_2 fertilisation, contingent on the assumption that downstream plant and ecosystem processes also affected indirectly by elevated $[CO_2]$ do not result in a compensating increased flux of carbon *out* of the system, for example due to increased plant mortality, soil respiration or carbon leaching. This sink would continue to exist until carbon outputs through vegetation and soil turnover came into balance with the increased inputs, a process that, even if changes in environmental conditions are halted, could take many decades or even centuries (Port et al., 2012).

The existence and strength of a "CO_2-caused" sink depends on a number of key factors. Firstly, are large enhancements of photosynthesis seen in early experiments in laboratories or open-top chambers realised in real world ecosystems across the globe? Second, to what processes or pools do plants allocate any additional carbon resulting from CO_2 fertilisation? Third, how long does carbon remain in those pools before they are turned over, either through tissue death or whole-plant mortality? Finally, what is the emergent ecosystem response arising as a sum of these effects, including the fate of turned over carbon? Whether or not CO_2 fertilisation results in a net terrestrial carbon sink therefore goes well beyond the question of whether or not primary production increases.

In this article, we assess the potential to improve TBM-based simulations of the contemporary and future terrestrial biospheric carbon cycle by reviewing how processes relating to vegetation growth and turnover under increasing [CO_2] are represented in these models in comparison to current knowledge. We restrict our scope to plant and vegetation processes, while recognising that soil carbon turnover is also critically important for biosphere carbon balance. Much of the focus is upon trees, as these reflect the major stores of vegetation biomass. In contrast to other reviews which have focused on synthesising results of ecosystem experiments (Norby et al., 1999; Ainsworth and Long 2005; Ainsworth and Rogers 2007; Leakey et al., 2009; Norby and Zak 2011), on comparisons of models to particular ecosystem experiments (Medlyn et al., 2015), or on modelling forest productivity (Hickler et al., 2015), we focus on the model development and data requirements in order to improve understanding and projections of the effect of increasing [CO_2] on the terrestrial vegetation carbon sink in TBMs.

We begin with an overview of relevant process representations in current TBMs (§2). We then discuss opportunities to improve TBM representations of processes related to the effects of elevated [CO_2] on photosynthesis (§3), allocation and turnover (§4), feedbacks of in-plant sink limitations on photosynthesis (§5), and direct effects of [CO_2] on turnover (§6), as well as indirect [CO_2]-induced feedbacks which influence the terrestrial carbon sink (§7). The core of the review focuses on processes relating to plant allocation of photosynthate, and its resultant effects on carbon turnover, which we contend to be the key area of uncertainty in understanding the effect of elevated [CO_2] on the terrestrial carbon sink. We define allocation here in its broadest sense, to encompass the division of carbon to all processes downstream of photosynthesis, including respiration and symbiotic associations, as well as the plant structural growth and allocation to non-structural carbohydrates. We limit the scope of our review to the effects of [CO_2] on the net flux of carbon to vegetation (the vegetation carbon sink) and resultant changes in vegetation carbon storage.

2. STATE OF THE ART IN GLOBAL TERRESTRIAL BIOSPHERE MODELLING

Fig. 1 shows an overview of carbon stocks and flows, and the governing processes, within a typical state-of-the-art TBM. Representations of photosynthesis in TBMs are rooted in data from controlled leaf-level experiments, often embodied in the Farquhar model (Farquhar et al., 1980). Initial enhancements in net primary productivity (NPP) in these models for a step-change increase in [CO_2] have been found to be broadly consistent with those observed in stand-scale Free-Air CO_2 Enrichment (FACE) experiments (Hickler et al., 2008; Piao et al., 2013; Zaehle et al., 2014), although these observed enhancements were not always maintained over multiple years, as we discuss further in §4. Fixed carbon not used for respiration (i.e. NPP) is assumed to be used for plant growth, and allocated to carbon compartments of the plant, typically fine roots, wood or leaves. Depending on the TBM, this allocation usually follows either fixed empirical coefficients, or functional balance rules whereby resource shortages (light, water, nutrients) lead to allocation shifts in an attempt to alleviate that shortage. Of these parsimonious approaches, allocation by functional balance has proven successful in capturing the nature of allocation shifts under elevated [CO_2] (De Kauwe et al., 2014). Photosynthesis is closely coupled to the water cycle through stomatal conductance and resultant effects on the leaf-internal CO_2 concentration (C_i), allowing indirect effects of elevated [CO_2] through increased water use efficiency to be simulated. Increasingly, TBMs are also explicitly representing nitrogen (N) dynamics in vegetation and soil, with photosynthesis being limited when nitrogen is in short supply, for instance through a limitation of the carboxylation capacity of Rubisco (V_{max}) (Yang et al., 2009; Zaehle and Friend 2010; Wang et al., 2010; Smith et al., 2014). Turnover of carbon and nitrogen stocks within vegetation is represented by a variety of processes relating to phenology and mortality, with the turned-over material being transferred to litter and soil stocks, from where they are eventually transferred to the atmosphere in the case of carbon, or partially leached or recycled into vegetation in the case of nitrogen. The rates and outcomes of these various processes relating to carbon and nitrogen uptake, allocation and turnover are closely intertwined with properties relating to ecosystem composition and canopy structure. The DGVM sub-group of TBMs also explicitly represent competition between different plant types, allowing them to emulate the functional consequences of changes in species composition brought about by changing environmental drivers.

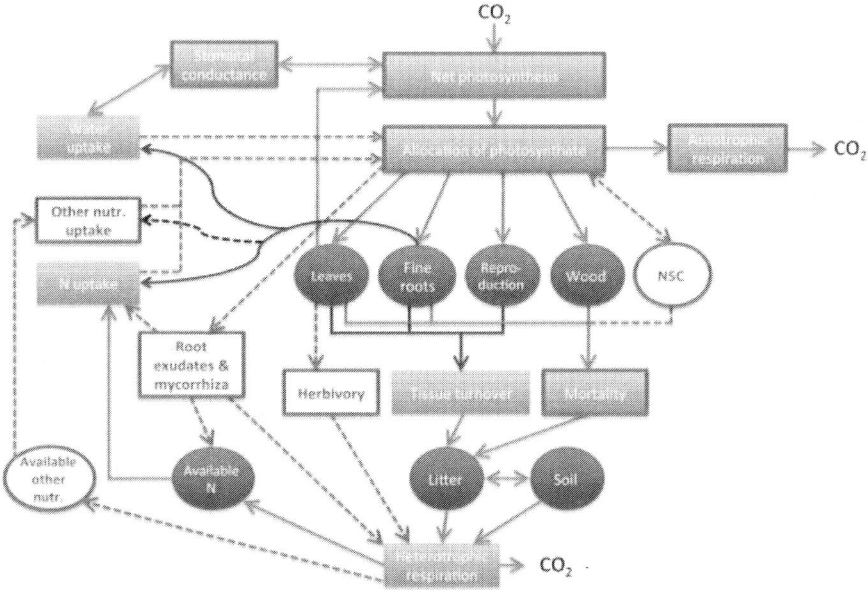

Figure 1. Schematic of stocks (ellipses) of carbon and nutrients in a typical TBM, and the processes which link them (rectangles). Blue shading indicates processes and pools typically represented, red borders those we argue are in need of update for the consideration of the effects of elevated [CO_2] on carbon cycling, and red borders and red text those we believe should be included but are generally missing currently. Shades of blue and red are only for clarity. Note that processes are shown from a TBM structural perspective, as opposed to an ecological perspective. Diagram is only indicative; process representations vary substantially between TBMs, and for clarity not all processes can be shown (for instance, vegetation dynamics are encapsulated within mortality, nutrient sources are excluded).

TBMs have proved successful in capturing important aspects of the global carbon cycle: modelled distributions and magnitudes of gross primary productivity (GPP) and NPP correspond with those derived from satellites and flux-tower observations (e.g. Piao et al., 2013; Anav et al., 2013), and models have been able to recreate the season cycle of [CO_2] in both hemispheres (Sitch et al., 2003; Forkel et al., 2016) as well as important aspects of the pattern of inter-annual variability in [CO_2] (Sitch et al., 2008). Nonetheless, the nature of the terrestrial carbon flux, as a relatively small residual of two very large fluxes (GPP and ecosystem respiration), means that even relatively small inaccuracies in these larger fluxes will cause very large uncertainties in the net terrestrial flux of carbon. This is reflected in the divergent trajectories of terrestrial carbon balance simulated by different models (Ciais et al., 2013; Sitch et al., 2015; Le Quéré et al., 2014).

3. ADVANCES IN UNDERSTANDING OF PHOTOSYNTHESIS

Some recent empirical evidence has suggested that the effect of elevated [CO_2] on photosynthesis could be underestimated by current modelling approaches. Sun et al. (2014) suggest that C_3 plants are currently much more strongly CO_2-limited than assumed in the photosynthesis algorithms used by TBMs, since these algorithms do not include the leaf mesophyll resistance, associated with diffusion of CO_2 from the sub-stomatal cavity to sites of carboxylation. As a consequence, the CO_2concentration in the models that is "seen" by the carbon-fixing enzymes under current atmospheric CO_2 levels is overestimated. As pointed out by Sun et al. (2014), this overestimation of CO_2 concentration in the chloroplast translates into an underestimation of the sensitivity of Rubisco to increasing [CO_2] since, at the same levels of stomatal conductance, temperature and photosynthetically active radiation, the carboxylation reaction by Rubisco is more sensitive to changes in CO_2 at low CO_2concentrations. When accounting for mesophyll resistance in an empirical approach within the TBM CLM4.5, Sun et al. simulated an enhanced effect of [CO_2] on global GPP of 0.05 Pg C a^{-1} $ppmv^{-1}$, compared to the standard formulation. This amounted to an additional 16% increase in global GPP over the period 1901–2010 in their simulations.

A second line of evidence also appearing to suggest a steeper response of GPP to [CO_2] than commonly assumed by models stems from the analysis of isotopomers (i.e. differences in stable isotope abundance at specific positions within a molecule) of deuterium in glucose and sucrose, which give an indication of the ratio of photorespiration (i.e. linked to the oxygenation reaction catalyzed by Rubisco) to carboxylation under different levels of [CO_2]. Such data assembled by Ehlers et al. (2015) from herbarium samples of different plant types collected between the late 19th and early 21st century indicate that this ratio decreased by ca. 25% since the late 19th century, demonstrating increased carboxylation over oxygenation in response to increasing [CO_2]. Using a simple model Ehlers et al. showed that this would result in a 35% increase in GPP, with all else being constant. This is substantially more than the 6–20% increases calculated by an ensemble of DGVMs over a similar period (Piao et al., 2013), or the 18% enhancement calculated for the TBM LPJ-GUESS following the same protocol but without changes in climate (authors, unpublished). Although the TBM simulations also integrate over other factors such as changes in temperature and leaf area index, these measurements appear to show some consistency with the results of Sun et al. (2014) and taken together appear to indicate an underestimation of GPP enhancements over the last century by global models. However, it should be cautioned that the herbarium samples used by Ehlers et al. (2015) are leaf, rather than canopy-scale, measurements and, further, are not necessarily representative of natural ecosystems globally, whilst Sun et al. (2014) rely on TBMs for upscaling the effects of their findings, meaning that the proposed effects on global GPP are therefore subject to the limitations in other aspects of these models discussed below. As we expand on in the following sections, such effects on GPP may only be realised in a small subset of plants or ecosystems.

4. ARE EFFECTS ON GROWTH AND CARBON STORAGE OVERESTIMATED?

Whilst understanding the presence (or absence), magnitude and continuity of the response of photosynthesis in (C_3) vegetation to [CO_2] is an important first step, from the perspective of ecosystem-level carbon balance a more fundamental question is whether enhanced leaf-level photosynthesis translates into enhanced growth, and whether this enhanced growth results in a net increase in carbon storage within the ecosystem. Correct allocation of plant carbon to different vegetation compartments and losses in the form of, e.g., respiration, exudates or volatile organic compounds (VOCs) represents a fundamental challenge for modellers. A wide range of empirical, allometric or evolution-based formulations exists (see Franklin et al., 2012 for a review), although these are usually applied only for allocation between plant compartments. Evolution-based approaches, which typically allocate carbon to maximise a fitness proxy, are shown to give realistic relative responses to environmental variability and climate change, but are included only to a limited extent in some large-scale models (e.g. Haverd et al., 2016). Empirical and allometric alternatives are the common choice in many models (§2). Likewise, allocation of carbon to purposes other than autotrophic respiration and growth, such as exudates or biogenic VOC emissions, are neglected in most TBMs. In this section we discuss allocation of carbon to growth (§4.1), then partitioning of that carbon within different growth components (§4.2) and then other carbon losses (§4.3).

4.1. Growth limitations

As noted above, most TBMs assume that, after subtraction of respiration costs from GPP, all remaining carbon is used for growth. This means that increases in GPP automatically translate into increases in growth unless compensated by a corresponding increase in respiration. Some authors argue that direct environmental limitations on growth may be more important than the availability of carbon substrate from photosynthesis, i.e. the strength of in-plant carbon sinks exerts a dominant control on carbon allocation and plant growth (e.g. Körner, 2003; Fatichi et al., 2014; Körner, 2015; Tardieu et al., 2011). These authors argue that, because environmental limitations typically affect growth processes before photosynthesis, TBMs currently overestimate increases in plant growth under elevated [CO_2]. Three sources of environmental limitation on growth have been debated most widely: water stress, low temperatures, and nutrient limitation.

Considerable evidence exists that low water potentials inhibit cell division due to low cell turgor (Muller et al., 2011; Tardieu et al., 2011) and, importantly, that the shape of this response differs from that for reduction of photosynthesis under low water potentials, such that growth in leaves and shoots typically becomes limited at higher water potentials than photosynthesis. There is also evidence of a decoupling between carbon availability and root growth under water stress (Muller et al., 2011; Fatichi et al., 2014). Furthermore, the capability of the plant to transport sugars may be reduced (Sevanto 2014), starving parts of the plant distal to the leaves of substrate for growth, although such responses are equivocal (Hartmann et al., 2015). Such a disconnect between growth and photosynthesis has been demonstrated for Mediterranean forests (Gea-Izquierdo et al., 2015), and is consistent with build-up of leaf sugars observed under water stress (Körner, 2003; Muller et al., 2011). It may help explain the relatively poor performance of ecosystems models at simulating net carbon exchange at water-limited sites (Morales et al., 2005). Fatichi et al. (2014) used a model analysis to demonstrate that implementing water limitations on growth might reduce accumulation of biomass under elevated [CO_2] by ca. 75% in a Swiss deciduous forest, although it remains to be proven whether this approach improves model performance in comparison to independent, e.g. experimental, observations of cause and effect. Using output from the TBM LPJ-GUESS, we find that a large fraction of global forests are located in areas that spend a substantial portion of the average year in water stress (Fig. 2). Based on output from this model, we therefore calculate that water limitations have the potential to influence growth conditions across a forested area which corresponds to 37% of the simulated global annual terrestrial C sink. Therefore if such limitations on growth are substantial and sustained, they could markedly reduce modelled estimates of the global C sink resulting from elevated [CO_2]. This effect may, however, be offset by reductions in water stress resulting from increased water use efficiency under elevated [CO_2] (Ainsworth and Rogers 2007; Leakey et al., 2009), although this is complicated by the unknown size of hydrological feedbacks with the atmosphere (Leuzinger et al., 2015). TBMs generally capture observed increases in water use efficiency under elevated [CO_2] (although see §5) (e.g. Medlyn et al., 2015); however, when it comes to growth limitations, data on tissue growth rates as a function of water stress are only available for a small selection of plants, but show substantial variation between species (Muller et al., 2011). Many more observations would be required to parameterise a model for the major global biomes. Therefore, the gathering of appropriate observations, and the testing of models including water-based growth limitations against ecosystem experiments under elevated [CO_2], is a pre-requisite for model advancement.

The rate of cell division is also a function of temperature, which can effectively halt growth both at very high and very low temperatures (Körner, 2003). Leuzinger et al. (2013) have shown that a sink-limited implementation of plant growth improves modelled biomass estimates for near-treeline boreal and alpine areas. However, sink limitations due to temperature are most likely to occur at the extremities of the growing season, when assimilation is in any case relatively low, and along biome boundaries determined by high and low temperature limits. Consideration of direct temperature limitation on growth may, for example, prove useful in projecting the advance of treelines under climate change, but is probably of limited relevance with regard to the effect of [CO_2] on the terrestrial carbon sink at the global scale.

Nutrient constraints under elevated [CO_2], underpinned by clear stoichiometric requirements for tissue formation (Elser et al., 2010), have been much discussed, with several notable reviews (Hungate et al., 2003; De Graaff et al., 2006; Reich et al., 2006a; Arneth et al., 2010a; Zaehle and Dalmonech 2011; Lenka and Lal 2012; Ciais et al., 2013; Medlyn et al., 2015). Although there can be some limited flexibility of carbon to nutrient

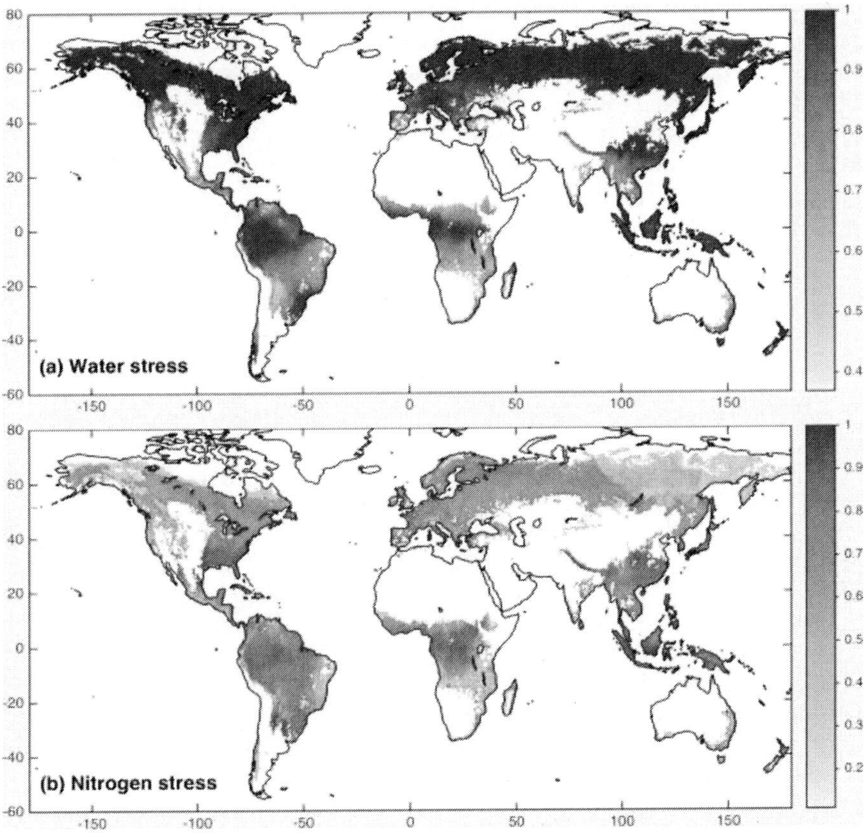

Figure 2. Estimated geographical distributions of limitations on growth in forested areas. (a) Fraction of time period 2001–2010 during which terrestrial vegetation is under water stress, as identified by a water scaling factor which varies between 0 and 1 according to the ratio of available water supply versus atmospheric demand, as modulated by plant stomata. A value of 1 indicates no water stress and all areas with a value of 0.95 or less are assumed to be substantially affected by water stress for at least part of the year. Output is from a simulation with the LPJ-GUESS dynamic global vegetation model carried out by Ahlström et al. (2012) using potential natural vegetation and forced using bias-corrected historical climate from the CCSM4 simulation carried out for the CMIP5 model intercomparison (Ciais et al., 2013). (b) Scaling factor from LPJ-GUESS for canopy carboxylation capacity (V_{max}) below its maximum rates (as determined by available light, temperature and leaf-internal CO_2 concentration) when available nitrogen is not sufficient to meet the V_{max} demand. A value of 1 indicates no nitrogen stress; lower values increasing nitrogen stress. Simulation as described in Smith et al. (2014) and results are shown for 1996–2005.

ratios within plants (e.g. Leakey et al., 2009), plant growth requires sufficient nutrient supply to sustain these ratios; Hungate et al. (2003) showed that some early TBM projections of terrestrial carbon uptake under elevated [CO_2] would require matching amounts of N far beyond reasonable projections of N availability. Experiments have shown that plants which were artificially fertilised with N showed much greater responses of plant biomass to elevated [CO_2] than those growing under native conditions (Oren et al., 2001; De Graaff et al., 2006; Reich et al., 2006a; McCarthy et al., 2010; Norby et al., 2010; Reich and Hobbie 2013), indicating a clear N limitation effect.

Diagnostic modelling of the assumed process by which elevated [CO_2] leads to a decline in the N concentration of plant litter, causing the microbial decomposers to immobilise nitrogen, and thereby reducing availability to the plants, suggests a limitation effect which progressively develops following an initial GPP and growth enhancement (Luo et al., 2004). The progressive nitrogen limitation hypothesis is supported by observations in FACE experiments in grassland (Reich et al., 2006b) and in an 11 year long FACE experiment in a plantation forest (ORNL; Norby et al., 2010). Initial but unsustained increases in annual biomass increment were also observed in a Swiss FACE experiment in mature trees (Körner et al., 2005), but in the only other closed-canopy forest FACE experiment (Duke) biomass increment remained elevated throughout the experiment (McCarthy et al., 2010). It is not clear whether the biomass increment at Duke would have persisted if the experiment had continued beyond the 10 years of elevated [CO_2] treatment (Norby and Zak 2011). We emphasise here the results for closed-canopy forests as the availability of nutrients can be strongly affected by the structure of the ecosystem (Körner, 2006; Norby and Zak 2011). In young ecosystems where canopy closure has not been achieved, individual plants have room to expand and access new resources. This is not the case in more mature ecosystems where canopy closure has been attained (Norby and Zak 2011). Large biomass increments in young trees (Norby et al., 1999) have relevance only for re-growing forests before they reach canopy closure (Körner, 2006), and do not give a clear picture of nutrient limitations likely to be present across wider ecosystems. Such differences in plant response clearly emphasise the importance, when parameterising models based on experimental findings, of considering ecosystem-level, rather than plant-level, responses.

Several TBMs have incorporated nitrogen limitation of productivity (and thus in these models, at least implicitly of growth), and results generally show N limitation, and a resultant decreased effect of elevated [CO_2] on modelled carbon sequestration, especially in the temperate and boreal zones (Fig. 2b; Zaehle and Friend 2010; Smith et al., 2014). Comparison of N-enabled TBMs against detailed ecosystem N measurements from two elevated CO_2 experiments in temperate forests suggested deficiencies in model ability to capture underlying processes at the stand scale (Zaehle et al., 2014). While Zaehle et al. (2014) were able to refute a number of common model assumptions, such as a high degree of stoichiometric flexibility in the allocation of C and N to new tissue, the experimental data were insufficient to constrain some assumptions, such as turnover rates of woody tissue and N losses through exudation (Medlyn et al., 2015). Further, TBMs lack constraint of the future ecosystem N supply (Smith et al., 2014), which requires mechanistic formulations of biological nitrogen fixation that can be applied at the global scale, as e.g. proposed by Fisher et al. (2010b). Also lacking are realistic large-scale parameterisations of N losses through leaching and gaseous emissions, which in forest ecosystems are strongly dependent on variations in soil water status, pH, texture and freeze-thaw processes, as well as microbial properties (Wu et al., 2010; Arneth et al., 2010b). So whilst progress has been made in modelling N limitations in TBMs in recent years, much work remains to be done to ensure that the models correctly simulate emergent ecosystem responses to [CO_2] and climate change.

Beyond nitrogen, other nutrients such as phosphorus and potassium may be very important in tropical regions (e.g. Lloyd et al., 2015; Sardans and Peñuelas 2015), where the relative effects of [CO_2] on photosynthesis are expected to be largest (Hickler et al., 2008), but phosphorus is considered in only a handful of TBMs (Wang et al., 2010; Goll et al., 2012; Yang et al., 2014), whilst to our knowledge potassium is explicitly considered in none. A recent study applying observed stoichiometric constraints to ESM simulations used in the IPCC 5th Assessment Report (Ciais et al., 2013) suggested that consideration of N and P availability would reduce global NPP by ca. 25% in 2100 (Wieder et al., 2015). In contrast to changes in nitrogen, for which limitation at ecosystem level is ultimately determined by the fixation rate of an abundantly available element, the finite plant-availability of other nutrients such as phosphorus imposes fundamental limitations on growth in ecosystems where they are in short supply. There is also evidence that limitations in non-nitrogen nutrients such as phosphorus and molybdenum may prevent enhancements of biological N fixation (van Groenigen et al., 2006). In contrast to water and temperature limitations, it is not yet clear whether GPP and growth are differentially affected by nutrient constraints, although stoichiometric arguments and observations of photosynthetic down-regulation mechanisms imply that growth is the controlling factor (§5).

Larger-scale support for strong limitations on stem growth comes from dendrological datasets which suggest that there has been no overall increase in stem growth in trees across many major biomes over the last century, despite a ca. 100 ppmv increase in atmospheric CO_2 mixing ratio

(Penuelas et al., 2011; Battipaglia et al., 2013; van der Sleen et al., 2015). However, these results stand in contrast to those of Cole et al. (2010) and McMahon et al. (2010) who found, based on forest inventory data, that forest growth rates in parts of the United States are increasing in excess of rates predicted from stand-age effects alone. The discrepancies between these two sources of data cannot easily be reconciled by the successional stage of the forest; McMahon et al. (2010) used a dataset with tree ages from 5 to 250 years. Additionally, Los (2013) attributed 40% of the global greening trend to the influence of increasing [CO_2] based upon satellite observations (although we note that greening differs from stem growth increment). While these growth increments cannot be definitively attributed to any one environmental driver, it nonetheless appears evident that forests differ greatly in their degree of stem growth limitation.

As discussed above, young forests without canopy closure are likely to respond very differently to elevated [CO_2] compared with mature forests, with arguments for little or no additional growth response in mature ecosystems focusing on the idea that these ecosystems are often making the maximum use of their available nutrient resources, i.e. they are nutrient-limited (see above), or upon arguments relating to ecosystem dynamics (e.g. Körner, 2009; Bugmann and Bigler 2011; see §6). A key objective of the next generation of forest FACE experiments is to evaluate whether the growth enhancements that were achieved in young forests or plantations in previous FACE experiments (even those with closed canopies) hold for mature old-growth ecosystems (Norby et al., 2016). However, the extent to which CO_2 enhancements of growth are limited by environmental conditions at the global scale remains an open question, which will require results from FACE, process-based ecosystem modelling, inventory and dendrological methods to be combined and reconciled.

Although the magnitude of sink-related growth limitations may be uncertain, their existence clearly is not. So does accounting for this require a paradigm shift in TBMs? This seems unlikely; the structure of TBMs allows for demand-side limitations to be included in the algorithms governing C allocation, as initial attempts have demonstrated (Leuzinger et al., 2013). The key aspect of the problem then becomes balancing supply and demand of C within the plant in order to retain mass balance. Short-term imbalances can be accounted for by the representation in TBMs of non-structural carbohydrates (NSC), and several TBMs include this pool of carbon in some form (e.g. Fisher et al., 2010a; Zaehle and Friend, 2010). The challenge here may be to correctly characterise the capacity and response rate of the NSC pool, as well as the strategy of the plant for allocating to NSC, and this is likely to differ greatly across species. The problem of parameterisation at the large scale is increased by difficulties in defining a baseline for absolute measurements of NSC (Quentin et al., 2015).

Alternatively, NSC dynamics may be informed not by observations of NSC per se, but by leaf- and root-phenology, which emerge from lags between growth and assimilation, buffered by NSC (Haverd et al., 2016). Longer-term imbalances can be resolved by characterising other plant C losses (§4.3) and by capturing individual-plant and whole-ecosystem responses to limitations on ecosystem water and nutrient resources (§5). Here, again, TBMs are well placed to adapt to better represent these processes, though data limitations remain an issue.

4.2. Allocation to different components of growth

Changes in allocation ratios, i.e. relative growth rates, between wood, leaves and fine roots are also of potentially critical importance for the terrestrial CO_2 sink. Assuming no change in turnover rates of individual tissues (see §6), then increases in allocation of carbon to a particular pool within the plant will lead to an increase in the standing biomass of that pool. The longer the residence time of carbon within that particular pool, the greater influence a given additional carbon input flux will have on pool size. Thus, if trees invest disproportionately in wood, which has a long carbon residence time (low turnover rate), a given increase in whole-plant growth will result in a greater increase in standing biomass than if allocation favours fast-turnover pools such as leaves or fine roots. Likewise the relative allocation between roots and leaves is important for determining inputs into soil carbon pools.

In a study of allocation responses to elevated [CO_2], De Kauwe et al. (2014) found increased allocation towards fine roots at the expense of wood and leaves at one forested FACE site (ORNL), but a small increase in wood allocation at the expense of leaves at another forested FACE site (Duke). These differences in allocation ratios had a profound influence on carbon storage; at Duke 88% of additional NPP generated by elevated [CO_2] remained at the end of the experiment, whereas at ORNL none of the additional NPP was reflected in the measured biomass at the end of the experiment, as it had been allocated to parts of the plant with rapid turnover rates. In comparing these results with predictions of ecosystem models including TBMs, De Kauwe et al. (2014) found that models with functional-balance approaches to allocation caught the general character of the observations, providing encouragement that such approaches may be suitable for modelling responses to elevated [CO_2]. Similarly, results from another forested FACE experiment found increases in photosynthesis, but no overall increase in stand biomass, indicating a shift in typical allocation patterns away from wood, although it was not clear where this carbon was allocated (Bader et al., 2010).

As well as changes at the plant scale, elevated [CO_2] can also modify allocation, and thus turnover of carbon at the ecosystem scale, if preferentially advantaged species have different allocation patterns compared to disadvantaged species. For instance, leguminous plants have been observed to be advantaged in some ecosystems under increased [CO_2] (Reich et al., 2006a). At its most extreme, such change can manifest as biome shifts. One of the most reported examples is the shrub encroachment and woody thickening observed in many semi-arid ecosystems around the world over recent decades (Liu et al., 2013; Mitchard and Flintrop 2013), along with a general increase in observations of vegetation cover (Donohue et al., 2013), which has been linked to CO_2-induced enhancement of photosynthesis and water-use efficiency. In another example, whilst many N-enabled TBMs also predict a decline in biospheric carbon storage when the interactive effects of climate change are considered (e.g. Zaehle et al., 2010; Fisher et al., 2010b), a study based on the LPJ-GUESS DGVM found that the combination of elevated [CO_2] with increasing N mineralisation rates in warming high-latitude soils leads to forest advance and increase in woody biomass (Wårlind et al., 2014). In the latter study, this provides a new transient sink for carbon in growing trees over the 21st century. This contrasts with the general expectation that N limitation will decrease growth and therefore carbon storage (§4.1). To understand the effect of elevated [CO_2] on the terrestrial C sink we therefore need to understand not only effects on carbon allocation and turnover at the plant scale, but also effects on vegetation dynamics at the ecosystem scale. Such understanding for forested ecosystems will not be obtained from ecosystem manipulation experiments, meaning that we must be confident in the ability of our models to correctly capture competitive dynamics between species with different functional traits, as well as having access to on-going monitoring data from unmanaged ecosystems to evaluate such predictions from models.

4.3. Additional losses

Root exudates and export to symbionts are prime candidates for additional plant carbon losses, not represented in most models. These processes are difficult to quantify experimentally. In total, as much as 60–70% of GPP has been observed to be transferred to the soil in studies of *Pinus sylvestris* and *Pseudotsuga menziesii* trees (Grayston et al., 1996), the vast majority of which is likely to be rapidly returned to the atmosphere through heterotrophic respiration. Separating the portion of the total plant-soil flux attributable to root exudates from that due to processes such as fine root turnover and growth respiration (which are often explicitly represented in TBMs) is challenging. Estimates of root exudate amount range from 1 to

10% of GPP under ambient conditions (Grayston et al., 1996). Experiments on trees under elevated [CO_2] have shown exudation to be enhanced by ca. 50% during the main growing season for a 200 ppmv increase in atmospheric CO_2 mixing ratio (Phillips et al., 2009; Phillips et al., 2011). One interpretation of this additional C exudation would be a simple overflow response to excessive carbon. Alternatively, plants may produce exudates to alleviate nutrient limitations, for instance through priming (§7). TBM estimates of plant N uptake may potentially be improved by accounting for such a response (Medlyn et al., 2015). The relation of exudate rate to [CO_2| is equivocal, however; Uselman et al. (2000) saw no such [CO_2]-mediated enhancement in an experiment in *Robinia pseudoacacia* grown in an artificial medium. Observations are required from a range of biomes, and particularly in real soils, both to assess the present-day carbon balance, and to understand the response of assimilate export to [CO_2].

A further possible carbon loss pathway is the synthesis of VOCs. Emissions of isoprene, the most globally emitted biogenic VOC, have been observed to constitute up to 10% of leaf GPP in some instances (Sharkey and Yeh 2001), although 1–2% might be a more common value globally (Guenther et al., 2012). There is evidence that emission rates of many VOCs are suppressed by elevated [CO_2] (Possell et al., 2004; Arneth et al., 2007), suggesting that VOCs are unlikely to be fundamental in balancing the global plant carbon budget.

It is questionable whether the carbon demands of the myriad processes that contribute to autotrophic respiration are adequately captured by the simple temperature or fixed-fraction functions currently employed in most TBMs. For instance, there is evidence that leaf respiration initially increases under water stress, possibly to repair damage inflicted by water deficit, a feature not captured by TBMs (Powell et al., 2013), and which might partially balance growth reductions under water stress. Evidence that plants are able to acclimate their respiration levels to temperature (see Smith and Dukes 2013 for an overview) is increasingly recognised and may be a standard feature in the next generations of TBMs. Evidence for a direct response of autotrophic respiration to elevated [CO_2] is, however, mixed (e.g. Ryan 1991; Bunce 2004), with strong increases observed for soybeans, but not in many other species examined (Leakey et al., 2009).

5. MULTI-SCALE FEEDBACKS ON PHOTOSYNTHESIS

Overwhelmingly the consensus from the last two decades of large-scale ecosystem manipulation experiments is that enhancements in assimilation are seen, but often to a lesser extent than in earlier, more artificial, experiments, and they were not always maintained over many years (Norby et al., 2010; Norby and Zak 2011), indicating feedbacks on assimilation rates. Reductions of photosynthesis to balance sources with sinks within plants may take many forms. Photosynthesis in trees and grasses is typically limited by carboxylation capacity at atmospheric CO_2 mixing ratios up to ca. 500 ppmv (Ainsworth and Rogers 2007). Both laboratory and field studies under elevated [CO_2] have shown that if source processes outstrip sinks, or transport of photosynthate away from the leaves is limited, then the build-up of leaf sugars leads to a direct down-regulation of photosynthesis (reviewed in Paul and Foyer 2001; Ainsworth and Long 2005; Ainsworth and Rogers 2007; Leakey et al., 2009). Such down-regulation reduces leaf-level photosynthetic capacity, either by reducing leaf Rubisco content and therefore carboxylation capacity (generally reflected by a reduction in leaf C:N ratio, although Rubisco activity has also been observed to be modified; Cen and Sage 2005), or by decreasing the maximum rate of electron transport. This is exemplified in observations that reductions in carboxylation capacity under elevated [CO_2] appear to be much greater under conditions of low N availability (Ainsworth and Rogers 2007; Leakey et al., 2009). Such reductions are not uniformly present across all species however. For instance, *Populus* trees have been observed to avoid down-regulation of leaf photosynthesis by transporting sugars away from leaves, combined with a large capacity to store non-structural carbohydrates (NSC), thereby maintaining the capacity of in-plant C sinks (Karnosky et al., 2003; Davey et al., 2006). Bader et al. (2010) also reported no down-regulation of photosynthesis under elevated [CO_2] in a FACE experiment in a mature forest stand in Switzerland.

The observations above demonstrate a close linkage between photosynthesis and the capacity of the plant to transport and utilise photosynthate. TBMs incorporating C–N interactions commonly parameterise a direct reduction in photosynthesis when N becomes limiting, in some cases by directly modelling leaf C:N ratios, which then results in less growth compared to a model with no N limitations, since less C is assimilated (e.g. Zaehle and Friend 2010; Smith et al., 2014). This assumes that plants allocate N such that photosynthesis is maximised within the constraints of the N limitation on structural growth. This parsimonious approach avoids the need to model leaf sugar concentration, which, given its strong coupling with phloem transport, presents huge data limitations (Fatichi et al., 2014) and currently seems an unrealistic target for global scale models. Effects of limitations in other resources needed for growth (water, nutrients other than N) could be parameterised in a similar manner. An alternative approach is to adopt a representation in which photosynthesis is regulated as a net result of the whole-plant carbon balance, i.e. down-regulation occurs when C assimilation exceeds the sum of all sinks and losses, including maximum allocation to NSC, and when there is nothing the plant can do with the extra C to attempt to alleviate any sink constraints. This alternative approach has the advantage that it accommodates a range of plant response strategies to changes in resource supply/demand arising under elevated [CO_2]. A disadvantage is that model accuracy in capturing photosynthetic down-regulation is then contingent on the currently limited observational basis on which to characterise the response of various sinks and losses of carbon within the plant (§4).

Another possible route to reducing photosynthesis is to reduce stomatal conductance, thereby reducing the leaf intercellular CO_2 concentration (C_i). Laboratory and ecosystem experiments often show that stomatal conductance is regulated such that C_i changes proportionally to [CO_2] (Drake et al., 1997; Ainsworth and Rogers 2007). This results in an increase of water use efficiency (WUE) under elevated [CO_2], as a given C_i can be maintained for less water loss. This increase in WUE has been widely observed (Leakey et al., 2009; Penuelas et al., 2011; De Kauwe et al., 2013; van der Sleen et al., 2015). Using a network of temperate and boreal forest sites where above-canopy fluxes for both CO_2 and water were both measured, Keenan et al. (2013) found a large increase in WUE over recent decades which they attributed to plant responses to increasing [CO_2]. The increase in WUE was so large, however, that their results suggested that trees must be closing their stomata to hold $C_i$relatively constant, i.e. they were prioritising limiting water loss over increasing GPP. The contrast to the previous studies finding proportionality of [CO_2] and C_i has sparked debate (e.g. Medlyn and De Kauwe, 2013). Nonetheless, if borne out by future research, one possible interpretation of the results of Keenan et al. (2013)would be that plants may be actively down-regulating photosynthesis in favour of maintaining turgor for growth. As TBMs currently assume proportionality of [CO_2] and C_i, large reductions in stomatal conductance as suggested by Keenan et al. (2013)could have profound implications for TBM simulations of terrestrial carbon cycling, affecting not only productivity, but many other processes through the resultant modification of soil water content. A decrease in C_i:[CO_2] ratio under elevated [CO_2] was also seen in tree ring data at three FACE sites (Battipaglia et al., 2013), whilst other studies of tree rings have revealed evidence of large increases in WUE as a result of rising [CO_2], without corresponding increases in growth, across a wide range of global ecosystems (Penuelas et al., 2011; van der Sleen et al., 2015). In contrast, syntheses of FACE experiments have found no such strong effect of increased [CO_2] on stomatal conductance at either leaf or stand scale (Leakey et al., 2009; Norby and Zak 2011). Possibly the discrepancy between results from FACE sites stems from the different sampling periods of the techniques involved (Leuzinger and Hättenschwiler, 2013). In comparing their historical samples to results from controlled [CO_2]

experiments, Ehlers et al. (2015) found evidence that there have been no adjustments of properties such as stomatal function or photosynthetic regulation in sugar beet in response to increased [CO_2], but as these results come from managed ecosystems, comparisons with natural ecosystems are non-trivial.

On longer time-scales there are other ways for plants to redistribute resources to balance supply and demand. Environmental limits may induce plants to reduce photosynthesis by reducing total leaf area, either through reduced leaf growth (§4.1), by investing in thicker leaves (Ainsworth and Long 2005), or by active changes in allocation patterns through changes in growth hormones (e.g. Ribeiro et al., 2012). As discussed in §4.2, changes in allocation patterns can have profound effects on vegetation carbon storage, and therefore the terrestrial C sink. Whilst some TBMs are able to capture allocation changes under increased [CO_2], changes in leaf mass per unit area (LMA) were recently highlighted as being a critical point for model development, with most TBMs treating LMA as a constant and none of those considered capturing the observed increase (Medlyn et al., 2015). Effective changes in leaf area at the ecosystem level may also accompany biome shifts (§4.2).

Overall, whilst none of the mechanisms moderating photosynthesis discussed above are clear-cut with respect to whether, how often, and in what species they occur, there is overwhelming evidence at the plant and ecosystem scales that plants and ecosystems regulate photosynthesis under increased [CO_2].

6. TURNOVER

Differences in the treatment of vegetation carbon turnover have been highlighted as the primary source of variation in TBM projections of vegetation C stocks for the 21st century (Friend et al., 2014), illustrating the huge uncertainty in our understanding of this process. Myriad factors influence ecosystem carbon turnover rates, including changes in drought-induced mortality, ecosystem disturbances, insect and pathogen outbreaks, and biome shifts. Changes in turnover will fundamentally impact upon the size of any CO_2-induced vegetation carbon sink (Müller et al., 2016). This topic is, however, worthy of multiple reviews by itself, and is a topic of increasing focus for the terrestrial modelling community (Running 2008; McDowell et al., 2011; Friend et al., 2014; Seidl et al., 2011; Johnson et al., 2016). Further, as discussed in §4.2, CO_2-induced shifts in allocation between leaves, stems and roots can alter turnover. In this section we restrict our focus to plant turnover processes which may be directly affected by increased [CO_2].

Bugmann and Bigler (2011) hypothesised that tree life cycles could be accelerated under elevated [CO_2], citing evidence that (i) trees in several elevated [CO_2] experiments invested little of the extra carbon gained in defence, thus making them more vulnerable to early mortality, (ii) earlier cone production was observed in trees subject to elevated [CO_2] at the Duke forest FACE experiment (Ladeau and Clark 2006), (iii) crops mature and die younger under elevated [CO_2] than under ambient conditions (Kimball et al., 1995), and (iv) growth increment in tree rings was negatively correlated with longevity in a study of three conifer species (Bigler and Veblen 2009). By parameterising tree longevity in a forest model based on observed correlation to growth rate in temperate tree species across multiple continents, they re-parameterised tree longevity in a forest model and showed that the simulated decrease in longevity tended to offset the increased growth. Furthermore, even without changes in longevity, individual tree growth enhancements of up to 30% could lead to a very small overall increase in stand biomass, mainly due to increased competition for light in the mature stand. These results again emphasise the dangers of assuming that observed tree-level relationships will hold true at the ecosystem scale, and if they turn out to be generally applicable for global forests, they would have profound effects for the size of any CO_2-driven terrestrial sink.

Lianas present an example of a community-level effect on vegetation turnover rates. Several studies now point towards increased growth of lianas in tropical forests under elevated [CO_2] (Körner, 2004; Phillips et al., 2002; Schnitzer and Bongers 2011). Lianas appear to be advantaged because they grow under light-limited rather than nutrient-limited conditions in the forest understorey, and light-limited plants benefit more than light-saturated plants from increased [CO_2] (Körner, 2009). Forest plots with lianas have been found to exhibit higher mortality of trees and branches compared to plots in which the lianas were cleared (van der Heijden et al., 2015). As with biome-type shifts, an increased portion of the canopy occupied by lianas also modifies the ratio of leaves to wood, in this case increasing it, resulting in a higher proportion of forest NPP being invested in short-lived tissues compared to stands in which lianas are not present.

Insect herbivory may be increased under elevated [CO_2]. This may occur through a reduction in investment in defence, as argued by Bugmann and Bigler (2011), and observed in soybean plants (Zavala et al., 2008). It may also occur through a reduction in leaf nutritional quality as a result of reduced nitrogen content, leading to herbivores increasing their leaf consumption to compensate. Evidence for the latter comes both from field studies (Stiling and Cornelissen 2007) and from palaeo-evidence showing that insect herbivory tends to be higher during periods of relatively high

[CO_2] and temperature (Currano et al., 2008). However, field observations under elevated [CO_2] have also reported a reduction in insect herbivory, apparently driven by lower densities of herbivores (Stiling et al., 2003; Hamilton et al., 2004; Knepp et al., 2005; Stiling and Cornelissen 2007). The reasons for these lower insect densities are not clear, but may relate to poorer nutrition slowing insect development, and therefore increasing their vulnerability to predators, or to insects preferentially choosing to feed on plants outside the experimental plots (Knepp et al., 2005). Elevated [CO_2] may also change the emission of defensive compounds such as VOCs, although the effects are difficult to generalise (Zavala et al., 2013), whilst decreased stomatal conductance can lead to a rise in leaf temperature which can increase insect metabolism and feeding rates (Zavala et al., 2013). Our limited understanding of such responses at the ecosystem scale makes including insect herbivory in TBMs a challenge, but given the large fraction of NPP consumed by herbivores, particularly in nutrient-rich ecosystems (Cebrian et al., 1998), their impacts could be of substantial importance for vegetation turnover rates, and thus for the terrestrial carbon sink (Metcalfe et al., 2014).

Interactions with pathogens have also been observed, but the responses are mixed, with susceptibility to some pathogens increased under increased [CO_2], but the effects of other pathogens being lower than under ambient [CO_2] (summarised in Reich et al., 2006a). Whilst, in theory, change in susceptibility to pathogens could profoundly affect vegetation carbon cycling, the variety of responses and the often close coupling of plant and pathogen species make ecosystem responses very difficult to predict; likely long-term impacts on the carbon cycle would be indirect through a change in the balance of ecosystem composition.

7. COMPLEX IMPACTS OF ELEVATED ATMOSPHERIC [CO_2] ON THE TERRESTRIAL CARBON CYCLE

7.1. Interactions with land-use change

Elevated [CO_2] also affects the terrestrial carbon sink through indirect feedback mechanisms other than climate change and the effects on photosynthesis and plant growth and turnover as discussed above. Land-use dynamics are important to consider, as the effect of increased [CO_2] on projections of crop (Rosenzweig et al., 2014; Challinor et al., 2014) and timber (Kirilenko and Sedjo 2007) yields, as well as water availability

(Elliott et al., 2014), are likely to affect future land-use patterns (Schmitz et al., 2014; Nelson et al., 2014) and thus modify land-use emissions, which currently offset the effects of elevated [CO_2] on terrestrial carbon storage (Le Quéré et al., 2014). Reductions in rotation period of forest plantations in response to accelerated growth under elevated [CO_2] stands to reduce the mean standing biomass at such managed sites (Körner 2006). The uncertainty in crop yield response to elevated [CO_2] has been identified as a major scientific challenge in crop modelling (Rosenzweig et al., 2014; Müller et al., 2016) but here additional experimental evidence is also needed to enhance process understanding. Most crops use the C_3photosynthetic pathways and are essentially subject to the same mechanisms as described above in sections 3–5, while C_4 crops such as maize (*Zea maize* L.), sorghum, sugar cane and millet species may be expected to show smaller, or different, responses to [CO_2]. Crop-specific allocation to yield organs represents an additional complication. There is evidence that stimulated photosynthesis under elevated [CO_2] does not increase crop yields proportionally, especially if nutrients are limiting (Leakey et al., 2009). However, Kimball (2016) argues that FACE experiments indeed show substantial increases in crop yields, especially when considering that the unsteady CO_2 concentrations in FACE experiments may reduce the effect (Bunce 2013). Also in crop production, elevated [CO_2] can lead to reduced consumptive water use in crop fields. This leads, especially in combination with enhanced crop yields, to higher crop water productivity and thus to a more efficient use of water resources (Deryng et al., 2016), although as in natural systems, hydrological feedbacks may prove important (Leuzinger et al., 2015).

7.2. Interactions with soil processes

The majority of carbon in the terrestrial biosphere is stored in the soils, where biomass from litterfall is converted to various forms of soil organic matter and eventually returned to the atmosphere as respired CO_2. This conversion of organic matter to carbon dioxide is driven by microorganisms that respond to temperature and moisture regimes, and to the chemical and structural composition of the mineral and organic soil compartments. The effect of elevated [CO_2] on soils was reviewed in Reich et al. (2006a) and De Graaff et al. (2006), who found that soil C content typically increased under elevated [CO_2] when additional N was added to the system, but not in ecosystems with low N availability, although measurement of effects on soil was confounded by long response times relative to experimental duration. Elevated [CO_2] can change the stoichiometry of the organic matter from litter by e.g. leading to lower nitrogen content of leaves and roots (Yang et al., 2011). This in turn changes the decomposition rate of organic matter (Zhang et al., 2008). It can also lead to a priming effect, i.e. the enhanced decomposition of soil organic matter by supplying easily degradable material in the form of root exudates under elevated CO_2 (Carney et al., 2007; Phillips et al., 2011), resulting in the break down of organic-mineral compounds that otherwise prevent mineralization of the associated organic matter (Keiluweit et al., 2015). Such priming effects can both reduce soil C content and liberate N, although Reich et al. (2006a) point out that any additional N made available by priming is likely to be negated in the longer-term by increased N immobilisation resulting from higher C inputs to the soil under increased [CO_2].

8. THE ROAD AHEAD: CLOSER INTEGRATION BETWEEN MEASUREMENTS AND MODELLING

What recommendations may be made for the improvement of TBMs when applied to modelling the effect of increasing [CO_2] on the terrestrial carbon sink? Advances in knowledge are sufficient for some of the processes discussed above to be implemented and tested in TBMs. For some processes and observed phenomena more consistent or comprehensive observations will be required in order to provide enough information to design robust parameterisations for implementation in models. A model is by definition a simplified representation of the real system it stands for, and TBM modelling generally strives for a reductionist approach, representing the biosphere and its component ecosystems in the least amount of detail necessary to reproduce salient features and dynamics. A model with high process complexity will only become more accurate if the functional forms, scaling and linkages among processes, drivers and descriptors of system state can be unequivocally constrained by observations, and this is rarely the case, even for empirically well-studied processes such as photosynthesis or transpiration. Overall the aim for global modellers must be to maintain model parsimony: only include new processes when they are well constrained and make a substantial difference to the output metric of interest. TBMs that are solely tuned to emerging properties such as net ecosystem exchange, rather than observations of individual processes, can be expected to fail in this context. Further, it will be important to ensure that when new process mechanisms are included, related parameterisations within the model are reassessed to ensure that any implicit "double counting" is avoided. We summarise recommendations below.

1)Within-plant carbon allocation, and its resultant effects on plant carbon turnover, is a key step governing any changes in the size of the vegetation carbon pool, and thus in the terrestrial C sink. Allocation responses may confound any direct association between primary productivity and the net terrestrial carbon flux. Inclusion of in-plant C sink constraints (beyond N availability, which is already included in some models), and their feedbacks to photosynthesis, is highly feasible within current TBM structure, but will require further work to parameterise typical growth responses to resource stress for different species globally, as well as refining N and P models and developing new parameterisations to test the importance of other nutrients. Inclusion of an NSC pool will likely be necessary to balance sources and sinks in the short term, as will the effects of photosynthetic down-regulation in the long-term. The HAVANA approach (Haverd et al., 2016) in which growth and NPP have separate time courses, but balance in the long term, provides a useful modelling framework.

2)It is not enough to understand response to [CO_2] at the single tree level; rather, whole ecosystem response, including effects of inter-plant and inter-species competition for resources, as well as large-scale plant-atmosphere interactions, must be considered. Some TBMs already aim to capture such effects. The challenge is to better evaluate and parameterise this aspect of the models, allowing confidence in the resultant emerging ecosystem responses.

3)Effects of [CO_2] on turnover rates at the tissue, plant and ecosystem scales have only been lightly investigated. Those changes relating to the partitioning of carbon among different pools within the plant are tightly tied to plant allocation strategies, and uncertainty should be reduced as our confidence in modelling plant allocation in TBMs increases. The veracity of possible acceleration of tree life cycles, and thus decreased tree longevity, must be a priority for future research.

4)Evidence suggests that plants may invest a large fraction of carbon through root exports to soil, and that these exports are likely to increase under nutrient deficiency induced by elevated [CO_2]. Below-ground dynamics, including root growth, exudates, symbioses and plant-soil interactions are not well observed in most empirical studies. Better quantification of this carbon export is crucial to understanding the whole-plant carbon balance, and effectively modelling plant allocation strategies.

5)The effect of mesophyll conductance on GPP is sufficiently constrained to be implemented in global models. Although we expect that it will be much more limited in its global influence than suggested by Sun et al. (2014), it is favourable nonetheless to incorporate the best knowledge if it does not substantially degrade model parsimony.

6)A large proportion of natural ecosystems has been cleared by man for use in agriculture, or appropriated for other uses such as forestry. Effects of elevated [CO_2] on the productivity of these managed ecosystems can feed back directly on the amount of land occupied by unmanaged vegetation, providing an indirect, but potentially very important, driver for the size of the terrestrial carbon sink. Enhanced TBMs which can also capture effects in managed ecosystems will be better able to feed into land-use assessments and provide overarching assessments of the effect of increased [CO_2] on the terrestrial C sink.

7)Although not reviewed in depth here, soil processes are key to the supply of nutrients and water, as well as representing the largest store of ecosystem carbon, and are often represented with very simple functions in TBMs. Further attention must be, and is being, paid to soil processes. The reader is directed to reviews by De Graaff et al. (2006) and Reich et al. (2006a), as well as recent modelling advances by e.g. Koven et al. (2013) and Todd-Brown et al. (2012).

ACKNOWLEDGEMENTS

The authors thank Anders Ahlström of Lund University, Sweden, for providing the data for Fig. 2a. TP and AA were funded by the European Commission's 7th Framework Programme, under Grant Agreement numbers 282672 (EMBRACE) and 603542(LUC4C). This work was supported, in part, by the German Federal Ministry of Education and Research (BMBF), through the Helmholtz Association and its research program ATMO. CM acknowledges financial support from the BMBF grant 01LN1317A (MACMIT). VH's contribution was supported by the Australian Climate Change Science Program. This is paper number 16 of the Birmingham Institute of Forest Research (BIFoR).

REFERENCES

1. Ahlström, A., et al., 2012. Robustness and uncertainty in terrestrial ecosystem carbon response to CMIP5 climate change projections. Environ. Res. Lett. 7 (4), 044008.
2. Ahlström, A., et al., 2015. The dominant role of semi-arid ecosystems in the trend and variability of the land CO2 sink. Science 6237, 895–899.
3. Ainsworth, E.A., Long, S.P., 2005. What have we learned from 15 years of free-air CO2 enrichment (FACE)? A meta-analytic review of the responses of photosynthesis, canopy properties and plant production to rising CO2. New Phytol. 165, 351–372.
4. Ainsworth, E.A., Rogers, A., 2007. The response of photosynthesis and stomatal conductance to rising [CO2]: mechanisms and environmental interactions. Plant Cell Environ. 30, 258–270.
5. Anav, A., et al., 2013. Evaluating the land and ocean components of the global carbon cycle in the CMIP5 earth system models. J. Clim. 26 (18), 6801–6843.
6. Arneth, A., et al., 2007. Process-based estimates of terrestrial ecosystem isoprene emissions: incorporating the effects of a direct CO2-isoprene interaction. Atmos. Chem. Phys. 7, 31–53.

7. Arneth, A., et al., 2010a. Terrestrial biogeochemical feedbacks in the climate system. Nat. Geosci. 3 (8), 525–532.
8. Arneth, A., et al., 2010b. From biota to chemistry and climate: towards a comprehensive description of trace gas exchange between the biosphere and atmosphere. Biogeosciences 7 (1), 121–149.
9. Bader, M.K., Siegwolf, R., Ko, C., 2010. Sustained enhancement of photosynthesis in mature deciduous forest trees after 8 years of free air CO2 enrichment. Planta 232, 1115–1125.
10. Battipaglia, G., et al., 2013. Elevated CO2 increases tree-level intrinsic water use efficiency: insights from carbon and oxygen isotope analyses in tree rings across three forest FACE sites. New Phytol. 197, 544–554.
11. Bigler, C., Veblen, T.T., 2009. Increased early growth rates decrease longevities of conifers in subalpine forests. Oikos 118 (February), 1130–1138.
12. Bugmann, H., Bigler, C., 2011. Will the CO2 fertilization effect in forests be offset by reduced tree longevity. Oecologia 165, 533–544.
13. Bunce, J.A., 2004. A comparison of the effects of carbon dioxide concentration and temperature on respiration: translocation and nitrate reduction in darkened soybean leaves. Ann. Bot. 93, 665–669.
14. Bunce, J.A., 2013. Effects of pulses of elevated carbon dioxide concentration on stomatal conductance and photosynthesis in wheat and rice. Physiol. Plant. 149, 214–221.
15. Carney, K.M., et al., 2007. Altered soil microbial community at elevated CO2 leads to loss of soil carbon. Proc. Natl. Acad. Sci. U. S. A. 104 (12), 4990–4995.
16. Cebrian, J., et al., 1998. The dependence of heterotrophic consumption and C accumulation on autotrophic nutrient content in ecosystems. Ecol. Lett. 1, 165–170.
17. Cen, Y.-P., Sage, R.F., 2005. The regulation of rubisco activity in response to variation in temperature and atmospheric CO2 partial pressure in sweet potato. Plant Physiol. 139, 979–990.
18. Challinor, a J., et al., 2014. A meta-analysis of crop yield under climate change and adaptation. Nat. Clim. Change 27 (April), 1–5.
19. Ciais, P. et al. 2013. Carbon and other biogeochemical cycles. Climate change 2013: the physical science basis. Contribution of Working Group I to the Fifth Assessment Report of the Intergovernmental Panel on Climate Change, pp. 465–570.
20. Cole, C.T., et al., 2010. Rising concentrations of atmospheric CO2 have increased growth in natural stands of quaking aspen (Populus tremuloides). Global Change Biol. 16, 2186–2197.
21. Currano, E.D., et al., 2008. Sharply increased insect herbivory during the Paleocene-Eocene Thermal Maximum. Proc. Natl. Acad. Sci. U. S. A. 105 (6), 1960–1964.
22. Davey, P.A., et al., 2006. Can fast growing plantation trees escape biochemical down-regulation of photosynthesis when grown throughout their complete production cycle in the open air under elevated carbon dioxide? Plant Cell Environ. 29, 1235–1244.
23. De Graaff, M.-A., et al., 2006. Interactions between plant growth and soil nutrient cycling under elevated CO2: a meta-analysis. Global Change Biol. 12, 2077–2091.
24. De Kauwe, M.G., et al., 2013. Forest water use and water use efficiency at elevated CO2: a model-data intercomparison at two contrasting temperate forest FACE sites. Global Change Biol. 19, 1759–1779.
25. De Kauwe, M.G., et al., 2014. Where does the carbon go? A model –data intercomparison of vegetation carbon allocation and turnover processes at two temperate forest free-air CO2 enrichment sites. New Phytol. 203, 883–899.
26. Deryng, D., et al., 2016. Regional disparities in the beneficial e ects of rising CO2 concentrations on crop water productivity. Nat. Clim. Change (advance online publication).
27. Donohue, R.J., et al., 2013. Impact of CO2 fertilization on maximum foliage cover across the globe's warm, arid environments. Geophys. Res. Lett. 40 (May), 3031–3035.
28. Drake, B.G., Gonzàlez-meler, M.A., Long, S.P., 1997. More efficient plants: a consequence of rising atmospheric CO2. Annu. Rev. Plant Physiol. Plant Mol. Biol. 48, 609–639.

29. Ehlers, I., et al., 2015. Detecting long-term metabolic shifts using isotopomers: CO2-driven suppression of photorespiration in C3 plants over the 20th century. Proc. Natl. Acad. Sci. U.S.A. 112 (51), 15585–15590.
30. Elliott, J., et al., 2014. Constraints and potentials of future irrigation water availability on agricultural production under climate change. Proc. Nat. Acad. Sci. U. S. A. 111, 3239–3244.
31. Elser, J.J., et al., 2010. Biological stoichiometry of plant production: metabolism: scaling and ecological response to global change. New Phytol. 186, 593–608.
32. Farquhar, G.D., von Caemmerer, S., Berry, J.A., 1980. A biochemical model of photosynthetic CO2 assimilation in leaves of C3 species. Planta 90, 78–90.
33. Fatichi, S., Leuzinger, S., Körner, C., 2014. Moving beyond photosynthesis: from carbon source to sink-driven vegetation modeling. New Phytol. 201, 1086–1095.
34. Fisher, J.B., et al., 2010a. Carbon cost of plant nitrogen acquisition: a mechanistic, globally applicable model of plant nitrogen uptake, retranslocation, and fixation. Global Biogeochem. Cycles 24, GB1014.
35. Fisher, R., et al., 2010b. Assessing uncertainties in a second-generation dynamic vegetation model caused by ecological scale limitations. New Phytol. 187 (3), 666–681.
36. Forkel, M., et al., 2016. Enhanced seasonal CO2 exchange caused by amplified plant productivity in northern ecosystem. Science 4971 (online first).
37. Franklin, O., et al., 2012. Modeling carbon allocation in trees: a search for principles. Tree Physiol. 32, 648–666.
38. Friedlingstein, P., Cox, P., Betts, R., 2006. Climate-carbon cycle feedback analysis: results from the C4MIP model intercomparison. J. Clim. 19, 3337–3353.
39. Friedlingstein, P., et al., 2014. Uncertainties in CMIP5 climate projections due to carbon cycle feedbacks. J. Clim. 27 (2), 511–526.
40. Friend, A.D., et al., 2014. Carbon residence time dominates uncertainty in terrestrial vegetation responses to future climate and atmospheric CO2. Proc. Natl. Acad. Sci. U. S. A. 111 (9), 3280–3285.
41. Gea-Izquierdo, G., et al., 2015. Modelling the climatic drivers determining photosynthesis and carbon allocation in evergreen Mediterranean forests using multiproxy long time series. Biogeosciences 12, 3695–3712.
42. Goll, D.S., et al., 2012. Nutrient limitation reduces land carbon uptake in simulations with a model of combined carbon: nitrogen and phosphorus cycling. Biogeosciences 9, 3547–3569.
43. Grayston, S.J., Vaughan, D., Jones, D., 1996. Rhizosphere carbon flow in trees: in comparison with annual plants: the importance of root exudation and its impact on microbial activity and nutrient availability. Appl. Soil Ecol. 5, 29–56.
44. van Groenigen, K., et al., 2006. Element interactions limit soil carbon storage. Proc. Natl. Acad. Sci. U. S. A. 103 (17).
45. Guenther, A.B., et al., 2012. The model of emissions of gases and aerosols from nature version 2.1 (MEGAN2.1): An extended and updated framework for modeling biogenic emissions. Geosci. Model Dev. 5, 1471–1492.
46. Hamilton, J.G., et al., 2004. Insect herbivory in an intact forest understory under experimental CO2 enrichment. Oecologia 138, 566–573.
47. Hartmann, H., McDowell, N.G., Trumbore, S., 2015. Allocation to carbon storage pools in Norway spruce saplings under drought and low CO2. Tree Physiol. 35, 243–252.
48. Haverd, V., et al., 2016. Coupling carbon allocation with leaf and root phenology predicts tree – grass partitioning along a savanna rainfall gradient. Biogeosciences 13, 761–779.
49. Hickler, T., et al., 2008. CO2 fertilization in temperate FACE experiments not representative of boreal and tropical forests. Global Change Biol. 14, 1531–1542.
50. Hickler, T., Rammig, A., Werner, C., 2015. Modelling CO2 impacts on forest productivity. Curr. For. Rep. 1 (2), 69–80.
51. Hungate, B.A., et al., 2003. Nitrogen and climate change. Science 302 (November), 1512–1513.

52. Johnson, M.O., et al., 2016. Variation in stem mortality rates determines patterns of aboveground biomass in Amazonian forests: implications for dynamic global vegetation models. Global Change Biol. (Accepted manuscript).
53. Jones, C., et al., 2013. Twenty-first-century compatible CO2 emissions and airborne fraction simulated by CMIP5 earth system models under four representative concentration pathways. J. Clim. 26 (13), 4398–4413.
54. Körner, C., 2003. Carbon limitation in trees. J. Ecol. 91, 4–17.
55. Körner, C., 2006. Plant CO2 responses: an issue of definition: time and resource supply. New Phytol. 172, 393–411.
56. Körner, C., 2015. Paradigm shift in plant growth control. Curr. Opin. Plant Biol. 25, 107–114.
57. Karnosky, D.F., et al., 2003. Tropospheric O3 moderates responses of temperate hardwood forests to elevated CO2: a synthesis of molecular to ecosystem results from the Aspen FACE project. Funct. Ecol. 17, 289–304.
58. Keenan, T.F., et al., 2013. Increase in forest water-use efficiency as atmospheric carbon dioxide concentrations rise. Nature 499, 324–327.
59. Keiluweit, M., et al., 2015. Mineral protection of soil carbon counteracted by root exudates. Nat. Clim. Change 5, 588–595.
60. Kimball, B.A., et al., 1995. Productivity and water use of wheat under free-air CO2 enrichment. Global Change Biol. 1, 429–442.
61. Kimball, B.A., 2016. Crop responses to elevated CO2 and interactions. Curr. Opin. Plant Biol. 31, 36–43.
62. Kirilenko, A.P., Sedjo, R.A., 2007. Climate change impacts on forestry. Proc. Natl. Acad. Sci. U. S. A. 104 (50), 19697–19702.
63. Knepp, R.G., et al., 2005. Elevated CO2 reduces leaf damage by insect herbivores in a forest community. New Phytol. 167, 207–218.
64. Körner, C., 2004. Through enhanced tree dynamics carbon dioxide enrichment may cause tropical forests to lose carbon. Philos. Trans. R. Soc. B: Biol. Sci. 359, 493–498.
65. Körner, C., 2009. Responses of humid tropical trees to rising CO2. Ann. Rev. Ecol. Evol. Syst. 40, 61–79.
66. Körner, C., et al., 2005. Carbon flux and growth in mature deciduous forest trees exposed to elevated CO2. Science 309, 1360–1362.
67. Koven, C.D., et al., 2013. The effect of vertically resolved soil biogeochemistry and alternate soil C and N models on C dynamics of CLM4. Biogeosciences 10 (11), 7109–7131.
68. Ladeau, S.L., Clark, J.S., 2006. Elevated CO2 and tree fecundity: the role of tree size, interannual variability: and population heterogeneity. Global Change Biol. 12, 822–833.
69. Le Quéré, C., et al., 2014. Global carbon budget 2013. Earth Syst. Sci. Data 6 (1), 235–263.
70. Quéré, et al., 2009. Trends in the sources and sinks of carbon dioxide. Nat. Geosci. 2 (12), 831–836.
71. Leakey, A.D.B., et al., 2009. Elevated CO2 effects on plant carbon, nitrogen, and water relations: six important lessons from FACE. J. Exp. Bot. 60 (10), 2859–2876.
72. Lenka, N.K., Lal, R., 2012. Soil-related constraints to the carbon dioxide fertilization effect. Crit. Rev. Plant Sci. 31, 342–357.
73. Leuzinger, S., Hättenschwiler, S., 2013. Beyond global change: lessons from 25 years of CO2 research. Oecologia 171, 639–651.
74. Leuzinger, S., et al., 2013. A sink-limited growth model improves biomass estimation along boreal and alpine tree lines. Global Ecol. Biogeogr. 22, 924–932.
75. Leuzinger, S., et al., 2015. The island effect in terrestrial global change experiments: a problem with no solution? AoB PLANTS (accepted manuscript).
76. Liu, F., et al., 2013. Woody plant encroachment into grasslands: spatial patterns of functional group distribution and community development. PLoS One 8 (12), e84364.
77. Lloyd, J., et al., 2015. Edaphic: structural and physiological contrasts across Amazon Basin forest –savanna ecotones suggest a role for potassium as a key modulator of tropical woody vegetation structure and function. Biogeosciences 12, 6529–6571.

78. Los, S.O., 2013. Analysis of trends in fused AVHRR and MODIS NDVI data for 1982–2006: indication for a CO2 fertilization effect in global vegetation. Global Biogeochem. Cycles 27, 318–330.
79. Luo, Y., et al., 2004. Progressive nitrogen limitation of ecosystem responses to rising atmospheric carbon dioxide. BioSci 54 (8), 731–739.
80. Müller, C., et al., 2016. Drivers and patterns of land biosphere carbon balance reversal. Environ. Res. Lett. 11 (4), 44002.
81. McCarthy, H.R., et al., 2010. Re-assessment of plant carbon dynamics at the Duke free-air CO2 enrichment site: interactions of atmospheric [CO2] with nitrogen and water availability over stand development. New Phytol. 185, 514–528.
82. McDowell, N.G., et al., 2011. The interdependence of mechanisms underlying climate-driven vegetation mortality. Trends Ecol. Evol. 26 (10), 523–532.
83. McMahon, S.M., Parker, G.G., Miller, D.R., 2010. Evidence for a recent increase in forest growth. Proc. Natl. Acad. Sci. U. S. A. 107 (8), 3611–3615.
84. Medlyn, B., De Kauwe, M., 2013. Carbon dioxide and water use in forests. Nature 499, 287–288.
85. Medlyn, B.E., et al., 2015. Using ecosystem experiments to improve vegetation models. Nat. Clim. Change 5 (6), 528–534.
86. Metcalfe, D.B., et al., 2014. Herbivory makes major contributions to ecosystem carbon and nutrient cycling in tropical forests. Ecol. Lett. 17, 324–332.
87. Mitchard, E.T.A., Flintrop, C.M., 2013. Africa's woodlands and savannas 1982–2006 woody encroachment and forest degradation in Sub-Saharan Africa's woodlands and savannas 1982–2006. Philos. Trans. R. Soc. B: Biol. Sci. 368 (July), 20120406.
88. Morales, P., et al., 2005. Comparing and evaluating process-based ecosystem model predictions of carbon and water fluxes in major European forest biomes. Global Change Biol. 11 (12), 2211–2233.
89. Muller, B., et al., 2011. Water deficits uncouple growth from photosynthesis, increase C content, and modify the relationships between C and growth in sink organs. J. Exp. Bot. 62 (6), 1715–1729.
90. Nelson, G.C., et al., 2014. Climate change effects on agriculture: economic responses to biophysical shocks. Proc. Natl. Acad. Sci. U. S. A. 111 (9), 3274–3279.
91. Norby, R.J., Zak, D.R., 2011. Ecological lessons from free-air CO2 enrichment (FACE) experiments. Ann. Rev. Ecol. Evol. Syst. 42, 181–203.
92. Norby, R.J., et al., 1999. Tree responses to rising CO2 in field experiments: implications for the future forest. Plant Cell Environ. 22, 683–714.
93. Norby, R.J., et al., 2010. CO2 enhancement of forest productivity constrained by limited nitrogen availability. Proc. Natl. Acad. Sci. U. S. A. 107 (45), 19368–19373.
94. Norby, R.J., et al., 2016. Model-data synthesis for the next generation of forest free-air CO2 enrichment (FACE) experiments. New Phytol. 209, 17–28.
95. Oren, R., et al., 2001. Soil fertility limits carbon sequestration by forest ecosystems in a CO2 -enriched atmosphere. Nature 411, 469–472.
96. Pan, Y., et al., 2011. A large and persistent carbon sink in the world's forests. Sci. (New York, N.Y.) 333 (July), 988–993.
97. Paul, M.J., Foyer, C.H., 2001. Sink regulation of photosynthesis. J. Exp. Bot. 52 (360), 1383–1400.
98. Penuelas, J., Canadell, J.G., Ogaya, R., 2011. Increased water-use efficiency during the 20th century did not translate into. Global Ecol. Biogeogr. 20, 597–608.
99. Phillips, O.L., et al., 2002. Increasing dominance of large lianas in Amazonian forests. Nature 418 (August), 770–774.
100. Phillips, R.P., Bernhardt, E.S., Schlesinger, W.H., 2009. Elevated CO2 increases root exudation from loblolly pine (Pinus taeda) seedlings as an N-mediated response. Tree Physiol. 29, 1513–1523.
101. Phillips, R.P., Finzi, A.C., Bernhardt, E.S., 2011. Enhanced root exudation induces microbial feedbacks to N cycling in a pine forest under long-term CO2 fumigation. Ecol. Lett. 14, 187–194.

102. Piao, S., et al., 2013. Evaluation of terrestrial carbon cycle models for their response to climate variability and to CO2 trends. Global Change Biol. 19 (7), 2117–2132.
103. Port, U., Brovkin, V., Claussen, M., 2012. The influence of vegetation dynamics on anthropogenic climate change. Earth Syst. Dyn. 3, 233–243.
104. Possell, M., et al., 2004. Interactive effects of elevated CO2 and soil fertility on isoprene emissions from Quercus robur. Global Change Biol. 10, 1835–1843.
105. Powell, T., et al., 2013. Confronting model predictions of carbon fluxes with measurements of Amazon forests subjected to experimental drought. New Phytol. 200 (2), 350–364.
106. Pugh, T.A.M., et al., 2015. Simulated carbon emissions from land-use change are substantially enhanced by accounting for agricultural management. Environ. Res. Lett. 10 (12), 124008.
107. Quentin, A.G., et al., 2015. Non-structural carbohydrates in woody plants compared among laboratories. Tree Physiol. (online first).
108. Reich, P.B., Hobbie, S.E., 2013. Decade-long soil nitrogen constraint on the CO2 fertilization of plant biomass. Nat. Clim. Change 3 (3), 278–282.
109. Reich, P.B., et al., 2006a. Carbon–nitrogen interactions in terrestrial ecosystems in response to rising atmospheric carbon dioxide. Annu. Rev. Ecol. Evol. Syst. 37, 611–636.
110. Reich, P.B., et al., 2006b. Nitrogen limitation constrains sustainability of ecosystem response to CO2. Nature 440 (April), 922–925.
111. Ribeiro, D.M., et al., 2012. Action of gibberellins on growth and metabolism of arabidopsis plants associated with high concentration of carbon dioxide. Plant Physiol. 160 (December), 1781–1794.
112. Rosenzweig, C., et al., 2014. Assessing agricultural risks of climate change in the 21st century in a global gridded crop model intercomparison. Proc. Natl. Acad. Sci. U. S. A. 111 (9), 3268–3273.
113. Running, S.W., 2008. Climate change Ecosystem disturbance, carbon, and climate. Science 321, 652–653.
114. Ryan, M.G., 1991. Effects of climate change on plant respiration. Ecol. Appl. 1 (2), 157–167.
115. Sardans, J., Penuelas, ˜ J., 2015. Potassium: a neglected nutrient in global change. Global Ecol. Biogeogr. 24, 261–275.
116. Schimel, D., Stephens, B.B., Fisher, J.B., 2015. Effect of increasing CO2 on the terrestrial carbon cycle. Proc. Natl. Acad. Sci. U. S. A. 112 (3), 436–441.
117. Schmitz, C., et al., 2014. Land-use change trajectories up to 2050: insights from a global agro-economic model comparison. Agric. Econ. 45, 69–84.
118. Schnitzer, S.A., Bongers, F., 2011. Increasing liana abundance and biomass in tropical forests: emerging patterns and putative mechanisms. Ecol. Lett. 14, 397–406.
119. Seidl, R., et al., 2011. Modelling natural disturbances in forest ecosystems: a review. Ecol. Modell. 222 (4), 903–924.
120. Sevanto, S., 2014. Phloem transport and drought. J. Exp. Bot. 65 (7), 1751–1759.
121. Sharkey, T.D., Yeh, S., 2001. Isoprene emission from plants. Annu. Rev. Plant Physiol. Plant Mol. Biol. 52, 407–436.
122. Sitch, S., et al., 2003. Evaluation of ecosystem dynamics, plant geography and terrestrial carbon cycling in the LPJ dynamic global vegetation model. Global Change Biol. 9 (2), 161–185.
123. Sitch, S., et al., 2008. Evaluation of the terrestrial carbon cycle, future plant geography and climate-carbon cycle feedbacks using five Dynamic Global Vegetation Models (DGVMs). Global Change Biol. 14, 2015–2039.
124. Sitch, S., et al., 2015. Recent trends and drivers of regional sources and sinks of carbon dioxide. Biogeosciences 12, 653–679.
125. Smith, N.G., Dukes, J.S., 2013. Plant respiration and photosynthesis in global-scale models: incorporating acclimation to temperature and CO2. Global Change Biol. 19, 45–63.
126. Smith, B., et al., 2014. Implications of incorporating N cycling and N limitations on primary production in an individual-based dynamic vegetation model. Biogeosciences 11, 2027–2054.

127. Smith, W.K., et al., 2016. Large divergence of satellite and Earth system model estimates of global terrestrial CO2 fertilization. Nat. Clim. Change 6, 306–310.
128. Stephens, B.B., et al., 2007. Weak northern and strong tropical land carbon uptake from vertical profiles of atmospheric CO2. Science 316, 1732–1735.
129. Stiling, P., Cornelissen, T., 2007. How does elevated carbon dioxide (CO2) affect plant-herbivore interactions? A field experiment and meta-analysis of CO2-mediated changes on plant chemistry and herbivore performance. Global Change Biol. 13, 1823–1842.
130. Stiling, P., et al., 2003. Elevated CO2 lowers relative and absolute herbivore density across all species of a scrub-oak forest. Oecologia 134, 82–87.
131. Sun, Y., et al., 2014. Impact of mesophyll diffusion on estimated global land CO2 fertilization. Proc. Natl. Acad. Sci. U. S. A. 111 (44), 1577415779.
132. Tardieu, F., Granier, C., Muller, B., 2011. Water deficit and growth. Co-ordinating processes without an orchestrator? Curr. Opin. Plant Biol. 14, 283–289.
133. Todd-Brown, K.E.O., et al., 2012. A framework for representing microbial decomposition in coupled climate models. Biogeochemistry 109, 19–33.
134. Uselman, S.M., Qualls, R.G., Thomas, R.B., 2000. Effects of increased atmospheric CO2, temperature: and soil N availability on root exudation of dissolved organic carbon by a N-fixing tree (Robinia pseudoacacia L.). Plant and Soil 222, 191–202.
135. van der Heijden, G., Powers, J.S., Schnitzer, S.A., 2015. Lianas reduce forest-level carbon accumulation and storage. Proc. Natl. Acad. Sci. U. S. A. (Early Edit). van der Sleen, P., et al., 2015. No growth stimulation of tropical trees by 150 years of CO2 fertilization but water-use efficiency increased. Nat. Geosci. 8 (January), 24–28.
136. Wårlind, D., et al., 2014. Nitrogen feedbacks increase future terrestrial ecosystem carbon uptake in an individual-based dynamic vegetation model. Biogeosciences 11, 6131–6146.
137. Wang, Y.P., Law, R.M., Pak, B., 2010. A global model of carbon: nitrogen and phosphorus cycles for the terrestrial biosphere. Biogeosciences 7, 2261–2282.
138. Wieder, W.R., et al., 2015. Future productivity and carbon storage limited by terrestrial nutrient availability. Nat. Geosci. 8, 441–444.
139. Wu, X., et al., 2010. Environmental controls over soil—atmosphere exchange of N2O, NO, and CO2 in a temperate Norway spruce forest. Global Biogeochem. Cycles 24, GB2012.
140. Yang, X., et al., 2009. Integration of nitrogen cycle dynamics into the Integrated Science Assessment Model for the study of terrestrial ecosystem responses to global change. Global Biogeochem. Cycles 23, GB4029.
141. Yang, Y., et al., 2011. Terrestrial C:N stoichiometry in response to elevated CO2 and N addition: a synthesis of two meta-analyses. Plant Soil 343, 393–400.
142. Yang, X., et al., 2014. The role of phosphorus dynamics in tropical forests –a modeling study using CLM-CNP. Biogeosciences 11, 1667–1681.
143. Zaehle, S., Dalmonech, D., 2011. Carbon-nitrogen interactions on land at global scales: current understanding in modelling climate biosphere feedbacks. Curr. Opin. Environ. Sustain. 3 (5), 311–320.
144. Zaehle, S., Friend, A.D., 2010. Carbon and nitrogen cycle dynamics in the O-CN land surface model: 1. Model description, site-scale evaluation, and sensitivity to parameter estimates. Global Biogeochem. Cycles 24, 1–13.
145. Zaehle, S., Friedlingstein, P., Friend, A.D., 2010. Terrestrial nitrogen feedbacks may accelerate future climate change. Geophys. Res. Lett. 37, 1–5.
146. Zaehle, S., et al., 2014. Evaluation of 11 terrestrial carbon –nitrogen cycle models against observations from two temperate free-air CO2 Enrichment studies. New Phytol. 202, 803–822.
147. Zavala, J.A., et al., 2008. Anthropogenic increase in carbon dioxide compromises plant defense against invasive insects. Proc. Natl. Acad. Sci. U. S. A. 105 (13), 5129–5133.
148. Zavala, J.A., Nabity, P.D., Delucia, E.H., 2013. An emerging understanding of mechanisms governing insect herbivory under elevated CO2. Annu. Rev. Entomol. 58, 79–97.
149. Zhang, D., et al., 2008. Rates of litter decomposition in terrestrial ecosystems: global patterns and controlling factors. J. Plant Ecol. 1 (2), 85–93.

Index